Mathematics for Industry

Volume 33

Aims & Scope

The meaning of "Mathematics for Industry" (sometimes abbreviated as MI or MfI) is different from that of "Mathematics in Industry" (or of "Industrial Mathematics"). The latter is restrictive: it tends to be identified with the actual mathematics that specifically arises in the daily management and operation of manufacturing. The former, however, denotes a new research field in mathematics that may serve as a foundation for creating future technologies. This concept was born from the integration and reorganization of pure and applied mathematics in the present day into a fluid and versatile form capable of stimulating awareness of the importance of mathematics in industry, as well as responding to the needs of industrial technologies. The history of this integration and reorganization indicates that this basic idea will someday find increasing utility. Mathematics can be a key technology in modern society.

The series aims to promote this trend by 1) providing comprehensive content on applications of mathematics, especially to industry technologies via various types of scientific research, 2) introducing basic, useful, necessary and crucial knowledge for several applications through concrete subjects, and 3) introducing new research results and developments for applications of mathematics in the real world. These points may provide the basis for opening a new mathematics-oriented technological world and even new research fields of mathematics.

To submit a proposal or request further information, please use the PDF Proposal Form or contact directly: Swati Meherishi, Executive Editor (swati.meherishi@springer.com).

Editor-in-Chief

Masato Wakayama, Kyushu University, Fukuoka, Japan

Series Editors

Robert S. Anderssen, Commonwealth Scientific and Industrial Research Organisation, Canberra, ACT, Australia
Yuliy Baryshnikov, Department of Mathematics, University of Illinois at Urbana-Champaign, Urbana, IL, USA
Heinz H. Bauschke, University of British Columbia, Vancouver, BC, Canada
Philip Broadbridge, School of Engineering and Mathematical Sciences, La Trobe University, Melbourne, VIC, Australia
Jin Cheng, Department of Mathematics, Fudan University, Shanghai, China
Monique Chyba, Department of Mathematics, University of Hawaii at Mānoa, Honolulu, HI, USA
Georges-Henri Cottet, Joseph Fourier University, Grenoble, Isère, France
José Alberto Cuminato, University of São Paulo, São Paulo, Brazil
Shin-ichiro Ei, Department of Mathematics, Hokkaido University, Sapporo, Japan
Yasuhide Fukumoto, Kyushu University, Nishi-ku, Fukuoka, Japan
Jonathan R. M. Hosking, IBM T.J. Watson Research Center, Scarsdale, NY, USA
Alejandro Jofré, University of Chile, Santiago, Chile
Masato Kimura, Faculty of Mathematics & Physics, Kanazawa University, Kanazawa, Japan
Kerry Landman, The University of Melbourne, Victoria, Australia
Robert McKibbin, Institute of Natural and Mathematical Sciences, Massey University, Palmerston North, Auckland, New Zealand
Andrea Parmeggiani, Dir Partenariat IRIS, University of Montpellier 2, Montpellier, Hérault, France
Jill Pipher, Department of Mathematics, Brown University, Providence, RI, USA
Konrad Polthier, Free University of Berlin, Berlin, Germany
Osamu Saeki, Institute of Mathematics for Industry, Kyushu University, Fukuoka, Japan
Wil Schilders, Department of Mathematics and Computer Science, Eindhoven University of Technology, Eindhoven, The Netherlands
Zuowei Shen, Department of Mathematics, National University of Singapore, Singapore, Singapore
Kim Chuan Toh, Department of Analytics and Operations, National University of Singapore, Singapore, Singapore
Evgeny Verbitskiy, Mathematical Institute, Leiden University, Leiden, The Netherlands
Nakahiro Yoshida, The University of Tokyo, Meguro-ku, Tokyo, Japan

More information about this series at http://www.springer.com/series/13254

Tsuyoshi Takagi · Masato Wakayama ·
Keisuke Tanaka · Noboru Kunihiro ·
Kazufumi Kimoto · Yasuhiko Ikematsu
Editors

International Symposium on Mathematics, Quantum Theory, and Cryptography

Proceedings of MQC 2019

 Springer

Editors
Tsuyoshi Takagi
Department of Mathematical Informatics
University of Tokyo
Tokyo, Japan

Masato Wakayama
Institute of Mathematics for Industry
Kyushu University
Fukuoka, Japan

Keisuke Tanaka
Department of Mathematical
and Computing Science
Tokyo Institute of Technology
Tokyo, Japan

Noboru Kunihiro
Department of Computer Science
University of Tsukuba
Ibaraki, Japan

Kazufumi Kimoto
University of the Ryukyus
Okinawa, Japan

Yasuhiko Ikematsu
Institute of Mathematics for Industry
Kyushu University
Fukuoka, Japan

ISSN 2198-350X   ISSN 2198-3518   (electronic)
Mathematics for Industry
ISBN 978-981-15-5193-2   ISBN 978-981-15-5191-8   (eBook)
https://doi.org/10.1007/978-981-15-5191-8

This Springer imprint is published by the registered company Springer Nature Singapore Pte Ltd.
The registered company address is: 152 Beach Road, #21-01/04 Gateway East, Singapore 189721, Singapore

Foreword

It is a great honor and pleasure for me to write some words for the book of extended abstracts of "International Symposium on Mathematics, Quantum Theory, and Cryptography (MQC 2019)".

I am currently supervising the CREST program "Modeling Methods allied with Modern Mathematics" funded by Japan Science and Technology Agency (JST). This program has 11 research teams, and Professor Tsuyoshi Takagi is directing one of them, the CREST CRYPTO-MATH team with the project titled "Mathematical Modelling for Next-Generation Cryptography". Hereby, we are pleased to support this symposium partly through the project of Professor Takagi. We are also happy to find speakers from several other teams of our CREST program.

Nowadays, it is a common understanding that cryptography is very important for sustaining society. And, as we all know, the modern cryptography is based on mathematics. Here "we" includes of course all the participants of this symposium, and I sincerely hope that "we" becomes most of the population partly through the activity of our program.

I am a geometer working on the structures on manifolds, but I gave from time to time lectures on the RSA cryptosystem to high school students. It was always easy to get the students excited about the beautiful mathematics used in the RSA cryptosystem.

I learned from the CREST CRYPTO-MATH team, however, that cryptography based on hardness of the integer factorization problem or the discrete logarithm problem faces a probable crisis because of advances in quantum computing. In fact, in these years there are already several companies planning to realize executing the quantum-based algorithm to attack the actual system of cryptography. They seem to demonstrating some part.

Of course, there are always questions on the cost and we should not overestimate or underestimate the probable effect which will happen in the next decade because of quantum computing. After all, it is really necessary to understand scientifically current theoretical achievement as well as current technical achievement. Here, I would like to share with all the participants from a vast area of research fields the fact that mathematics is the key for understanding.

As I learned that this symposium deals with all technical aspects of mathematical cryptography secure in the era of quantum computers, I sincerely hope that the participants would share the achievement from multiple aspects and would have the advantage to progress their research from this base. I strongly believe these research efforts will help people to enjoy a safer and sustainable society, not only at the national level, but also in the global prospective as well.

I hope to see a lot of exciting presentations as well as extensive and fruitful discussions where this book of extended abstracts would help, which will contribute to the success of this symposium.

Fukuoka, Japan   Takashi Tsuboi
September 2019

Preface

MQC 2019, the International Symposium on Mathematics, Quantum Theory, and Cryptography, was held at the IMI auditorium of Kyushu University in Fukuoka, Japan, during September 25–27, 2019. The symposium was organized by the CREST CRYPTO-MATH Project: "Mathematical Modelling for Next-Generation Cryptography", which was supported by Japan Science and Technology Agency (JST) to construct mathematical modeling of next-generation cryptography using wide-range mathematical theories. This symposium was held to mainly express the culmination of our project for these five years.

The symposium introduced new mathematical results in order to strengthen information security, simultaneously making fresh insights and developing the respective areas of mathematics. The symposium consists of 3 keynote addresses and 16 invited talks. The keynote addresses were given by Daniel Braak (Max Planck Institute), Johannes Buchmann (Technische Universitat Darmstadt), and Kouichi Semba (National Institute of Information and Communications Technology, NICT).

These proceedings consist of the papers/surveys selected from the talks of MQC 2019. Original research papers/surveys on all technical aspects of mathematical cryptography secure in the era of quantum computers were solicited. The topics include: (1) Mathematics and quantum theory for the next-generation cryptography such as number theory, algebraic geometry, lattice theory, representation theory, multivariate polynomial theory, quantum computation, mathematical physics, and probability theory; (2) Cryptosystems that have the potential to be safe against quantum computers such as hash-based signature schemes, lattice-based cryptosystems, multivariate cryptosystems, and quantum cryptographic schemes. There were 13 papers selected for publication. In addition, these proceedings contain 5 resumes corresponding to the remaining talks.

Many people contributed to the success of MQC 2019. We are very grateful to all of the Program Committee members as well as the external reviewers for their fruitful comments and discussions on their areas of expertise. We would also like to thank the students who supported to hold MQC 2019 smoothly.

Finally, we would like to express our gratitude to our partners and sponsors: JST CREST (Grant Number JPMJCR14D6), Kyushu University, Tokyo Institute of Technology, The University of Tokyo, and Advanced Innovation powered by Mathematics Platform (AIMaP).

Fukuoka, Japan
September 2019

Tsuyoshi Takagi
Masato Wakayama
Keisuke Tanaka
Noboru Kunihiro
Kazufumi Kimoto
Yasuhiko Ikematsu

The original version of the book was revised: Incorrect Part FM names in the website for second part FM now updated with correct title. The correction to the book is available at https://doi.org/10.1007/978-981-15-5191-8_19

Contents

About the Editors

Prof. Tsuyoshi Takagi received his B.Sc. and M.Sc. degrees in mathematics from Nagoya University in 1993 and 1995, respectively. He was engaged in research on network security at NTT Laboratories from 1995 to 2001. He received his Ph.D. from Technical University of Darmstadt in 2001. He is currently a Professor in the Graduate School of Information Science and Technology at University of Tokyo. His current research interests are information security and cryptography. He received DOCOMO Mobile Science Award in 2013, IEICE Achievement Award in 2013, and JSPS Prize in 2014. Dr. Takagi was a Program Chair of the 7th International Conference on Post-Quantum Cryptography, PQCrypto 2016.

Prof. Masato Wakayama is a Professor of Mathematics, Vice President at Tokyo University of Science (TUS) and Principal Fellow at Center for Research and Development Strategy, Japan Science and Technology Agency (CRDS/JST). He is also Professor Emeritus at Kyushu University. He obtained Ph.D. from Hiroshima University in 1985. His research interests include Representation Theory, Number Theory and Mathematical Physics, and has published over 100 referred research papers. He has contributed his experience and expertise to both academic works and university administration. His current appointments include Chair of Asia Pacific Consortium of Mathematics for Industry (2014–). He is the Editor-in-Chief of the Springer series "Mathematics for Industry". e-mail: wakayama@rs.tus.ac.jp

Prof. Keisuke Tanaka is a Professor in the School of Computing at Tokyo Institute of Technology. He received his B.S. from Yamanashi University in 1992, and his M.S. and Ph.D. in Computer Science from Japan Advanced Institute of Science and Technology in 1994 and 1997, respectively. Before joining Tokyo Institute of Technology, he was Research Engineer at NTT Information Platform Laboratories. His research interests include theory of cryptography, cryptocurrency and blockchain technology, and cybersecurity.

Prof. Noboru Kunihiro received his B.E., M.E. and Ph.D. in Mathematical Engineering and Information Physics from the University of Tokyo in 1994, 1996 and 2001, respectively. He has been a professor of University of Tsukuba since 2019. He was a researcher of NTT Communication Science Laboratories from 1996 to 2002. He was an associate professor of the University of Electro-Communications from 2002 to 2008. He was an associate professor of the University of Tokyo from 2008 to 2019. His research interest includes cryptography, information security.

Prof. Kazufumi Kimoto received his Ph.D. in Mathematics from Kyushu University in 2003. He was an assistant professor of the University of the Ryukyus from October 2003 to December 2010. He was an associate professor of the University of the Ryukyus from January 2011 to March 2015. He has been a professor of the University of the Ryukyus since April 2015. His research interest includes representation theory, number theory, and combinatorics.

Prof. Yasuhiko Ikematsu received his Ph.D. in Mathematics in 2016 from Kyushu University. He was a research fellow at the Institute of Mathematics for Industry, Kyushu University from 2016 to 2018 and in Department of Mathematical Informatics, University of Tokyo from April to December in 2018. He is currently an assistant professor in Institute of Mathematics for Industry, Kyushu University. His research interests include number theory and multivariate cryptography.

Keynote

Sustainable Cryptography

Johannes Buchmann

Abstract Cryptography is a fundamental tool for cybersecurity and privacy which must be protected for long periods of time. However, the security of most cryptographic algorithms relies on complexity assumptions that may become invalid over time. In this talk I discuss how sustainable cybersecurity and privacy can be achieved in this situation.

J. Buchmann (✉)
Technical University of Darmstadt, Hochschulstr. 10, 64289 Darmstadt, Germany
e-mail: johannes.buchmann@tu-darmstadt.de

© The Author(s) 2021
T. Takagi et al. (eds.), *International Symposium on Mathematics,
Quantum Theory, and Cryptography*, Mathematics for Industry 33,
https://doi.org/10.1007/978-981-15-5191-8_1

What Kind of Insight Provide Analytical Solutions of Quantum Models?

Daniel Braak

Abstract There are several concepts of what constitutes the analytical solution of a quantum model, as opposed to the mere "numerically exact" one. This applies even if one considers only the determination of the discrete spectrum of the corresponding Hamiltonian, setting aside such important questions as the asymptotic dynamics for long times. In the simplest case, the spectrum can be given in closed form, the eigenvalues E_j, $j = 0, \ldots, N \leq \infty$ read $E_j = f(j, \{p_k\})$, where f is a known function of the label $j \in \mathbb{N}_0$ and the $\{p_k\}$ are a set of numbers parameterizing the Hamilton operator. This kind of solution exists only in cases where the classical limit of the model is Liouville-integrable. Some quantum-mechanical many-body systems allow the determination of the spectrum in terms of auxiliary parameters $[\{k_j\}, \{n_l\}]$ as $E(\{n_l\}) = f(\{k_j(\{n_l\})\})$ where the $\{k_j(\{n_l\})\}$ satisfy a coupled set of transcendental equations, following from a certain ansatz for the eigenfunctions. These systems (integrable in the sense of Yang-Baxter (Eckle 2019)) may have a Hilbert space dimension growing exponentially with the system size L, i.e., $N \sim e^L$. The simple enumeration of the energies with the label j is replaced by the multi-index $\{n_l\}$. Although no priori knowledge about the spectrum is available, its statistical properties can be computed exactly (Berry and Tabor 1977). Other integrable and also non-integrable models exist where N depends polynomially on L and the energies E_j are the zeroes of an analytically computable transcendental function, the so-called G-function $G(E, \{p_k\})$ (Braak 2013a, 2016), which is proportional to the spectral determinant. Although no closed formula for E_j as function of the index j exists, detailed qualitative insight into the distribution of the eigenvalues can be obtained (Braak 2013b). Possible applications of these concepts to information compression and cryptography are outlined.

D. Braak (✉)
Max Planck Institute for Solid State Research, Heisenbergstraße 1, 70569 Stuttgart, Germany
e-mail: d.braak@fkf.mpg.de

© The Author(s) 2021

T. Takagi et al. (eds.), *International Symposium on Mathematics,
Quantum Theory, and Cryptography*, Mathematics for Industry 33,
https://doi.org/10.1007/978-981-15-5191-8_2

5

References

M.V. Berry, M. Tabor, Proc. R. Soc. Lond. A **356**, 375 (1977)
D. Braak, Ann. Phys. (Berlin) **525**, L23 (2013a)
D. Braak, J. Phys. A: Math. Theor. **46**, 175301 (2013b)
D. Braak, in *Applications + Practical Conceptualization + Mathematics = Fruitful Innovation*, ed. by R. Anderssen (Springer, Berlin, 2016)
H.-P. Eckle, *Models of Quantum Matter: A First Course on Integrability and the Bethe Ansatz* (Oxford University Press, Oxford, 2019)

Emerging Ultrastrong Coupling Between Light and Matter Observed in Circuit Quantum Electrodynamics

Kouichi Semba

Abstract The strength of the coupling between an atom and a single electromagnetic field mode is defined as the ratio of the vacuum Rabi frequency to the Larmor frequency, and is determined by a small dimensionless physical constant, the fine structure constant $\alpha = Z_{vac}/2R_K$. On the other hand, the quantum circuit including Josephson junctions behaving as artificial atoms and it can be coupled to the electromagnetic field with arbitrary strength (Devoret et al. 2007). Therefore, the circuit quantum electrodynamics (circuit QED) is extremely suitable for studying much stronger light-matter interaction.

We have used a Josephson junction atom, a flux qubit, harmonic oscillator coupled system. This circuit is well described by the Hamiltonian shown in Eq. (1).

$$\mathcal{H}_{\text{total}} = -\frac{\hbar}{2}(\Delta\sigma_x + \varepsilon\sigma_z) + \hbar\omega_0(\hat{a}^\dagger\hat{a} + \frac{1}{2}) + \hbar g\sigma_z(\hat{a} + \hat{a}^\dagger). \tag{1}$$

The first, second, and third terms represent the energy of the qubit, the energy of the harmonic oscillator, and the interaction energy, respectively. If the coupling strength g becomes as large as the atomic and cavity frequencies (Δ and ω_o, respectively), the energy eigenstates including the ground state are predicted to be highly entangled (Hepp and Lieb 1973; Ashhab and Nori 2010). We have experimentally achieved this deep strong coupling using a superconducting-flux-qubit LC-oscillator system (Yoshihara et al. 2017). By carefully designing a superconducting persistent-current qubit interacting with an LC harmonic oscillator that has a large zero-point fluctuation current via a large shared Josephson inductance, we have realized circuits with $\frac{g}{\omega_o}$ ranging from 0.72 to 1.34 and $\frac{g}{\Delta} \gg 1$. From the transmission spectroscopy, we have observed unconventional transition spectra and selection rules which can be interpreted using predicted energy levels which are well described by Schrödinger-cat-like entangled states between persistent-current states and displaced vacuum or Fock states of the oscillator (Yoshihara et al. 2017). By using two-tone spectroscopy, the energies of the six lowest levels of each circuit have been determined. We have

K. Semba (✉)
National Institute of Information and Communications Technology,
4-2-1 Nukui-Kitamachi, Koganei, Tokyo 184-8795, Japan
e-mail: semba@nict.go.jp

© The Author(s) 2021
T. Takagi et al. (eds.), *International Symposium on Mathematics,*
Quantum Theory, and Cryptography, Mathematics for Industry 33,
https://doi.org/10.1007/978-981-15-5191-8_3

observed huge light shifts, i.e., Lamb shifts, qubit energy shift due to coupling to vacuum field, that exceed 90% of the bare qubit frequencies and Stark shifts, inversions of the qubits' ground and excited states when there are only a few photons in the oscillator (Yoshihara et al. 2018). We have also observed collective coupling between an engineered 4300 ensemble of flux qubits and a superconducting resonator (Kakuyanagi et al. 2016), and considered the condition for observing generation of superradiant ground state in the presence of parameter fluctuations (Ashhab and Semba 2017).

References

S. Ashhab, F. Nori, Phys. Rev. A **81**, 042311 (2010)
S. Ashhab, K. Semba, Phys. Rev. A **95**, 053833 (2017)
M. Devoret, S. Girvin, R. Schoelkopf, Ann. Phys. (Leipzig) **16**(10–11), 767–779 (2007)
K. Hepp, E.H. Lieb, Ann. Phys. (NY) **76**, 360–404 (1973)
K. Kakuyanagi, Y. Matsuzaki, C. Deprez, H. Toida, K. Semba, H. Yamaguchi, W.J. Munro, S. Saito, Phys. Rev. Lett. **117**, 210503 (2016)
F. Yoshihara, T. Fuse, S. Ashhab, K. Kakuyanagi, S. Saito, K. Semba, Nat. Phys. **13**, 44 (2017)
F. Yoshihara, T. Fuse, Z. Ao, S. Ashhab, K. Kakuyanagi, S. Saito, T. Aoki, K. Koshino, K. Semba, Phys. Rev. Lett. **120**, 183601 (2018)

Summary

Verified Numerical Computations and Related Applications

Shin'ichi Oishi

Abstract The author has been engaged in the study of numerical computations with result verification starting from 1990.

Summary

The author has been engaged in the study of numerical computations with result verification starting from 1990. As a result, the following results have been obtained:

1. We have proposed a concept of error-free transformations for calculating not only approximate values of numerical evaluations of certain arithmetic expressions consisting of additions, subtractions and multiplications, but also exact error of such numerical evaluations. Using this concept, we have established the way of getting numerical solutions for various problems in numerical linear algebra with required accuracy. Especially, we have established the verified numerical methods for the following problems:

   a. Finite dimensional linear equations including extremely ill-conditioned problems.
   b. Matrix eigenvalue problems.

2. We have proposed various verified numerical methods for various problems including

   a. Calculation of ill-conditioned definite integrals.
   b. Boundary value problems for nonlinear differential equations based on invention of methods for eigenvalue evaluation of associated linearized problems.

In this talk, we will review some of these results and will mention possible applications for cryptography.

S. Oishi (✉)
Waseda University, 3-4-1 Okubo, Tokyo, Shinjuku-ku 169-8555, Japan
e-mail: oishi@waseda.jp

© The Author(s) 2021

T. Takagi et al. (eds.), *International Symposium on Mathematics,
Quantum Theory, and Cryptography*, Mathematics for Industry 33,
https://doi.org/10.1007/978-981-15-5191-8_4

A Review of Secret Key Distribution Based on Bounded Observability

Jun Muramatsu

Abstract Secret key distribution is a technique for a sender and a receiver to share a secret key, which is not known by any eavesdropper, when they share no common secret information in advance. By using this technique, the sender and the receiver can transmit a message securely in the sense that the message remains secret from any eavesdropper. We introduced a secret key distribution based on the Bounded Observability (Muramatsu et al. 2010, 2013, 2015), which provides a necessary and sufficient condition for the possibility of secret key distribution. This condition describes limits on the information obtained by observation of a random object, and models the practical difficulty of completely observing random physical phenomena.

Keywords Secret key distribution · Information-theoretic security · Secret key agreement · Bounded observability

References

J. Muramatsu, K. Yoshimura, P. Davis, Information theoretic security based on bounded observability, in *Information Theoretic Security*, vol. 5973. Lecture Notes on Computer Science. (Springer, Berlin, 2010), pp. 128–139

J. Muramatsu, K. Yoshimura, P. Davis, A. Uchida, T. Harayama, Bounded Observability ni moto-duku himitu-kagi-haisou (in Japanese). Bull. Jpn Soc. Ind. Appl. Math. (Ouyou-suuri) **23**(1), 11–20 (2013)

J. Muramatsu, K. Yoshimura, P. Davis, A. Uchida, T. Harayama, Secure key distribution based on bounded observability. Proc. IEEE **103**(10), 1762–1780 (2015)

J. Muramatsu (✉)
NTT Communication Science Laboratories, NTT Corporation, 2-4, Seika-cho, Soraku-gun, Hikaridai, Kyoto 619-0237, Japan
e-mail: jun.muramatsu@ieee.org

© The Author(s) 2021
T. Takagi et al. (eds.), *International Symposium on Mathematics, Quantum Theory, and Cryptography*, Mathematics for Industry 33,
https://doi.org/10.1007/978-981-15-5191-8_5

Quantum Computing and Information Theory

Quantum Random Numbers Generated by a Cloud Superconducting Quantum Computer

Kentaro Tamura and Yutaka Shikano

Abstract A cloud quantum computer is similar to a random number generator in that its physical mechanism is inaccessible to its users. In this respect, a cloud quantum computer is a black box. In both devices, its users decide the device condition from the output. A framework to achieve this exists in the field of random number generation in the form of statistical tests for random number generators. In the present study, we generated random numbers on a 20-qubit cloud quantum computer and evaluated the condition and stability of its qubits using statistical tests for random number generators. As a result, we observed that some qubits were more biased than others. Statistical tests for random number generators may provide a simple indicator of qubit condition and stability, enabling users to decide for themselves which qubits inside a cloud quantum computer to use.

Keywords Cloud quantum computer · Random number generator · NIST SP 800-22 · Stability

1 Introduction

Given a coin with an unknown probability distribution, there are two approaches to decide whether the coin is fair (Tamura and Shikano 2019). The first approach is to examine the coin itself; one expects an evenly shaped coin to yield fair results. The

K. Tamura (✉)
Department of Applied Physics and Physico-Informatics, Keio University,
3-14-1 Hiyoshi, Kohoku, Yokohama 223-8522, Japan
e-mail: cicero@keio.jp

Y. Shikano
Quantum Computing Center, Keio University, 3-14-1 Hiyoshi, Kohoku,
Yokohama 223-8522, Japan
e-mail: yutaka.shikano@keio.jp

Institute for Quantum Studies, Chapman University, 1 University Dr., Orange,
CA 92866, USA

© The Author(s) 2021
T. Takagi et al. (eds.), *International Symposium on Mathematics,
Quantum Theory, and Cryptography*, Mathematics for Industry 33,
https://doi.org/10.1007/978-981-15-5191-8_6

second approach is to actually toss the coin a number of times to see if the output is sound. In this approach, the coin is treated as a black box. A random number generator is similar to a coin in that it is expected to produce unbiased and independent 0s and 1s. Unlike a coin, however, the physical mechanism of a random number generator is often inaccessible to its users. Therefore, users rely on statistical tests to decide the fairness of the device from its output.

Random number generators play an important role in cryptography, particularly in the context of key generation. For example, the security of the RSA cryptosystem is based on keys that are determined by random choices of two large prime numbers (Boneh 1999). If the choices of prime numbers are not random, an adversary could predict future keys and hence compromise the security of the system. Randomness in cryptography derives from what is called the seed. The seed is provided by physical random number generators (Schindler and Killmann 2003; Ugajin et al. 2017). It is required that the physical mechanism of a physical random number generator remains a black box for the seed to be unpredictable. Given that the measurement outcomes are theoretically unpredictable in quantum mechanics, random number generators based on quantum phenomena are a promising source of unpredictability (Pironio et al. 2010; Ma et al. 2016; Herrero-Collantes and Garcia-Escartin 2017).

Cloud quantum computers are quantum computers that are accessed online (Srivastava et al. 2016; Gibney 2017; Castelvecchi 2017; Xin et al. 2018; Yamamoto et al. 2019; National Academies of Sciences, Engineering, and Medicine 2019). In order to use a cloud quantum computer, users are required to send programs specifying the quantum circuit to be executed and the number of times the circuit should be run (LaRose 2019). When a user's turn arrives, the quantum computer executes the program and returns the results (Preskill 2018). A similarity between random number generators and cloud quantum computers is that its users do not have direct access to the physical mechanism of the device. So, as far as the users are concerned, both random number generators and cloud quantum computers are black boxes. In the field of random number generation, much research has been done on how to characterize the device from its output. This leads to the creation of statistical tests for random number generators. The present study aims to introduce the idea of statistical tests for random number generators to the field of cloud quantum computing. This aim is supported by three points. Firstly, the cloud quantum computer is a black box to its users, which is also the case with random number generators. Secondly, quantum computers become random number generators when given certain programs. Finally, the cloud quantum computer lacks a simple benchmark that would enable its users to decide the condition of the device.

The rest of this article is organized as follows. In Sect. 2, statistical tests for random number generators are generally explained. In Sect. 3, a group of statistical tests called the NIST SP 800-22 is reviewed. In Sect. 4, we present the results of the statistical analysis of random number samples obtained from the cloud quantum computer, IBM 20Q Poughkeepsie, and the test results of the eight statistical tests from the NIST SP 800-22. Finally, Sect. 5 is devoted to the conclusion. In the appendix, a measure of uniformity often employed in the field of cryptography, the min-entropy, is explained.

2 Statistical Tests for Random Number Generators

Statistical tests for random number generators are necessary to confirm that a random number generator is suitable for use in encryption processes (Demirhan and Bitirim 2016). Random number generators used in this context are required to have unpredictability. This means that given any subset of a sequence produced by the device, no adversary can predict the rest of the sequence, including the output from the past. Statistical tests aim to detect random number generators that produce sequences with a significant bias and/or correlation.

When subjected to statistical tests, a random number generator is considered a black box. This means that the only information available is its output. Under the null hypothesis that the generator is unbiased and independent, one expects its output to have certain characteristics. The characteristics of the output are quantified by the test statistic, whose probability distribution is known. From the test statistic, the probability that a true random number generator produces an output with a worse test statistic value is calculated. This probability is called the p-value. If the p-value is below the level of significance α, the generator fails the test, and the null hypothesis that the generator is unbiased and independent is rejected. Since statistical tests for random number generators merely rule out significantly biased and/or correlated generators, these tests do not verify that a device is the ideal random number generator. Nevertheless, a generator that passes the tests is more reliable than a generator that doesn't. This is why statistical tests are usually organized in the form of test suites, so as to be comprehensive. Some well known test suites are the NIST SP 800-22 (Bassham 2010), TestU01 (L'ecuyer and Simard 2007), and the Dieharder test.

Because statistical tests are designed to check for statistical anomalies under the hypothesis that the generator is unbiased, a biased random number generator would naturally fail the tests. This can be a problem when testing quantum random number generators, as they can be biased and unpredictable at the same time. Given that statistically faulty generators can still be unpredictable, the framework of statistical tests fails to capture the essence of randomness: unpredictability. There have been attempts to assure the presence of unpredictability by exploiting quantum inequalities, but they have not reached the point of replacing statistical tests altogether.

3 NIST SP 800-22

The NIST SP 800-22 is a series of statistical tests for cryptographic random number generators provided by the National Institute of Standards and Technology (Bassham 2010). Random number generators for cryptographic purposes are required to have unpredictability, which is not strictly necessary in other applications such as simulation and modeling, but is a crucial element of randomness. The test suite contains 16 tests, each test with a different test statistic to characterize deviations of binary sequences from randomness. The entire testing procedure of the NIST SP 800-22 is divided into three steps. The first step is to subject all samples to the 16 tests. For each sample, each test returns the probability that the sample is obtained from an unbiased and independent RNG. This probability, which is called the p-value, is then compared

Table 1 The minimum length n required for each test in order to obtain meaningful results. The tests not employed in the present study are shaded in gray. Note that the tests will be referred to by their test number in Sect. 4

Test number	Test name	Minimum length
1	Frequency	$n \geq 100$
2	Frequency within a block	$n \geq 100$
3	Runs	$n \geq 100$
4	Longest run of ones	$n \geq 128$
	Binary matrix rank	$n \geq 38912$
5	DFT	$n \geq 1000$
	Non-overlapping T. M.	$n \geq 8m - 8$
	Overlapping T. M.	$n \geq 10^6$
	Maurer's Universal	$n > 387840$
	Linear complexity	$n \geq 10^6$
	Serial	$n > 16$
6	Approximate entropy	$n > 64$
7	Cumulative sums (forward)	$n \geq 100$
8	Cumulative sums (backward)	$n \geq 100$
	Random excursions	$n \geq 10^6$
	Random excursions variant	$n \geq 10^6$

to the level of significance $\alpha = 0.01$. If the p-value is under the level of significance, the sample fails the test. The second step involves the proportion of passed samples for each test. Under the level of significance $\alpha = 0.01$, 1% of samples obtained from an unbiased and independent RNG is expected to fail each test. If the proportion of passed samples is too high or too low, the RNG fails the test. Finally, p-value uniformity is checked for each test. Suppose one tested 100 binary samples. This yields 100 p-values per test. If the samples are independent, the p-values should be uniformly distributed for all tests. The distribution of p-values is checked via the chi-squared test.

In the following sections, eight tests from the NIST SP 800-22 are explained (Table 1). The input sequence will be denoted by $\varepsilon = \varepsilon_1 \varepsilon_2 \cdots \varepsilon_n$, and the ith element by ε_i.

3.1 Frequency Test

The frequency test aims to test whether a sequence contains a reasonable proportion of 0s and 1s. If the probability of obtaining the sequence from an independent and unbiased random number generator is lower than 1%, it follows that the random number generator is not "independent and unbiased". The minimum sample length required for this test is 100.

Test Description

1. Convert the sequence into ± 1 using the formula: $X_i = 2\varepsilon_i - 1$.
2. Add the elements of X together to obtain S_n.

3. Compute test statistic: $s_{obs} = |S_n|/\sqrt{n}$.
4. Compute p-value $= \mathrm{erfc}(s_{obs}/\sqrt{2})$ using complementary error function shown as

$$\mathrm{erfc}(z) = \frac{2}{\sqrt{\pi}} \int_z^\infty e^{-u^2} du. \tag{1}$$

5. Compare p-value to 0.01. If p-value ≥ 0.01, then the sequence passes the test. Otherwise, the sequence fails.

Example: $\varepsilon = 1001100010$, length $n = 10$.
1. $1, 0, 0, 1, 1, 0, 0, 0, 1, 0 \rightarrow +1, -1, -1, +1, +1, -1, -1, -1, +1, -1$.
2. $S_{10} = 1 - 1 - 1 + 1 + 1 - 1 - 1 - 1 + 1 - 1 = -2$.
3. $s_{obs} = |-2|/\sqrt{10} \approx 0.632455$.
4. P-value $= \mathrm{erfc}(s_{obs}/\sqrt{2}) \approx 0.527089$.
5. P-value $= 0.527089 > 0.01 \rightarrow$ the sequence passes the test.

This test is equivalent to testing the histogram for bias. Because the test only considers the proportion of 1s, sequences such as 0000011111 or 0101010101 would pass the test. Failing this test means that the sample is overall biased.

3.2 Frequency Test Within a Block

Firstly, the sequence is divided into N blocks of size M. The frequency test is then applied to the respective blocks. As a result, one obtains N p-values. The second part of this test aims to check whether the variance of the p-values is by chance or not. This is called the chi-squared (χ^2) test. For meaningful results, a sample with a length of at least 100 is required. The following is the test description.

Test Description
1. Divide the sequence into $N = \lfloor \frac{n}{M} \rfloor$ non-overlapping blocks of size M.
2. Determine the proportion of 1s in each block using

$$\pi_i = \frac{\sum_{j=1}^M \varepsilon_{(i-1)M+j}}{M}. \tag{2}$$

3. Compute χ^2 statistic $\chi^2_{obs} = 4M \sum_{i=1}^N \left(\pi_i - \frac{1}{2}\right)^2$.

4. Compute p-value $= 1 - \text{igamc}\left(\frac{N}{2}, \frac{\chi^2_{obs}}{2}\right)$. Note that igamc stands for the incomplete gamma function.

$$\Gamma(z) = \int_0^\infty t^{z-1} e^{-t} \tag{3}$$

$$\text{igamc}(a, x) \equiv \frac{1}{\Gamma(a)} \int_0^x e^{-t} t^{(a-1)} dt \tag{4}$$

5. Compare p-value to 0.01. If p-value ≥ 0.01, then the sequence passes the test. Otherwise, the sequence fails.

Example: $\varepsilon = 1001100010$, length: $n = 10$.
1. If $M = 3$, then $N = 3$ and the blocks are 100, 110, 001. The final 0 is discarded.
2. $\pi_1 = 1/3, \pi_2 = 2/3, \pi_3 = 1/3$.
3. $\chi^2_{obs} = 4M \sum_{i=1}^N (\pi_i - \frac{1}{2})^2$.
4. $\chi^2_{obs} = 4 \times 3 \times \left\{ (\frac{1}{3} - \frac{1}{2})^2 + (\frac{2}{3} - \frac{1}{2})^2 + (\frac{1}{3} - \frac{1}{2})^2 \right\} = 1$.
5. P-value $= 1 - \text{igamc}(\frac{3}{2}, \frac{1}{2}) = 0.801252$.
6. P-value $= 0.801252 > 0.01 \rightarrow$ the sequence passes the test.

This test divides the sequence into blocks and checks each block for bias. Depending on the block size, samples such as 001100110011 or 101010101010 could pass the test. Failing this test means that certain sections of the sequence are biased.

3.3 Runs Test

The proportion of 0s and 1s does not suffice to identify a random sequence. A run, which is an uninterrupted sequence of identical bits, is also a factor to be taken into account. The runs test determines whether the lengths and oscillation of runs in a sequence are as expected from a random sequence. A minimum sample length of 100 is required for this test. The following is the test description.

Test Description

1. Compute proportion of ones $\pi = \left(\sum_j \varepsilon_j \right) / n$.
2. If the sequence passes frequency test, proceed to next step. Otherwise, the p-value of this test is 0.
3. Compute test statistic $V_n(\text{obs}) = \sum_{k=1}^{n-1} (\varepsilon_k \oplus \varepsilon_{k+1}) + 1$, where $\oplus$ stands for the XOR operation.

4. Compute p-value $= \text{erfc}\left(\frac{|V_n(\text{obs})-2n\pi(1-\pi)|}{2\sqrt{2n\pi}(1-\pi)}\right)$.
5. Compare p-value to 0.01. If p-value ≥ 0.01, then the sequence passes the test. Otherwise, the sequence fails.

Example: $\varepsilon = 1010110001$, length $n = 10$.
1. $\pi = \frac{5}{10} = 0.5$.
2. $|\pi - 0.5| = 0 < \frac{2}{\sqrt{n}} = \frac{2}{\sqrt{10}} = 0.63 \rightarrow$ test is applicable.
3. $V_{10}(\text{obs}) = (1 + 1 + 1 + 1 + 0 + 1 + 0 + 0 + 1) + 1 = 7$.
4. P-value $= \text{erfc}\left(\frac{|7-2\times10\times0.5\times(1-0.5)|}{2\times\sqrt{2\times10}\times0.5\times(1-0.5)}\right) = 0.21$.
5. P-value $= 0.21 \geq 0.01$, so sequence passes the test.

3.4 The Longest Run of Ones Within a Block Test

This test determines whether the longest runs of ones $111\cdots$ within blocks of size M is consistent with what would be expected in a random sequence. The possible values of M for this test are limited to three values, namely, 8, 128, and 10,000, depending on the length of the sequence to be tested.

Table 2 Choices of M for the longest runs of ones within a block test

Minimum length n	M
128	8
6,272	128
750,000	10000

Table 3 Classifications of each block

Classes v_i	$M \geq 8$	$M \geq 128$	$M \geq 100000$
v_0	≤ 1	≤ 4	≤ 10
v_1	2	5	11
v_2	3	6	12
v_3	≥ 4	7	13
v_4		8	14
v_5		≥ 9	15
v_6			≥ 16

Test Description

1. Divide the sequence into blocks of size M. The choices of M and N are determined in regard to the length of the sequence. N denotes the number of blocks, and the elements exceeding the number of blocks are discarded. The possible choices of n and M provided by NIST are shown in Table 2.
2. Classify each block into the following categories regarding M and the length of the longest run in each block. See Table 3.
3. Compute $\chi^2(\text{obs}) = \sum_{i=0}^{K} \frac{(v_i - N\pi_i)^2}{N\pi_i}$. Note that K, N, and π_i are determined by M. See Tables 4 and 5.
4. Compute p-value $= 1 - \text{igamc}\left(\frac{K}{2}, \frac{\chi^2(\text{obs})}{2}\right)$.
5. Compare p-value to 0.01. If p-value ≥ 0.01, then the sequence passes the test. Otherwise, the sequence fails.

Example: $n = 10000$

1. $M = 128$ and $N = 49$. The remaining 3728 elements are discarded.
2. The counts for the longest run of ones are $v_0 = 6, v_1 = 10, v_2 = 10, v_3 = 7, v_4 = 7$, and $v_5 = 9$.

Table 4 Values of K and N corresponding to M

M	K	N
8	3	16
128	5	49
10000	6	75

Table 5 Values of π_i corresponding to K and M

Classes	π_i		
	$K = 3, M = 8$	$K = 5, M = 128$	$K = 6, M = 10000$
v_0	0.2148	0.1174	0.0882
v_1	0.3672	0.2430	0.2092
v_2	0.2305	0.2493	0.2483
v_3	0.1875	0.1752	0.1933
v_4		0.1027	0.1208
v_5		0.1124	0.0675
v_6			0.0727

3.

$$\chi^2(\text{obs}) = \frac{(6 - 49 \times 0.1174)^2}{49 \times 0.1174} + \frac{(10 - 49 \times 0.2430)^2}{49 \times 0.2430}$$

$$+ \frac{(10 - 49 \times 0.2493)^2}{49 \times 0.2493} + \frac{(7 - 49 \times 0.1752)^2}{49 \times 0.1752}$$

$$+ \frac{(7 - 49 \times 0.1027)^2}{49 \times 0.1027} + \frac{(9 - 49 \times 0.1124)^2}{49 \times 0.1124}$$

$$= 3.994459.$$

4. P-value $= 1 - \text{igamc}\left(\frac{5}{2}, \frac{3.994459}{2}\right) = 0.550214$.
5. P-value $= 0.550214 \geq 0.01$, so the sequence passes the test.

3.5 Discrete Fourier Transform Test

This test checks for periodic patterns in the sequence by performing a discrete Fourier transform (DFT). The minimum sample length required for this test is 1000. The following is the test description.

Test Description

1. Convert the sequence ε of 0s and 1s into a sequence X of -1s and $+1$s.
2. Apply a DFT on X: $S = DFT(X)$. This should yield a sequence of complex variables representing the periodic components of the sequence of bits at different frequencies.
3. Compute $M = \text{modulus}(S') \equiv |S'|$, where S' is the first $\frac{n}{2}$ element of S. This produces a sequence of peak heights.
4. Compute $T = \sqrt{\left(\log_e \frac{1}{0.05}\right)}$. This is the 95 % peak height threshold value. 95 % of the values obtained by the test should not exceed T for a random sequence.
5. Compute $N(\text{ideal}) = \frac{0.95n}{2}$, which is the expected theoretical number of peaks that are less than T.
6. Compute $N(\text{obs})$, which is the actual number of peaks in M that are less than T.
7. Compute $d = \frac{N(\text{ideal}) - N(\text{obs})}{\sqrt{n \cdot 0.95 \cdot 0.05 \cdot \frac{1}{4}}}$.
8. Compute p-value $= \text{erfc}\left(\frac{|d|}{\sqrt{2}}\right)$.
9. Compare p-value to 0.01. If p-value ≥ 0.01, then the sequence passes the test. Otherwise, the sequence fails.

This test checks for periodic features. Samples with periodic features may look like 0110011001100110 or 010010100101001 among various other possibilities. Failing this test suggests that the sample has periodic patterns. It is noted that the probability distribution of the test statistic d should be rectified as it does not converge to the standard normal distribution (Hamano 2005).

Example: $\varepsilon = 1001010011$, length $n = 10$.
1. $X = 2\varepsilon_1 - 1, 2\varepsilon_2 - 1, \ldots, 2\varepsilon_n - 1 = 1, -1, -1, 1, -1, 1, -1, -1, 1, 1$.
2. $N(\text{ideal}) = 4.75$.
3. $N(\text{obs}) = 4$.
4. $d = \dfrac{(4.75-4)}{\sqrt{10 \cdot 0.95 \cdot 0.05 \cdot \frac{1}{4}}} = 2.147410$.
5. P-value $= \text{erfc}\left(\dfrac{|2.147410|}{\sqrt{2}}\right) = 0.031761$.
6. P-value $= 0.031761 \geq 0.01$, so the sequence passes the test.

3.6 Approximate Entropy Test

The approximate entropy test compares the frequency of m-bit overlapping patterns with that of $(m + 1)$-bit patterns in the sequence. It checks whether the relation of two frequencies is what is expected from an unbiased and independent RNG. The level of significance is $\alpha = 0.01$. This test can be applied to samples with lengths equal to or larger than 64. The test description is below.

Test Description
1. Append the first $m - 1$ bits of the sequence to the end of the sequence.
2. Divide the sequence into overlapping blocks with a length of m.
3. There are 2^m possible m-bit blocks. Count how many of each possible m-bit block there are in the sequence.
4. Compute $\frac{\text{count}}{n} \log_e(\frac{\text{count}}{n})$ for each count.
5. Compute the sum of all counts φ_m.
6. Replace m with $m + 1$ and repeat steps 1 through 5 to obtain φ_{m+1}.
7. Calculate test statistic obs $= 2n(\log_e(n) - (\varphi_m - \varphi_{m+1}))$.
8. Derive p-value $= 1 - \text{igamc}(2^{(m-1)}, \text{obs}/2)$.
9. Compare p-value with level of significance $\alpha = 0.01$. If p-value ≥ 0.01, the result is pass. Otherwise, the sequence fails the test.

Example: $\varepsilon = 1011010010$, length $n = 10$, $m = 3$.
1. $\varepsilon = 1011010010 \rightarrow 101101001010$.
2. $101101001010 \rightarrow 101, 011, 110, 101, 010, 100, 001, 010, 101, 010$.
3. "000" : 0, "001" : 1,
   "010" : 3, "011" : 1, "100" : 1, "101" : 3, "110" : 1, "111" : 0.
4. "000" : 0, "001" : $0.1\log_e(0.1)$, "010" : $0.3\log_e(0.3)$,
   "011" : $0.1\log_e(0.1)$, "100" : $0.1\log_e(0.1)$, "101" : $0.3\log_e(0.3)$,
   "110" : $0.1\log_e(0.1)$, "111" : 0.
5. $\varphi_3 = -1.643418$
6. $\varphi_{3+1} = -2.025326$.
7. obs $= 2 \times 10 \times (\log_e(10) - (-1.643418 - (-2.025326))) = 6.224774$.
8. P-value $= 1 - \text{igamc}(2^{(3-1)}, 6.224774/2) = 0.622069$.
9. P-value $= 0.622069 \geq 0.01$. The sequence passes the test.

The approximate entropy test checks for correlation between the number of m-bit patterns and $(m + 1)$-bit patterns in the sequence. The difference between the number of possible m-bit patterns and the number of possible $(m + 1)$-bit patterns in the sequence is computed, and if this difference is too small or too large, the two patterns are correlated.

3.7 Cumulative Sums Test

The cumulative sums test is basically a random walk test. It checks how far from 0 the sum of the sequence in terms of ± 1 reaches. For a sequence that contains uniform and independent 0s and 1s, the sum should be close to 0. This test requires a minimum sample length of 100.

Test Description
1. Convert 0 to -1 and 1 to $+1$.
2. In forward mode, compute the sum of the first i elements of X. In backward mode, compute the sum of the last i elements of X.
3. Find the maximum value z of the sums.
4. Compute the following p-value. Φ is the cumulative distribution function for the standard normal distribution.

$$\text{P-value} = 1 - \sum_{k=(\frac{-n}{z}+1)/4}^{(\frac{n}{z}-1)/4} \left[\Phi\left(\frac{(4k + 1)z}{\sqrt{n}}\right) - \Phi\left(\frac{(4k - 1)z}{\sqrt{n}}\right) \right]$$

$$+ \sum_{k=(\frac{-n}{z}-3)/4}^{(\frac{n}{z}-1)/4} \left[\Phi\left(\frac{(4k+3)z}{\sqrt{n}}\right) - \Phi\left(\frac{(4k+1)z}{\sqrt{n}}\right) \right]. \quad (5)$$

5. Compare p-value to $\alpha = 0.01$. If p-value ≥ 0.01, the result is pass. Otherwise, the sequence fails the test.

Example: $\varepsilon = 1011010010$, length $n = 10$.
1. $\varepsilon = 1011010010 \to X = 1, -1, 1, 1, -1, 1, -1, -1, 1, -1$.
2. Forward mode: $S_1 = 1$, $S_2 = 1 + (-1) = 0$, $S_3 = 1 + (-1) + 1 = 2$,
   $S_4 = 1 + (-1) + 1 + 1$, $S_5 = 1 + (-1) + 1 + 1 + (-1) = 1$,
   $S_6 = 1 + (-1) + 1 + 1 + (-1) + 1 = 2$, $S_7 = 1 + (-1) + 1 + 1 + (-1) + 1 + (-1) = 1$, $S_8 = 1 + (-1) + 1 + 1 + (-1) + 1 + (-1) + 1 = 2$,
   $S_9 = 1 + (-1) + 1 + 1 + (-1) + 1 + (-1) + 1 + (-1) = 1$.
3. In forward mode, the maximum value is $z = 2$.
4. P-value = 0.941740 for both forward and backward.
5. P-value = 0.941740 ≥ 0.01. The sequence passes the test.

Once the p-value has been calculated for all tests and samples, the proportion of samples that passed the test is computed for each test. Let us consider a case where 1000 samples were subjected to each of the 15 tests. This results in 1000 p-values per test. For example, if 950 out of 1000 samples passed the frequency test, the proportion of passed samples is 0.95. If the proportion of passed samples falls within the following range for all 15 tests, the samples pass the second step of the NIST SP 800-22. The acceptable range of proportion is calculated with

$$(1-\alpha) \pm 3\sqrt{\frac{\alpha(1-\alpha)}{m}}, \quad (6)$$

where α stands for the level of significance and m the sample size. It is noted that it is controversial whether the coefficient should be 3. A suggestion that the coefficient should be 2.6 exists (Marek et al. 2015). In the case of the current example, Eq. (6) can be calculated using $\alpha = 0.01$ and $m = 1000$ as

$$(1-0.01) \pm 3\sqrt{\frac{0.01(1-0.01)}{1000}} = 0.99 \pm 0.0094. \quad (7)$$

From the fact that 0.95 is not within the acceptable range, it follows that the samples fail the frequency test. The same process is done with all 16 tests, and unless the samples pass all tests, the result is that the hypothesis that the RNG is unbiased and independent is rejected.

The final step of the NIST SP 800-22 is to evaluate the p-value uniformity of each test. In order to perform the chi-squared (χ^2) test, the p-value is divided into 10 regions: $[k, k + 0.1)$ for $k = 0, 1, \ldots, 9$. The test statistic is given by

$$\chi^2 = \sum_{i=1}^{10} \frac{(\text{number of samples in } i\text{th region} - \text{sample size}/10)^2}{\text{sample size}/10}. \tag{8}$$

When the number of samples in each region is 2, 8, 10, 13, 17, 17, 13, 10, 8, 2, the test statistic (8) is calculated as $\chi^2 = 25.200000$. From χ^2, the p-value is

$$\text{p-value} = \text{igamc}\left(\frac{9}{2}, \frac{\chi^2}{2}\right). \tag{9}$$

Therefore, in the current example where $\chi^2 = 25.200000$, the p-value is 0.002758. The level of significance for the p-value uniformity is $\alpha = 0.0001$. So when the p-value is 0.002758, it follows that the p-value distribution is uniform. The p-value uniformity test requires at least 55 samples. As mentioned before, it is remarked that passing the NIST SP 800-22 does not ensure a sequence to be truly random (Kim et al. 2020; Fan et al. 2014; Haramoto and Matsumoto 2019).

4 Quantum Random Number Generation on the Cloud Quantum Computer

According to quantum mechanics, the measurement outcomes of the superposition state $(|0\rangle + |1\rangle)/\sqrt{2}$ along the computational basis ideally form random number sequences. This means that the resulting sequences are expected to pass the statistical tests for RNGs explained previously. Here, the computational basis, $|0\rangle$ and $|1\rangle$, spans the two-dimensional Hilbert space. In a quantum computer, the desired state $(|0\rangle + |1\rangle)/\sqrt{2}$ is generated from the initial state $|0\rangle$ by applying the Hadamard gate to a single quantum bit (qubit). Note that in this process, the initial state is always the same. Unlike classical random number generators and pseudorandom number generators that require random seeds to produce independent sequences, quantum random number generators are capable of producing independent sequences with the same seed. This reduces the risk of the output of a random number generator being predicted from the seed, because all possible outputs come from the same seed.

In the present study, the cloud superconducting quantum computer, IBM 20Q Poughkeepsie, was used. The device was given the circuit in Fig. 1a and was repeatedly instructed to execute the circuit 8192 times without interruption from 2019/05/09 11:24:27 GMT. Because the quantum computer has multiple users across the globe, interruption between jobs occur (Aleksandrowicz et al. 2019). 8192 is the maximum number of uninterrupted executions (shots) available. Running the circuit with 8192 shots yields a binary sequence with a length of 8192 per qubit. This process was

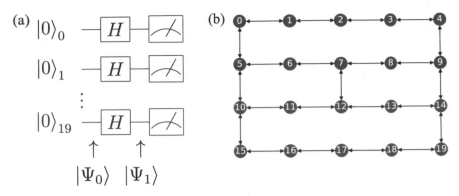

Fig. 1 **a**: QRNG quantum circuit using the Hadamard gate. **b**: Device topology of IBM 20Q Poughkeepsie provided by Qiskit

Table 6 The correspondence between calibration start/end time and time of job sent. All dates and times are in GMT

	Start time (GMT)	End time (GMT)
1	2019/05/08 23:34:19	2019/05/09 05:10:24
2	2019/05/09 21:58:54	2019/05/10 06:23:42
3	2019/05/10 23:07:22	2019/05/11 02:48:12
4	2019/05/11 20:59:21	2019/05/11 23:33:42
5	2019/05/12 20:50:41	2019/05/12 23:24:58

automatically repeated across calibrations. The device goes through calibration once in a day as seen in Table 6.

As a result, 579 samples were obtained from the IBM 20Q Poughkeepsie device. Note that each qubit produced 579 samples, each with a length of 8192. The samples were subjected to the eight tests from the NIST SP 800-22, which are: the frequency test, frequency within a block test, runs test, longest runs within a block test, DFT test, approximate entropy test, and the cumulative sums test (forward, backward). The p-value of each test corresponding to the respective samples was computed. For each test, the proportion of passed samples was checked. The acceptable range of the proportion of passed samples for 579 samples under the level of significance $\alpha = 0.01$ is >0.977595.

By constantly running the IBM 20Q Poughkeepsie device for five days, we obtained 579 samples for each of the 20 qubits. In theory, these samples should qualify as the output of an ideal random number generator. In random number generation, the output sequences are checked for two properties: bias and patterns. When the sequences show signs of bias or patterns, the device is not in ideal condition. The same logic applies to the cloud quantum computer. We also simulated the same quantum circuit on the simulator with the obtained noise parameters such as the T1 and T2 time, the coherent error, the single-qubit error, and the readout error, all of

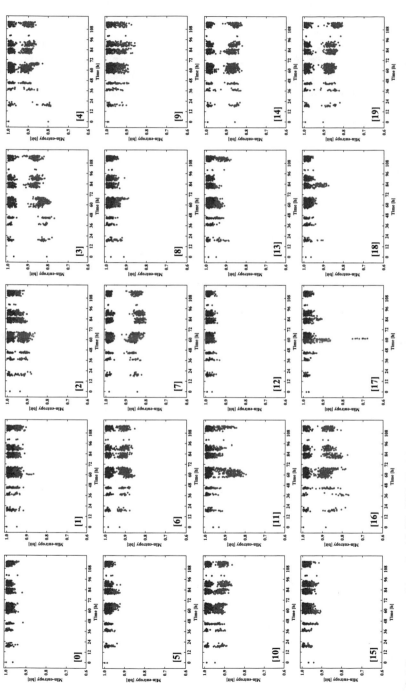

Fig. 2 Min-entropy transition of qubit [0]~[19]. The blue plots are the experimental results and the red plots the noisy simulation results. The figure has been rotated 90°. The horizontal axis ranges from 2019/05/09 11:24 GMT to 2019/05/14 07:54 GMT

which are updated. The simulator is referred to as the noisy simulator in the following. The noisy simulator program was also provided by IBM (Aleksandrowicz et al. 2019).

In the present section, the random number output of each qubit inside the IBM 20Q Poughkeepsie device is analyzed. The qubits that are connected by arrows in Fig. 1b represent the pairs of qubits on which the controlled NOT gate can operate. The controlled NOT gate is a two-qubit gate.

The min-entropy, whose definition and properties are seen in the Appendix, was computed for each qubit from the 579 samples. This resulted in 579 min-entropy transition plots for 20 qubits. Figure 2 is organized to form the topology of the IBM 20Q Poughkeepsie. The min-entropy takes values from 0 to 1 depending on the highest probability of the probability distribution. When the probability distribution is uniform, the min-entropy is 1. Figure 2 shows how each qubit has a unique tendency for min-entropy. Qubit [17], for example, shows a sudden drop in min-entropy at around 60 h. This does not occur in simulation. A sudden drop in min-entropy suggests that the measurement results can vary depending on when the cloud quantum computer executes a circuit. Overall, the noisy simulator tends to have a higher min-entropy compared to the actual device. According to Aleksandrowicz et al. (2019), the readout error that IBM provides does not reflect the asymmetry between the error output 1 on the state $|0\rangle$ and the error output 0 on the state $|1\rangle$. The discrepancy between the min-entropy of the actual device and the simulator suggests that readout asymmetry exists.

Next, the samples were checked for bias. Each qubit produced 579 samples with a length of 8192, which form 4,743,168-bit sequences when chronologically connected. Figure 2 demonstrates the proportion of 1s in the entire sequence output by each qubit. Under the level of significance $\alpha = 0.01$, the proportion of 1s of a 4,743,168-bit sequence should fall between the red lines. The result is that none of the qubits produced acceptable proportions of 1s as seen in Fig. 3. Furthermore, Fig. 4 shows that the actual device failed to pass the eight statistical tests, which indicates that the current quantum computing device does not have the statistical properties of a uniform random number generator.

The problem with histograms as seen in Fig. 3 is that they fail to detect certain anomalies. For example, a sequence consisting of all 0s for the former half and all 1s for the latter half yields a perfect histogram. However, such a sequence is clearly not random. To compensate for this flaw, we focused on the transition of the number of 1s in the sequence. Ideally, the number of 1s in a random number sequence should always be roughly half of the sequence length. The difference between the ideal number of 1s and the observed number of 1s for the 4,743,168-bit sequence of each qubit is examined in Fig. 5. Note that here, too, the figures are aligned topologically. Figure 5 shows the stability of each qubit in terms of the proportion of 1s in its output; a linear plot suggests that the qubit is being stably operated. While qubit[7] is more biased than qubit[17] overall, the line representing qubit[7] shows more stability than that of qubit[17]. Furthermore, the noisy simulator does not capture the trend of the qubits. Therefore, the discrepancy between the output of the actual device and the

Fig. 3 The proportion of 1s of qubit[0]~[19]. The acceptable range under the level of significance $\alpha = 0.01$ is between the two dotted lines. The blue bars are the experimental results and the red plots the noisy simulation results

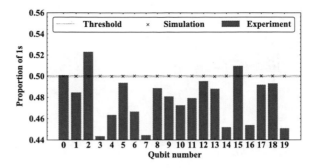

noisy simulator may not only be a result of readout asymmetry, but also time-varying parameters.

5   Conclusion

We characterized the qubits in a cloud quantum computer by using statistical tests for random number generators to provide a potential indicator of the device's condition. The IBM 20Q Poughkeepsie device was repeatedly run for a period of five days, and 579 samples with a length of 8192 were obtained for each of the 20 qubits. For comparison, the noise parameters obtained in the experiment were used to run the noisy simulator. Samples from both the actual device and the simulator were statistically analyzed for bias and patterns. To evaluate the uniformity of each sample, the min-entropy was computed. The transition of min-entropy showed that the qubits have unique characteristics. We identified a sudden drop of min-entropy in qubit [17]. The histogram of the proportion of 1s in the 4,743,168-bit sequences produced by each qubit revealed that, overall, none of the qubits produced acceptable proportions of 1s. However, we evaluated each qubit's stability from the time-series data of the proportion of 1s and found that qubits [0] and [12] were relatively stable. Finally, eight tests from the NIST SP 800-22 were applied to the 529 samples of the 20 qubits. None of the qubits cleared the standards of the test suite. However, the test results showed that qubits [0] and [12] were the closest to the ideal in terms of the proportion of passed samples for each test.

As is the case with random number generators, a cloud quantum computer is a black box to its users. Therefore, users are required to decide for themselves when to use a cloud quantum computer and which qubits to choose. Statistical tests for random number generators are a potential candidate for a simple indicator of qubit condition and stability inside a cloud quantum computer (Shikano et al. 2020).

Acknowledgements The authors thank Hidetoshi Okutomi, Atsushi Iwasaki, Shumpei Uno, and Rudy Raymond for valuable discussions. This work is partially supported by JSPS KAKENHI (Grant Nos. 17K05082 and 19H05156) and 2019 IMI Joint Use Research Program Short-term Joint

K. Tamura and Y. Shikano

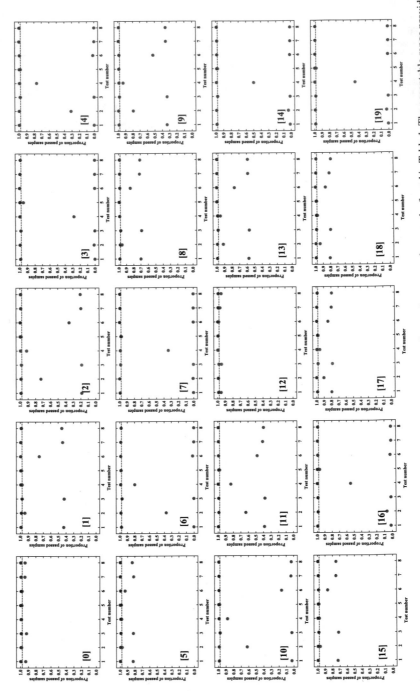

Fig. 4 The proportion of passed samples for each test. The test names corresponding to the test numbers can be found in Table 1. The acceptable range provided by the NIST is above the red line marking the proportion 0.977595. The blue plots are the experimental results and the red plots the noisy simulation results. The figure has been rotated 90°

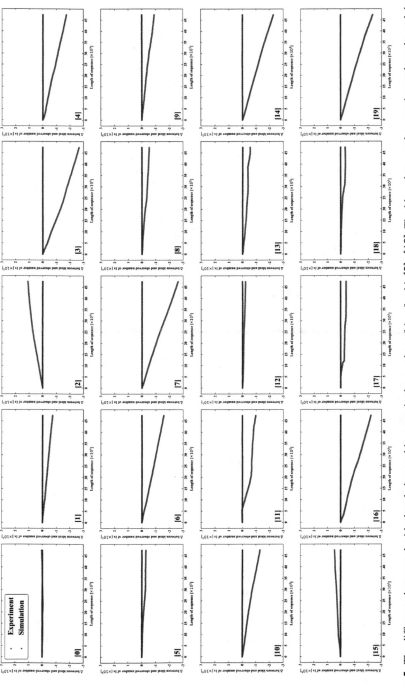

Fig. 5 The difference between the ideal and observed increase in the number of 1s of qubit [0]~[19]. The blue plots are the experimental results and the red plots the noisy simulation results. The figure has been rotated 90°

Fig. 6 Relation between
Shannon's entropy and
min-entropy

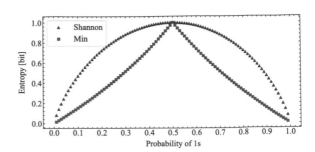

Research "Mathematics for quantum walks as quantum simulators". The results presented in this paper were obtained in part using an IBM Q quantum computing system as part of the IBM Q Network. The views expressed are those of the authors and do not reflect the official policy or position of IBM or the IBM Q team.

Appendix: Min-entropy

Among various entropy measures for uniformity, the min-entropy is often used in the context of cryptography. The min-entropy for a random variable X is defined as follows:

$$H_{\infty}(X) = -\log_2\left(\max_{x \in \{0,1\}} Pr[X = x]\right). \tag{10}$$

On the other hand, Shannon's entropy, which is also a measure for uniformity, is defined as follows:

$$H_{sh}(X) = -\sum_{x \in \{0,1\}} Pr[X = x]\log_2 Pr[X = x]. \tag{11}$$

Both measures (10) and (11) take values ranging from 0 to 1 for a random variable on $\{0, 1\}$. The reason why the min-entropy is more appropriate in the context of cryptography is that it is more sensitive than Shannon's entropy. This is apparent from Fig. 6. Figure 6 compares the min-entropy and Shannon's entropy corresponding to the probability of X yielding 1. The min-entropy provides a clearer distinction of probability distributions close to uniform than Shannon's entropy.

The min-entropy also indicates the probability that an adversary with knowledge of the probability distribution of X predicts the outcome of X correctly (Zhang et al. 2016). Here, the adversary predicts the value that appears with the highest probability. For this reason, the min-entropy considers the maximum probability of X.

References

G. Aleksandrowicz et al., version: 0.10.1 (2019). https://doi.org/10.5281/zenodo.2562110

L.E. Bassham et al., NIST Special Publication 800-22rev1a (2010)

D. Boneh, Notices Amer. Math. Soc. **46**, 203 (1999)

D. Castelvecchi, Nature **543**, 159 (2017)

H. Demirhan, N. Bitirim, J. Statisticians, Stat. Actuar. Sci. **9**, 1 (2016)

L. Fan, H. Chen, S. Gao, in *Information Security Applications, WISA 2013*, Lecture Notes in Computer Science, vol. 8267 ed. by Y. Kim, H. Lee, A. Perrig (Springer, Cham, 2014), p. 52

E. Gibney, Nature **541**, 447 (2017)

K. Hamano, IEICE Trans. Fundam. **88**, 67 (2005)

H. Haramoto, M. Matsumoto, Math. Comput. Simul. **161**, 66 (2019)

M. Herrero-Collantes, J.C. Garcia-Escartin, Rev. Mod. Phys. **89**, 015004 (2017)

S.-J. Kim, K. Umeno, A. Hasegawa (2020), arXiv:nlin/0401040

R. LaRose, Quantum **3**, 130 (2019)

P. L'ecuyer, R. Simard, ACM Trans. Math. Softw. (TOMS) **33**, 22 (2007)

X. Ma, X. Yuan, Z. Cao, B. Qi, Z. Zhang, npj Quantum Inf. **2**, 16021 (2016)

S. Marek, R. Zdeněk, M. Vashek, M. Kinga, S. Alin, Rom. J. Inf. Sci. Tech. **18**, 1 (2015)

National Academies of Sciences, *Engineering, and Medicine, Quantum Computing: Progress and Prospects* (The National Academies Press, Washington DC, 2019)

S. Pironio et al., Nature **464**, 1021 (2010)

J. Preskill, Quantum **2**, 79 (2018)

W. Schindler, W. Killmann, in *Cryptographic Hardware and Embedded Systems - CHES 2002*, vol. 2523, Lecture Notes in Computer Science, ed. by B.S. Kaliski, K. Koç, C. Paar (Springer, Berlin, 2003), p. 431

R. Srivastava, I. Choi, T. Cook, *The Commercial Prospects for Quantum Computing* (Networked Quantum Information Technologies, 2016)

K. Tamura, Y. Shikano, in *Proceedings of Workshop on Quantum Computing and Quantum Information*, ed. by M. Hirvensalo, A. Yakaryilmaz, TUCS Lecture Notes, vol. 30 (2019), http://urn.fi/URN:ISBN:978-952-12-3840-6

K. Ugajin et al., Opt. Exp. **25**, 6511 (2017)

T. Xin et al., Sci. Bull. **63**, 17 (2018)

Y. Yamamoto, M. Sasaki, H. Takesue, Quant. Sci. Tech. **4**, 020502 (2019)

X. Zhang, Y. Nie, H. Liang, J. Zhang, in IEEE-NPSS Real Time Conference (RT). Padua **1**, (2016)

Y. Shikano, K. Tamura, R. Raymond, EPTCS **315**, 18–25 (2020)

Quantum Factoring Algorithm: Resource Estimation and Survey of Experiments

Noboru Kunihiro

Abstract It is known that Shor's algorithm can break many cryptosystems such as RSA encryption, provided that large-scale quantum computers are realized. Thus far, several experiments for the factorization of the small composites such as 15 and 21 have been conducted using small-scale quantum computers. In this study, we investigate the details of quantum circuits used in several factoring experiments. We then indicate that some of the circuits have been constructed under the condition that the order of an element modulo a target composite is known in advance. Because the order must be unknown in the experiments, they are inappropriate for designing the quantum circuit of Shor's factoring algorithm. We also indicate that the circuits used in the other experiments are constructed by relying considerably on the target composite number to be factorized.

Keywords RSA · Quantum computer · Shor's quantum factoring algorithm · Oversimplified Shor's algorithm · Physical experiment

1 Introduction

It is crucial to evaluate the security of cryptosystems in order to securely use cryptographic technology. The security of RSA cryptosystems (Rivest et al. 1977), which are currently used widely, is based on the difficulty of factoring problem, and the evaluating the difficulty of the factoring problem is essential. Based on the security analysis, a 2048-bit composite number is widely used as a standard at present. It is known that prime factorization is possible in quantum polynomial time on the

N. Kunihiro (✉)
Faculty of Engineering, Information and Systems, University of Tsukuba, 1-1-1 Tennodai, Tsukuba, Ibaraki 305-8573, Japan
e-mail: kunihiro@cs.tsukuba.ac.jp

© The Author(s) 2021
T. Takagi et al. (eds.), *International Symposium on Mathematics,*
Quantum Theory, and Cryptography, Mathematics for Industry 33,
https://doi.org/10.1007/978-981-15-5191-8_7

39

bit length of the composite number using the Shor's algorithm (Shor 1997). Hence, almost all the currently used public-key cryptosystems will be broken if large-scale quantum computers are realized. Therefore, to prepare for the realization of quantum computers, quantum-resistant cryptography is researched actively at present (NIST 2020).

From the theoretical viewpoint, it has been evaluated how much resources are needed for the prime factorization of composite number of the currently used sizes (1024-bit, 2048-bit) (Häner 2017; Kunihiro 2005). However, from the experimental viewpoint, several experiments have been performed for the prime factorization of small composite numbers such as 15 and 21 (Lucero et al. 2012; Martin-Lopez et al. 2012; Monz et al. 2016; Politi 2009; Vandersypen 2001). In addition, commercial services for small-scale quantum computers such as IBM Q (2020) are beginning to be launched, and it is expected that the Noisy Intermediate-Scale Quantum (NISQ) technology might be available in the near future (Preskill 2018).

This paper presents a detailed survey of actual quantum experiments for prime factorization based on Shor's algorithm (Lucero et al. 2012; Martin-Lopez et al. 2012; Monz et al. 2016; Politi 2009; Vandersypen 2001). We give a detailed explanation of the circuits used in the experiments. We also indicate that some of them are problematic because they use a secret information in the circuit construction.

2 Outline of Shor's Quantum Factoring Algorithm (Shor 1997)

2.1 Quantum Computation

This subsection provides the basic facts about quantum gates (Nielsen and Chuang 2000). For the other information about quantum gates and circuits, refer to Nielsen and Chuang (2000).

We first explain a quantum bit, or *qubit*. A qubit has two possible states $|0\rangle$ and $|1\rangle$. We represent a single-qubit state as $\alpha |0\rangle + \beta |1\rangle$, where $\alpha, \beta \in \mathbb{C}$ and $|\alpha|^2 + |\beta|^2 = 1$. The gate that maps this state into $\alpha |1\rangle + \beta |0\rangle$ is called the NOT gate. The following matrix form is convenient for representing the NOT gate. Let a matrix X be

$$X = \begin{pmatrix} 0 & 1 \\ 1 & 0 \end{pmatrix}.$$

Suppose that the quantum state $\alpha |0\rangle + \beta |1\rangle$ is written in the vector form as

$$\begin{pmatrix} \alpha \\ \beta \end{pmatrix},$$

where the first entry corresponds to the amplitude for $|0\rangle$ and the second entry to the amplitude for $|1\rangle$. The corresponding output from the NOT gate is given by

$$X \begin{pmatrix} \alpha \\ \beta \end{pmatrix} = \begin{pmatrix} \beta \\ \alpha \end{pmatrix}.$$

The quantum gates on a single qubit can be described, in general, using 2×2 matrices. Furthermore, the matrix must be unitary. In fact, $X^\dagger X = I$ should hold, where $X^\dagger$ denotes the adjoint of X and I an identity matrix.

We then show the other important single-qubit gates, namely, the Z and H gates, in addition to the NOT gate. The matrix forms for the Z and H gates are given as follows.

$$Z = \begin{pmatrix} 1 & 0 \\ 0 & -1 \end{pmatrix}, \quad H = \frac{1}{\sqrt{2}} \begin{pmatrix} 1 & 1 \\ 1 & -1 \end{pmatrix}$$

The H gate is usually referred to as the Hadamard gate. The Hadamard gate turns the state $|0\rangle$ into $(|0\rangle + |1\rangle)/\sqrt{2}$ and the state $|1\rangle$ into $(|0\rangle - |1\rangle)/\sqrt{2}$ because

$$H \begin{pmatrix} 1 \\ 0 \end{pmatrix} = \begin{pmatrix} 1/\sqrt{2} \\ 1/\sqrt{2} \end{pmatrix} \quad \text{and} \quad H \begin{pmatrix} 0 \\ 1 \end{pmatrix} = \begin{pmatrix} 1/\sqrt{2} \\ -1/\sqrt{2} \end{pmatrix}.$$

Furthermore, employing the Hadamard gate, we can construct the flat superposition from the state $|0\rangle$.

We now discuss multiple-qubit gates. The first gate is the Controlled-NOT (C-NOT) gate, which has two input qubits. The action of the C-NOT gate can be described as

$$|0\rangle\,|0\rangle \rightarrow |0\rangle\,|0\rangle, \ |0\rangle\,|1\rangle \rightarrow |0\rangle\,|1\rangle, |1\rangle\,|0\rangle \rightarrow |1\rangle\,|1\rangle, \text{ and } |1\rangle\,|1\rangle \rightarrow |1\rangle\,|0\rangle.$$

Equivalently, we can describe the action as

$$|a\rangle|b\rangle \rightarrow |a\rangle|b \oplus a\rangle,$$

where $\oplus$ denotes the exclusive OR.

The second one is the Toffoli gate, which has three input qubits. The action of the Toffoli gate can be described as

$$|a\rangle|b\rangle|c\rangle \rightarrow |a\rangle|b\rangle|c \oplus (a \wedge b)\rangle,$$

where $\wedge$ denotes the logical operator AND. The first two qubits are the *control* qubits and the third one is the *target* qubit.

We can consider the *generalized* version of the Toffoli gate as follows.

$$|c_1\rangle|c_2\rangle \cdots |c_n\rangle|t\rangle \rightarrow |c_1\rangle|c_2\rangle \cdots |c_n\rangle|t \oplus (c_1 \wedge c_2 \wedge \cdots \wedge c_n)\rangle.$$

In this case, the first n qubits are the *control* qubits, and the last qubit is the target qubit. It is well known that the generalized Toffoli gate can be decomposed into several Toffoli gates (Nielsen and Chuang 2000).

We then explain the controlled circuit. We denote a unitary operation by U. The action of the control-U circuit (C-U circuit) is described as

$$|0\rangle |x\rangle \ \rightarrow \ |0\rangle |x\rangle, \quad |1\rangle |x\rangle \rightarrow \ |1\rangle U|x\rangle.$$

Or, equivalently, the action can be described as

$$|c\rangle |x\rangle \rightarrow |c\rangle U^c |x\rangle.$$

We explain the Quantum Fourier Transformation (QFT). The QFT on a basis $|0\rangle, |1\rangle, \ldots, |N-1\rangle$ is defined to be a linear operation with the following action on the states:

$$|j\rangle \rightarrow \frac{1}{\sqrt{N}} \sum_{k=0}^{N-1} \exp\left(\frac{2\pi i j k}{N}\right) |k\rangle.$$

The circuit for the QFT is constructed with the Hadamard gates and the controlled rotation gates. For the details, see the Sect. 5 in Nielsen and Chuang (2000). The inverse QFT is defined to be the *inverse* operation of QFT.

2.2 Shor's Quantum Factoring Algorithm

Let N denote a target composite to be factored, and n denote a bit length of N. To simplify the discussion, hereafter, we assume that p are q are distinct prime integers and that N is the product of p and q. Let a denote a positive integer coprime to N. The final goal of Shor's algorithm is to find the prime factors p and q. However, before doing so, the algorithm will find a positive integer r such that a^r mod $N = 1$ as a subgoal. This positive integer r is called an order. If we know the order r, we can easily find the prime factors p and q of N with high probability.

We will now explain Shor's factoring algorithm in detail. Letting $m = 2n$, we first prepare the initialized state as follows:

$$\underbrace{|0\rangle}_{m\text{-qubit}} \underbrace{|1\rangle}_{n\text{-qubit}},$$

where the first register (referred to as the control register in Martin-Lopez et al. 2012 or the period register in Monz et al. 2016) is of m qubits and the second register (referred to as the work register in Martin-Lopez et al. 2012 or the computational register in Monz et al. 2016) is of n qubits. We may use ancilla in the calculation if required. Applying the Hadamard gate to the first register, we obtain the flat super-

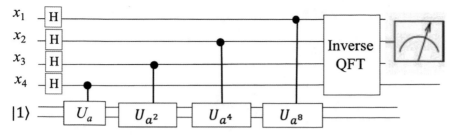

Fig. 1 Shor's quantum factoring algorithm for the case of $m = 4$

position as follows:

$$\frac{1}{2^{m/2}} \sum_{x=0}^{2^m-1} \underbrace{|x\rangle}_{m\text{-qubit}} \underbrace{|1\rangle}_{n\text{-qubit}} .$$

Subsequently, we apply the modular exponentiation to this superposition to obtain the following state:

$$\frac{1}{2^{m/2}} \sum_{x=0}^{2^m-1} \underbrace{|x\rangle}_{m\text{-qubit}} \underbrace{|a^x \bmod N\rangle}_{n\text{-qubit}} .$$

We then apply the inverse of the Quantum Fourier Transformation to this state. At the last step, we obtain some value by measuring the first register. Using the measured value, we calculate the order r with the help of the continued fraction algorithm and then we find the prime factors of N by classical computers.

Here, the modular exponentiation is operated by sequentially applying C–U_a, C–U_{a^2}, C–U_{a^4}, C–$U_{a^{2^j}}$, and C–$U_{a^{2^{m-1}}}$ circuits, as shown in Fig. 1. Note that the action of the U_b operator is described as $|x\rangle \rightarrow |bx \bmod N\rangle$.

Suppose that we can find the order r of a modulo N. For simplicity, let us assume r to be even. By computing $\gcd(a^{r/2} - 1 \bmod N, N)$, we can find the prime factors of N with high probability.

Hereafter, we do not discuss the part of the Hadamard transformation and the part of the inverse of Quantum Fourier Transformation because the circuit complexity of both these parts can be ignored compared with that of the modular exponentiation part. Hereafter, we focus on the discussion of the resources necessary for modular exponentiation.

Table 1 Number of qubits and elementary gates (Kunihiro 2005)

	The number of qubits	The number of gates
R-ADD	$3n + 2$	$270n^3 + O(n^2)$
GT-ADD	$2n + 4 \rightarrow 2n + 2$ (Häner 2017)	$\frac{16}{3}n^5 + O(n^4)$
Q-ADD	$2n + 3 \rightarrow 2n + 2$ (Takahashi and Kunihiro 2006)	$97n^4 + O(n^4)$

2.3 Circuit Construction and Resource Estimation for Shor's Quantum Factoring Algorithm

The modular exponentiation can be executed by performing $O(n^3)$ gate operations for the standard construction of circuit. Kunihiro gave three construction types for modular exponentiation (Kunihiro 2005). These constructions adopt different types of addition circuits. In Kunihiro (2005), the number of qubits and the number of gates for Shor's factoring circuit were evaluated precisely. It was also shown that $3n + 2$ qubits and $270n^3 + O(n^2)$ Toffoli gates are required for modular exponentiation if the addition circuit similar to the classical addition is adapted. This result implies that we require 6146 qubits and 3.04×10^{12} Toffoli gates for factoring a 2048-bit composite. Table 1 presents the resource estimation of n-bit composite for quantum factoring. Table 2 shows those of 768-bit composite and 2048-bit composite. Note that the current world record for factoring is 768-bit composite (Kleinjung 2010) and the current recommendation of RSA composite is with 2048-bit.

In addition to the classical addition-based circuits (referred to as R-ADD in Table 1), (Kunihiro 2005) also gave a resource estimation, which was derived from both the circuits based on the Generalized Toffoli gate and circuits based on the Quantum Addition (referred to as GT-ADD and Q-ADD in Table 1, respectively). The circuits based on the Generalized Toffoli gate require $2n + 4$ qubits and $\frac{16}{3}n^5$ Toffoli gate and those based on the Quantum Addition requires $2n + 3$ qubits and $20n^4$ C–NOT gates and $37n^4$ single-qubit gates. Takahashi and Kunihiro proposed the circuit construction that works even for $2n + 2$ qubits for the necessary qubits (Takahashi and Kunihiro 2006). Häner et al. also presented a similar result (Häner 2017).

The resource estimation for solving the elliptic curve discrete logarithm problem was presented in Roetteler et al. (2017), and further improvement is provided in Kurama and Kunihiro (2019).

2.4 Survey of Quantum Experiments for Factoring

In 2001, a research group of IBM performed an experiment for factoring 15 by implementing Shor's algorithm by using Nuclear Magnetic Resonance (NMR) (Van-

Table 2 Number of qubits and elementary gates for 768 and 2048 bits (Kunihiro 2005)

	World record ($n = 768$)		Recommended ($n = 2048$)	
	Qubits	# of gates	Qubits	# of gates
R-ADD	2306	1.22×10^{11}	6146	3.04×10^{12}
GT-ADD	1540		4100	
Q-ADD	1539	8.68×10^{11}	4099	1.22×10^{13}

Table 3 Summary of quantum experiments for factoring

Device	Research group	Year	Target	# of qubits	Embedding of order information
NMR	IBM (Vandersypen 2001)	2001	15	6	✓
Photonic chip	U. of Bristol (Politi 2009)	2009	15	4	✗(used)
Superconductivity	UCSB (Lucero et al. 2012)	2012	15	3	✓
Ion trap	U. Innsbruck (Monz et al. 2016)	2016	15	6	✓
Photon	U. of Bristol (Martin-Lopez et al. 2012)	2012	21	$1 + \log_2 3$	✗(used)

dersypen 2001). Since the group's pioneering work, several experiments based on Shor's algorithm have been conducted. Table 3 summarizes five of these experiments, of which four experiments dealt with the factorization of 15, and the fifth one with the factorization of 21.

Because the bit length of composite 15 is 4, it requires at least 14 qubits with standard construction based on the usual addition (R-ADD) and 10 qubits with the construction based on Takahashi and Kunihiro (2006) to factorize 15. As can be seen, all of the experiments employed fewer qubits than those in the above-mentioned construction for general composites. We can say that the circuits for factoring are customized to factor the target composites such as 15 and 21, and are not based on the general construction. In Sect. 3, we describe the detailed circuits without using the order information based on Lucero et al. (2012), Monz et al. (2016), and Vandersypen (2001). Though their circuits do not use any secret information, they are applicable to specific composite such as $2^n - 1$ for an even integer n, which are never used for RSA composite. In Sect. 4, we describe the detailed circuits by using the order information based on Martin-Lopez et al. (2012) and Politi (2009). These circuit constructions are inappropriate since the order information must be secret.

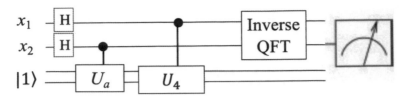

Fig. 2 Shor's factoring algorithm for $N = 15$

3   Quantum Circuits Without Using the Order Information

Before describing the details of each quantum circuits for factoring 15, we explain a common strategy for factoring 15. The positive integers relatively coprime to 15 are given by 2, 4, 7, 8, 11, 13, and 14. Their order modulo 15 are given by 4, 2, 4, 4, 2, 4, and 2, respectively. Clearly, the elements with order 4 are 2, 7, 8, and 13. In many cases, we consider using them as a. Note that $a^2 \bmod 15 = 4$ for $a = 2, 7, 8$, and 13.

For the element a with the order 4, $a^{2^k} \bmod 15$ is always 1 for integers $k \geq 2$. Hence, $U_{a^{2^k}}$ for $k \geq 2$ becomes an identity operation and they can be ignored in the calculation. On the basis of the above-mentioned observation, it is sufficient to implement C–U_a and C–$U_{a^2 \bmod 15}$ circuits for the modular exponentiation. Here, $a^2 \bmod 15 = 4$ and the necessary operation can be simplified into C–U_a and C–U_4. Hence, while constructing the quantum circuits, it is sufficient to consider a multiplication circuit by employing a as $a = 2, 4, 7, 8$, and 13. From the above-mentioned discussion, the general form for factoring $N = 15$ is given by Fig. 2 under the condition that the element of order 4 element is used.

3.1   Quantum Factoring Experiment Shown in Vandersypen (2001)

The literature (Vandersypen 2001) shows an experiment of factoring $N = 15$ using NMR. The experiment uses $a = 7$ as a chosen element. The order of 7 modulo 15 is given by 4.

As described previously, it is sufficient to construct multiplication circuits with 7 and 4. The multiplication circuit with 4 will be constructed by using the following strategy. Here, we denote a 4-bit nonnegative integer by $(y_3 y_2 y_1 y_0)_2$. By multiplying it with 4, we have $(y_3 y_2 y_1 y_0 00)_2$. By calculating the residue by 15, we have $(y_1 y_0 y_3 y_2)_2$. In summary, the multiplication of $(y_3 y_2 y_1 y_0)_2$ by 4 modulo 15 is given by $(y_1 y_0 y_3 y_2)_2$. It is sufficient to construct a circuit transferring $|y_3 y_2 y_1 y_0\rangle$ into $|y_1 y_0 y_3 y_2\rangle$ instead of directly implementing the multiplication circuit. From the above-mentioned discussion, it is sufficient to swap the first and the third qubits and swap the second and the fourth qubits for multiplication with 4 and taking modulo

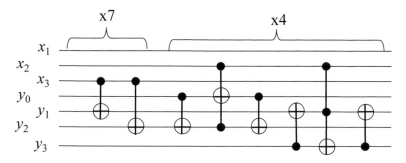

Fig. 3 Quantum Circuit for Factoring 15 in Vandersypen (2001)

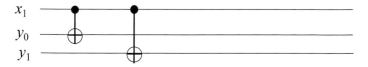

Fig. 4 Experiment for $a = 4$ and $N = 15$ in Lucero et al. (2012)

15. The swap operation can be executed without using ancilla qubits. Furthermore, the controlled–SWAP can be divided into one Toffoli gate and two C–NOT gates.

Subsequently, we explain the multiplication circuit with 7. Their shown circuit does not directly implement the multiplication with 7. We can easily verify that it is sufficient that $|0\rangle |1\rangle$ is mapped to $|0\rangle |1\rangle$ and $|1\rangle |1\rangle$ is mapped to $|1\rangle |7\rangle$ for multiplication with 7 in this situation. This operation can be executed via controlled-addition with 6. In this experiment, the controlled-addition with 6 is implemented by using two controlled-NOT gates.

On the basis of the above-mentioned idea, the authors of Vandersypen (2001) implemented the circuit as depicted in Fig. 3. Note that no ancilla qubit was used in applying U_a and U_4, and consequently only six qubits were involved in the implementation.

3.2 Quantum Factoring Experiment Shown in Lucero et al. (2012)

This experiment involves the factorization of 15 and uses $a = 4$ as the chosen element. Note that the order of 4 is 2. Hence, it is sufficient to implement U_4 for the experiment. In the circuit shown in Lucero et al. (2012), the circuit for multiplication with 4 is not implemented directly. It is sufficient to implement the circuit that transforms $|0\rangle |1\rangle \rightarrow |0\rangle |1\rangle$ and $|1\rangle |1\rangle \rightarrow |1\rangle |4\rangle$. This operation can be executed via controlled-addition with 3. In this experiment, the controlled-addition with 3 is implemented by using two C-NOT gates. Summing up the above discussion, the authors in Lucero et al. (2012) presented the circuit depicted in Fig. 4.

Note that no ancilla qubit was used in applying U_4 and consequently only three qubits were involved in the implementation.

3.3 Quantum Factoring Experiment Shown in Monz et al. (2016)

The authors presented the circuits not only for $a = 7$ but also for several other a's in the experiments. Concretely, the authors showed the circuit for $a = 2, 7, 8, 11$, and 13, and $a^2 \mod 15 = 4$ for these a's. Hence, it is sufficient to construct the U_a circuit and U_4 circuits. As shown in Sect. 3.1, the U_4 circuit can be constructed using SWAP. In Monz et al. (2016), the authors showed that the multiplication circuit U_a can also be constructed using SWAP and NOT gate.

We first present the multiplication circuit for $a = 2$. We denote the binary representation of a by $(a_3a_2a_1a_0)_2$ as previously. The double of a modulo 15 is given by $(a_2a_1a_0a_3)_2$ in the binary representation. The state $|a_2a_1a_0a_3\rangle$ can be obtained from $|a_3a_2a_1a_0\rangle$ using the following three sequential SWAP operations: SWAP between the first and second qubits, SWAP between the second and third qubits, and then SWAP between the third and fourth qubits. We can verify its correctness by following transition: $|a_3a_2a_1a_0\rangle \rightarrow |a_2a_3a_1a_0\rangle \rightarrow |a_2a_0a_3a_0\rangle \rightarrow |a_2a_0a_0a_3\rangle$.

We then consider the multiplication circuit for $a = 8$. The multiplication of a with 8 is given by $(a_0a_3a_2a_1)_2$ in the binary representation. The state $|a_0a_3a_2a_1\rangle$ can be obtained from $|a_3a_2a_1a_0\rangle$ using the following three sequential SWAP operations: SWAP between the third and fourth qubits, SWAP between the second and third qubits, and then SWAP between the first and second qubits.

We, thus, know that we can implement the multiplication with 2, 4, and 8 by using only the SWAP circuit.

We then implement the multiplication with $a = 7, 11$, and 13; the values of $15 - a$ for them are given by $a = 8, 4$, and 2, respectively. To construct the multiplication circuits with 7, 11, and 13, we will use the above-mentioned property. For the multiplication with $a = 13$, we first apply the multiplication with 2, and we then apply the NOT gate for all of the four qubits. Figure 5 depicts the concrete multiplication circuit with them. We can also obtain the multiplication circuits for $a = 7, 11$ in a similar manner.

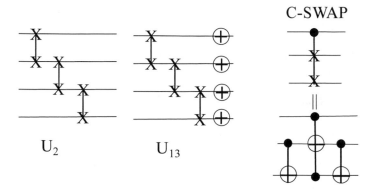

Fig. 5 Unitary operations U_2, U_{13} and the circuit for C–SWAP

4 Quantum Circuits with Explicitly Using the Order information

This section presents two experiments that explicitly use the order information. We want to emphasize that these experiments are inappropriate for employing in factoring algorithms because the purpose of Shor's algorithm is to find the order of a given element.

4.1 Quantum Factoring Experiment of $N = 15$ Shown in Politi (2009)

The authors of Politi (2009) conducted an experiment that factorized 15 with an element $a = 7$. The order of $a = 7$ is given by 4. Because the order is 4, the only four values, namely, 1, 7, 4, and 13 can appear in the second register, and the authors utilized this property. The authors represented these four values by using two bits. Concretely speaking, they adopted the following encoding: $1 \rightarrow 0(= 00)_2, 7 \rightarrow 1(= 01)_2, 4 \rightarrow 2(= 10)_2, 13 \rightarrow 3(= 11)_2$.

As described previously, it is sufficient to implement the multiplication circuits with 7 and 4. The multiplication with 7 corresponds to the addition with $+1$ under the encoding and the multiplication with 4 corresponds to addition with $+2$. These operations can be implemented using only one C–NOT gate. Summing up the above-mentioned discussion, the entire circuit is depicted in Fig. 6.

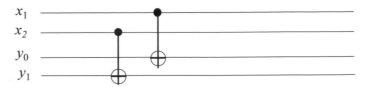

Fig. 6 Quantum circuit for $N = 15$ in Politi (2009)

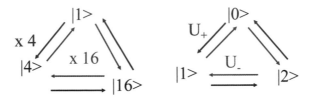

Fig. 7 Unitary operations U_+ and U_-

4.2 Quantum Factoring Experiment of $N = 21$ Shown in Martin-Lopez et al. (2012)

The target of this experiment is 21. In this experiment, a is set to $a = 4$. Because $a^3 \bmod 21 = 1$, the order of a modulo 21 is given by 3. Note that the purpose of Shor's algorithm is to obtain the order 3. The only three elements, namely, 1, 4, and 16 can appear in the second register.

It is sufficient to construct the quantum circuits $U_{4^{2^k} \bmod 21}$ for $k = 0, 1, 2, \ldots$ for the modular exponentiation. Note that $4^{2^k} \bmod 21 = 4$ for even k and $4^{2^k} \bmod 21 = 16$ for odd k. Then, it is sufficient to apply the unitary operation U_4 for even k and U_{16} for odd k.

In the experiment of Martin-Lopez et al. (2012), the following encoding is adapted as in the case of $N = 15$.

$$1 \to 0, \quad 4 \to 1, \quad 16 \to 2$$

We consider the multiplication with 4 and 16 under the aforementioned encoding. The multiplication with 4 is mapped into addition with $+1$, and the multiplication with 16 is mapped into addition with $+2$ or, equivalently, -1.

The experiment in Martin-Lopez et al. (2012) utilized a *qutrit*, which takes three quantum states instead of qubits, as the second register. We denote the unitary operations by

$$U_+ : |x\rangle \mapsto |x + 1 \bmod 3\rangle, \quad U_- : |x\rangle \mapsto |x - 1 \bmod 3\rangle.$$

The operations U_+ and U_- act on the quantum states as depicted in Fig. 7.

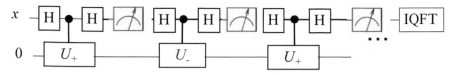

Fig. 8 Quantum circuit for $N = 21$ in Martin-Lopez et al. (2012)

Using the above-mentioned notation, Fig. 8 depicts the quantum circuit for factoring $N = 21$ described in Martin-Lopez et al. (2012). Here, in the circuit construction, the so-called qubit-recycling technique is employed to reduce the number of qubits. For the details of the qubit-recycling technique, refer to Martin-Lopez et al. (2012).

4.3 Oversimplified Shor's Algorithm (Smolin et al. 2013)

As described previously, the purpose of Shor's algorithm is to find the order of a given element. Hence, the circuit that explicitly utilizes the order information is inappropriate for (even the simplified version of) Shor's factoring algorithm. If we can use the order information, we can, in principle, factorize any large composite. We will explain the details of this fact by following the description provided in Smolin et al. (2013).

The modular exponentiation part in Shor's algorithm constructs the quantum superposition as follows:

$$\frac{1}{2^{m/2}} \sum_{x=0}^{2^m-1} |x\rangle |a^x \bmod N\rangle$$

from the flat superposition $\frac{1}{2^{m/2}} \sum_{x=0}^{2^m-1} |x\rangle |1\rangle$.

However, the circuits described in this section constructs the quantum superposition as follows:

$$\frac{1}{2^{m/2}} \sum_{x=0}^{2^m-1} |x\rangle |x \bmod r\rangle$$

from the flat superposition $\frac{1}{2^{m/2}} \sum_{x=0}^{2^m-1} |x\rangle |0\rangle$.

In this discussion, the following encoding is employed:

$$a^x \bmod N \mapsto x \bmod r.$$

This encoding includes the encodings described in Sects. 4.1 ($r = 4$) and 4.2 ($r = 3$) as a special case. This discussion is mathematically correct, but, it is inappropriate from the computational viewpoint because finding the order r is strongly believed to be infeasible in the classical polynomial time.

Fig. 9 Oversimplified
factoring algorithm

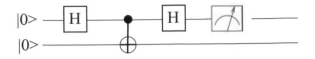

This circuit is constructed on the basis of the knowledge of the order r. Under this encoding, the operation $U_{a^{2j}}$ is transformed into the addition operation with $2^j \bmod r$. Assume that $r = 4$. The unitary operation $U_{a^{2j}}$ for $j = 0$ corresponds to the addition with 1; that for $j = 1$ corresponds to the addition with 2; that for $j \geq 2$ corresponds to an identity operation. Next, we assume that $r = 3$. The unitary operation $U_{a^{2j}}$ for even j corresponds to the addition with 1; that for odd j corresponds to the addition with 2 or, equivalently, -1. Note that all the addition is performed under the modulo 3.

To indicate that this kind of circuit that explicitly utilizes the order information is meaningless for the implementations of Shor's factoring algorithm, Smolin et al. (2013) presented the factoring circuits by using an element with order 2. Because the order r is 2, it is sufficient to construct the superposition as follows:

$$\frac{1}{\sqrt{2}} \sum_{x=0}^{1} |x\rangle |0\rangle \mapsto \frac{1}{\sqrt{2}} \sum_{x=0}^{1} |x\rangle |x\rangle = \frac{1}{\sqrt{2}} (|00\rangle + |11\rangle).$$

Figure 9 depicts the entire circuit described in Smolin et al. (2013).

We can find the element with order 2 for a large composite N using the following algorithm.

Input: $k \in \mathbb{Z}$
Output: a $2k$-bit composite N and an element a with order 2 modulo N
  Step1: Find two distinct k-bit primes p and q. Compute $N = pq$.
  Step2: Find a such that $a = +1 \bmod p$ and $a = -1 \bmod q$. Concretely, perform the following procedures to compute a.

Step2-1: Calculate $\bar{q} = q^{-1} \bmod p$.
Step2-2: Calculate $a = -1 + 2\bar{q}q$.

Furthermore, we provide a SageMath (2020) code for the above-mentioned algorithm with 2048-bit RSA.

```
1  k=1024
2  p=random_prime(2^k-1, false, 2^(k-1))
3  q=random_prime(2^k-1, false, 2^(k-1))
4  N=p*q
5  a= crt(1, -1, p, q)
```

We can easily verify that it holds that $a = +1 \bmod p$ and $a = -1 \bmod q$. Because $a^2 \equiv 1 \pmod{p}$ and $a^2 \equiv 1 \pmod{q}$, we have $a^2 \equiv 1 \pmod{N}$, and the order of a is a divisor of 2, implying that the order is 1 or 2. Because $a \not\equiv 1 \pmod{N}$, we

Table 4 Level of quantum experiments for factoring

Research Group	Year	Target	Level
IBM (Vandersypen 2001)	2001	15	Level 2
U. of Bristol (Politi 2009)	2009	15	Level 1
UCSB (Lucero et al. 2012)	2012	15	Level 2
U. Innsbruck (Monz et al. 2016)	2016	15	Level 2
U. of Bristol (Martin-Lopez et al. 2012)	2012	21	Level 1

can assert that the order of a is exactly 2. Furthermore, as $\gcd(a^{2/2} - 1, N) = p$, we can find a prime factor p of N.

In Smolin et al. (2013), the authors presented the prime factorization of a 20, 000-bit composite, showing that this kind of oversimplification is meaningless for the implementation of Shor's factoring algorithm.

5 Summary and Concluding Remarks

We reviewed the resource estimation of quantum factoring based on Shor's algorithm. We then presented a survey of the state-of-the-art circuit construction. We also indicated some of them as inappropriate for factoring circuits because the order information was embedded in the circuits (Sect. 4). The others considerably utilized the property of the target composite, and hence, they have no extensibility to the general composite (Sect. 3).

More experiments on factoring based on Shor's algorithm will be conducted using various devices. As we mentioned in this paper, we have to carefully analyze the circuit construction.

Based on the current status of quantum experiments for factoring, we introduce the following three levels of circuit construction for quantum factoring.

Level 1 Quantum factoring: The order information is embedded in the circuit. The experiment under Level 1 cannot be considered as a quantum experiment for factoring.

Level 2 Quantum factoring: The circuit relies considerably on the property of a target composite. The experiment under Level 2 can be considered as a quantum experiment for factoring, meaning that the compiled version of the circuits is acceptable. However, we cannot apply this circuit construction to the general composite, and hence, this circuit construction has no scalability.

Level 3 Quantum factoring: The circuit does not use any specific property of the target composite. The circuit under Level 3 is desirable.

Table 4 presents the levels for quantum factoring circuits shown in this paper. As can be seen, there is no experiment with Level 3.

Acknowledgements This research was partially supported by JST CREST Grant Number JPMJCR14D6, Japan and JSPS KAKENHI Grant Number JP16H02780. The authors thank Dr. Tetsuya Izu, who gave the information about quantum factoring circuits. They also thank Prof. Naoki Yamamoto and Prof. Yutaka Shikano for helpful discussions.

References

T. Häner, M. Roetteler, K.M. Svore, Factoring using $2n + 2$ qubits with Toffoli based modular multiplication. Quantum Inf. Comput. **17**(7&8), 673–684 (2017)

IBM Q, https://www.research.ibm.com/ibm-q/

T. Kleinjung, K. Aoki, J. Franke, A. Lenstra, E. Thome, J. Bos, P. Gaudry, A. Kruppa, P. Montgomery, D.A. Osvik, H. te Riele, A. Timofeev, P. Zimmermann, Factorization of a 768-bit RSA modulus in *Proceedings of CRYPTO2010*, LNCS 6223 (2010), pp. 333–350

N. Kunihiro, Exact analyses of computational time for factoring in quantum computers, in *IEICE Transactions on Fundamentals of Electronics, Communications and Computer Sciences*, vol. E88-A, No. 1 (2005), pp. 105–111

R. Kurama, N. Kunihiro, New quantum algorithms for modular inverse and its application on the elliptic curve discrete logarithm problem, in *The Poster Presentation of AQIS2019* (2019)

E. Lucero, R. Barends, Y. Chen, J. Kelly, M. Mariantoni, A. Megrant, P. O'Malley, D. Sank, A. Vainsencher, J. Wenner, T. White, Y. Yin, A.N. Cleland, J.M. Martinis, Computing prime factors with a Josephson phase qubit quantum processor. Nat. Phys. **8**, 719–723 (2012)

E. Martin-Lopez, A. Laing, T. Lawson, R. Alvarez, X.-Q. Zhou, J.L. O'Brien, Experimental realisation of Shor's quantum factoring algorithm using qubit recycling. Nat. Photonics **6**, 773–776 (2012)

T. Monz, D. Nigg, E.A. Martinez, M.F. Brandl, P. Schindler, R. Rines, S.X. Wang, I.L. Chuang, R. Blatt, Realization of a scalable Shor algorithm. Science **351**(6277), 1068–1070 (2016)

M.A. Nielsen, I.L. Chuang, *Quantum Computation and Quantum Information* (Cambridge University Express, Cambridge, 2000)

NIST, Post-Quantum Cryptography, https://csrc.nist.gov/projects/post-quantum-cryptography

A. Politi, J.C.F. Matthews, J.L. O'Brien, Shor's quantum factoring algorithm on a photonic chip. Science **325**(5945), 1221 (2009)

J. Preskill, Quantum computing in the NISQ era and beyond. Quantum **2**, 79 (2018)

R.L. Rivest, A. Shamir, L.M. Adelman, A method for obtaining digital signature and public-key cryptosystems, MIT-LCS-TM-082 (1977)

M. Roetteler, M. Naehrig, K.M. Svore, K. Lauter, Quantum resource estimates for computing elliptic curve discrete logarithms, in *Proceedings of ASIACRYPT 2017*, LNCS 10625 (2017), pp. 241–270

SageMath – Open-Source Mathematical Software System, http://www.sagemath.org/

P. Shor, Polynomial-time algorithms for prime factorization and discrete logarithms on a quantum computer. SIAM J. Comput. **26**(5), 1484–1509 (1997)

J.A. Smolin, G. Smith, A. Vargo, Oversimplifying quantum factoring. Nature **499**, 163–165 (2013)

Y. Takahashi, N. Kunihiro, A quantum circuit for Shor's factoring algorithm using $2n + 2$ qubits. Quantum Inf. Comput. **6**(2), 184–192 (2006)

L.M.K. Vandersypen, M. Steffen, G. Breyta, C.S. Yannoni, M.H. Sherwood, I.L. Chuang, Experimental realization of Shor's quantum factoring algorithm using nuclear magnetic resonance. Nature **414**, 883–887 (2001)

Towards Constructing Fully Homomorphic Encryption without Ciphertext Noise from Group Theory

Koji Nuida

Abstract In CRYPTO 2008, 1 year earlier than Gentry's pioneering "bootstrapping" technique for the first fully homomorphic encryption (FHE) scheme, Ostrovsky and Skeith III had suggested a completely different approach towards achieving FHE. They showed that the NAND operator can be realized in some *non-commutative* groups; consequently, homomorphically encrypting the elements of the group will yield an FHE scheme, without ciphertext noise to be bootstrapped. However, no observations on how to homomorphically encrypt the group elements were presented in their paper, and there have been no follow-up studies in the literature. The aim of this paper is to exhibit more clearly what is sufficient and what seems to be effective for constructing FHE schemes based on their approach. First, we prove that it is sufficient to find a surjective homomorphism $\pi : \widetilde{G} \to G$ between finite groups for which bit operators are realized in G and the elements of the kernel of π are indistinguishable from the general elements of $\widetilde{G}$. Secondly, we propose new methodologies to realize bit operators in some groups G. Thirdly, we give an observation that a naive approach using matrix groups would never yield secure FHE due to an attack utilizing the "linearity" of the construction. Then we propose an idea to avoid such "linearity" by using combinatorial group theory. Concretely realizing FHE schemes based on our proposed framework is left as a future research topic.

Keywords Fully homomorphic encryption · Non-commutative group · Combinatorial group theory

K. Nuida (✉)
Graduate School of Information Science and Technology,
The University of Tokyo, Tokyo, Japan
e-mail: nuida@mist.i.u-tokyo.ac.jp

National Institute of Advanced Industrial Science and Technology (AIST), Tokyo, Japan

© The Author(s) 2021
T. Takagi et al. (eds.), *International Symposium on Mathematics,
Quantum Theory, and Cryptography*, Mathematics for Industry 33,
https://doi.org/10.1007/978-981-15-5191-8_8

57

1 Introduction

Until the pioneering work by Gentry (2009) in 2009, it had been a long-standing open problem to construct *fully homomorphic encryption* (*FHE*) that enables arbitrary "computation on encrypted data" via "homomorphic" operations on the ciphertexts. After Gentry's work, studies of FHE to improve the efficiency (e.g. Chillotti et al. 2016; Ducas and Micciancio 2015; Gentry et al. 2012; Stehlé and Steinfeld 2010) and to give various frameworks of construction (e.g. Brakerski and Vaikuntanathan 2011; Cheon and Stehlé 2015; van Dijk et al. 2010; Gentry and Halevi 2011; Nuida and Kurosawa 2015) have been one of the main research topics in cryptology (see, e.g. Silverberg 2013 for a survey). Here we emphasize that all the previous FHE schemes in the literature rely on Gentry's "bootstrapping" framework. Namely, ciphertexts for these FHE schemes involve "noise" terms to conceal plaintexts, and the noise is increased by homomorphic operations and will finally collapse the ciphertext; hence the increased noise must be cancelled before the collapse. The bootstrapping, which is the additional procedure for noise cancellation, is a major bottleneck for efficiency improvement, makes the syntax of FHE less analogical to the classical homomorphic encryption, and causes somewhat unclear treatments regarding so-called circular security.

On the other hand, in 2008 (1 year earlier than Gentry 2009), Ostrovsky and Skeith III (2008) had suggested a completely different, group-theoretic approach towards achieving FHE. Namely, they showed that the **NAND** operator (which is sufficient for constructing arbitrary bit operators) can be realized (in a certain suitable sense) in some *non-commutative* groups. Consequently, if the elements of the underlying group can be homomorphically encrypted, then it will yield an FHE scheme where the ciphertexts involve no noise terms; hence, the bootstrapping procedure will no longer be required. However, no observations on how to homomorphically encrypt the group elements were presented in their paper and, to the author's best knowledge, there have been no follow-up studies in the literature based on their approach. The aim of this paper is to exhibit more clearly what is sufficient and what seems to be effective for constructing "noise-free" FHE schemes based on their approach.

1.1 Our Contributions

In Sect. 3, we revisit the approach towards constructing FHE suggested in Ostrovsky and Skeith (2008). We give a formalization of "realizations of bit operators in groups" in a slightly generalized manner (e.g. our formalization can also handle probabilistic realizations of bit operators, which were not considered in Ostrovsky and Skeith 2008). Then we reduce the problem of "homomorphically encrypting the elements of a group G" to finding a surjective homomorphism $\pi : \widetilde{G} \to G$ from another finite group $\widetilde{G}$ (which plays the role of the ciphertext space) satisfying certain conditions and prove that the resulting FHE scheme is CPA-secure if the elements of the kernel of π (ker π) are indistinguishable from the general elements of $\widetilde{G}$ even when a certain generating set of ker π is publicly given. This clarifies the problem to be solved from a group-theoretic viewpoint.

In Sect. 4, we propose new methodologies to realize bit operators in some groups, which are different from the previous methodology in Ostrovsky and Skeith (2008) analogous to Barrington's theorem (Barrington 1986) (recalled in Sect. 4.1). Our result enlarges the possibility of the underlying group G to find a suitable construction.

Finally, in Sect. 5, we give some observations and discussions on how to find a suitable homomorphism $\pi: \widetilde{G} \rightarrow G$. In Sect. 5.2, we give an observation that a naive approach to construct the group $\widetilde{G}$ by using embedding of a matrix group G into a larger matrix group and then taking its random conjugate would never yield a secure FHE scheme, due to the existence of a kind of "linear" constraint that separates the elements of ker π from general elements of $\widetilde{G}$ (where the "linearity" causes that such a constraint does not disappear even by taking random conjugate). This observation shows an importance of finding a homomorphism $\pi: \widetilde{G} \rightarrow G$ onto a given underlying group G without linear constraints for elements of ker π. Towards constructing such a homomorphism π, in Sect. 5.3, we propose another approach using combinatorial group theory, i.e. the properties of presentations of groups in terms of generators and fundamental relations. Then, in Sect. 5.4, we discuss several problems to be resolved in order to realize our proposed approach, many of which would be of independent interest from mathematical viewpoints.

2   Preliminaries

Let $a \leftarrow X$ mean that a random variable X takes a value a. Let $a \leftarrow_R X$ mean that an element a is chosen uniformly at random from a finite set X. The *statistical distance* between two probability distributions X, Y over a finite set A is defined by $\Delta(X, Y) = (1/2) \sum_{z \in A} |\Pr[z \leftarrow X] - \Pr[z \leftarrow Y]|$. For $\varepsilon \geq 0$, we say that X is ε-close to Y, if $\Delta(X, Y) \leq \varepsilon$. We say that a function $\varepsilon = \varepsilon(\lambda) \geq 0$ is *negligible*, if $\varepsilon = \lambda^{-\omega(1)}$. We say that $\varepsilon \in [0, 1]$ is *overwhelming*, if $1 - \varepsilon$ is negligible; and ε is *noticeable*, if there exist integers $n \geq 1$ and $\lambda_0 > 0$ for which we have $\varepsilon > \lambda^{-n}$ for every $\lambda > \lambda_0$.

A *public-key encryption (PKE)* scheme consists of the following three algorithms. The *key generation algorithm* $\mathsf{Gen}(1^\lambda)$ outputs a pair of a public key pk and a secret key sk. The *encryption algorithm* $\mathsf{Enc}(m) = \mathsf{Enc}_{\mathsf{pk}}(m)$ outputs a ciphertext for a plaintext m. The *decryption algorithm* $\mathsf{Dec}(c) = \mathsf{Dec}_{\mathsf{sk}}(c)$ for a ciphertext c outputs either a plaintext or a "failure" symbol $\perp$. The *correctness* of a PKE scheme means that, for any plaintext m, the probability $\Pr[\mathsf{Dec}_{\mathsf{sk}}(\mathsf{Enc}_{\mathsf{pk}}(m)) \neq m]$ (taken over the internal randomness for the algorithms) is negligible.

For a finite set M, we say that a set $\mathcal{F}$ of operators on M is *functionally complete*, if any (multivariate) function with inputs and outputs in M can be computed by combining operators in $\mathcal{F}$. We say that a PKE scheme with plaintext space M is a *fully homomorphic encryption (FHE)* scheme, if there exist a functionally complete set $\mathcal{F}$ of operators on $\mathcal{M}$ and an efficient *homomorphic evaluation algorithm* Eval with the property that, for each, say n-ary operator $f \in \mathcal{F}$ ($f: \mathcal{M}^n \to \mathcal{M}$) and for given ciphertexts c_i for plaintexts m_i ($i = 1, \ldots, n$), the algorithm $\mathsf{Eval}_{\mathsf{pk}}(f; c_1, \ldots, c_n)$ outputs a ciphertext for plaintext $f(m_1, \ldots, m_n) \in \mathcal{M}$ with overwhelming probability.

We say that a PKE scheme with plaintext space $\mathcal{M}$ is *CPA-secure*, if for any probabilistic polynomial-time (PPT) adversary $\mathcal{A}$, the *advantage* $\mathsf{Adv}_{\mathcal{A}}(\lambda) = |\Pr[b = b^*] - 1/2|$ of $\mathcal{A}$ is negligible, where $\Pr[b = b^*]$ is the probability that $b = b^*$ holds in the following game:

$$(\mathsf{pk}, \mathsf{sk}) \leftarrow \mathsf{Gen}(1^\lambda)\,;\ (m_0, m_1, \mathsf{st}) \leftarrow \mathcal{A}(\mathsf{submit}, 1^\lambda, \mathsf{pk})\,;$$

$$b^* \leftarrow_R \{0, 1\}\,;\ c^* \leftarrow \mathsf{Enc}_{\mathsf{pk}}(m_{b^*})\ :\ b \leftarrow \mathcal{A}(\mathsf{guess}, 1^\lambda, \mathsf{pk}, \mathsf{st}, c^*)\ .$$

The reader may refer to a textbook of group theory (e.g. Robinson 1996) for definitions and basic facts for groups mentioned without explicit references.

3 Our Framework for FHE

In this section, we describe our framework towards constructing FHE free from ciphertext noise. This can be seen as formalizing a framework suggested in Khamsemanan et al. (2016) and Ostrovsky and Skeith (2008).

3.1 Group-Theoretic Realization of Functions

Roughly speaking, a group-theoretic realization of a function in a group is emulating the function "by using the group operators only". To formalize it, we prepare some definitions. Let $w = w(x_1, \ldots, x_n)$ be a sequence of finite length over alphabet $\{x_1, x_1^{-1}, \ldots, x_n, x_n^{-1}\}$, called a *group word* with variables $x_1, \ldots, x_n$. Then one can *substitute* given elements $g_1, \ldots, g_n$ of a group into the variables $x_1, \ldots, x_n$ in $w(x_1, \ldots, x_n)$ to yield an element of the same group, denoted by $w(g_1, \ldots, g_n)$.

Then we define a group-theoretic realization of functions as follows. In comparison to a similar definition in Khamsemanan et al. (2016) that was deterministic with a single component, our formulation here also covers probabilistic situations with multiple components.

Definition 1 Let G be a group and $\mathcal{M}$ be a set. Let $\mathcal{F}$ be a set of functions of the form $f: \mathcal{M}^{\ell_f} \to \mathcal{M}$ with $\ell_f \geq 1$. We define a *group-theoretic realization* (or simply a *realization*) *of $\mathcal{F}$ in G* to be a collection of the following objects:

- a polynomially bounded integer $n \geq 1$, which we call the *degree* of the realization;

- non-empty and mutually disjoint subsets $X_m \subseteq G^n$ for all $m \in \mathcal{M}$;
- for $f \in \mathcal{F}$, a collection $\vec{w}_f(\vec{x}_1, \ldots, \vec{x}_{\ell_f}, \vec{y})$ of n group words $w_{f,i}(\vec{x}_1, \ldots, \vec{x}_{\ell_f}, \vec{y})$ $(i = 1, \ldots, n)$ of polynomially bounded lengths, where $\vec{x}_j = (x_{j,1}, \ldots, x_{j,n})$ for $j = 1, \ldots, \ell_f$ and $\vec{y} = (y_1, \ldots, y_k)$;
- a collection $\vec{r} = (r_1, \ldots, r_k)$ of polynomial-time samplable random variables on G;

satisfying the following condition, where negl is some negligible value: For any $f \in \mathcal{F}$, any $m_1, \ldots, m_{\ell_f} \in \mathcal{M}$, and any $\vec{g}_i = (g_{i,1}, \ldots, g_{i,n}) \in X_{m_i}$ $(i = 1, \ldots, \ell_f)$, the probability $\Pr[\vec{w}_f(\vec{g}_1, \ldots, \vec{g}_{\ell_f}, r_1, \ldots, r_k) \notin X_{f(m_1,\ldots,m_{\ell_f})}]$ taken over the random choices of values of $r_1, \ldots, r_k \in G$ is not larger than negl.

For each $f \in \mathcal{F}$, we denote by $\mathcal{A}_f$ an algorithm that, for given inputs $\vec{g}_1, \ldots, \vec{g}_{\ell_f} \in G^n$, outputs $\vec{w}_f(\vec{g}_1, \ldots, \vec{g}_{\ell_f}, r_1, \ldots, r_k) \in G^n$ where the values of random variables $r_1, \ldots, r_k$ are sampled according to the specified distributions.

We note that, in the formulation above, some of the random variables r_h may take a constant value in G. When all the random variables appearing in a realization are constant, we call the realization *deterministic*, or else call it *probabilistic*.

3.2 Lift of Realization of Functions

Given a group homomorphism $\widetilde{G} \to G$ and a realization of functions in the target group G, the notion of a "lift" of the realization up to the source group $\widetilde{G}$ defined below plays a role of homomorphic operations in our proposed framework for FHE. We note that such a notion was not introduced in the previous work (Khamsemanan et al. 2016; Ostrovsky and Skeith 2008).

Definition 2 We suppose that a set $\mathcal{F}$ of functions on $\mathcal{M}$ has a realization in a group G as in Definition 1. Let $\pi \colon \widetilde{G} \to G$ be a surjective group homomorphism. We define a *lift* of the realization up to $\widetilde{G}$ to be a collection of polynomial-time samplable random variables $\widetilde{r}_1, \ldots, \widetilde{r}_k$ on $\widetilde{G}$ with the property that each value $\pi(\widetilde{r}_h) \in G$ has the same probability distribution as r_h. Then for each $f \in \mathcal{F}$, we denote by $\widetilde{\mathcal{A}}_f$ an algorithm that outputs $\vec{w}_f(\widetilde{\vec{g}}_1, \ldots, \widetilde{\vec{g}}_{\ell_f}, \widetilde{r}_1, \ldots, \widetilde{r}_k) \in (\widetilde{G})^n$ for given inputs $\widetilde{\vec{g}}_1, \ldots, \widetilde{\vec{g}}_{\ell_f} \in (\widetilde{G})^n$ where the values of random variables $\widetilde{r}_1, \ldots, \widetilde{r}_k$ are sampled according to the specified distributions.

In the following, we also write as π the map $(\widetilde{G})^n \to G^n$ with $\pi(\widetilde{g}_1, \ldots, \widetilde{g}_n) = (\pi(\widetilde{g}_1), \ldots, \pi(\widetilde{g}_n))$.

Lemma 1 *In the situation of Definition 2, let $f \in \mathcal{F}$, $m_1, \ldots, m_{\ell_f} \in \mathcal{M}$, and let $\widetilde{\vec{g}}_i \in (\widetilde{G})^n$ satisfy $\pi(\widetilde{\vec{g}}_i) \in X_{m_i}$ for each $i = 1, \ldots, \ell_f$. Then the probability $\Pr[\pi(\widetilde{\mathcal{A}}_f(\widetilde{\vec{g}}_1, \ldots, \widetilde{\vec{g}}_{\ell_f})) \notin X_{f(m_1,\ldots,m_{\ell_f})}]$ is bounded by the same negligible value* negl *as in Definition 1.*

Proof As $\pi : \widetilde{G} \to G$ is a group homomorphism, we have

$$\pi(w_{f,i}(\vec{\widetilde{g}}_1, \ldots, \vec{\widetilde{g}}_{\ell_f}, \widetilde{r}_1, \ldots, \widetilde{r}_k)) = w_{f,i}(\pi(\vec{\widetilde{g}}_1), \ldots, \pi(\vec{\widetilde{g}}_{\ell_f}), \pi(\widetilde{r}_1), \ldots, \pi(\widetilde{r}_k))$$

for any $i = 1, \ldots, \ell_f$ and any values of the random variables $\widetilde{r}_h$. By Definition 1, the claim follows from the fact that the probability distribution for each $\pi(\widetilde{r}_h)$ is identical to r_h. □

3.3 The Proposed Framework

Based on the definitions above, here we describe our proposed framework for constructing FHE:

$\mathsf{Gen}(1^\lambda)$: Choose the following objects according to the security parameter λ, where $\mathcal{M}$ is the set of plaintexts and $\mathcal{F}$ is a functionally complete set of operators on $\mathcal{M}$:

- a group-theoretic realization (of some degree n) of $\mathcal{F}$ on a group G;
- a surjective group homomorphism $\pi : \widetilde{G} \to G$ and a lift of the realization of $\mathcal{F}$ up to $\widetilde{G}$;
- a polynomial-time samplable random variable r_{ker} on the kernel $\ker \pi$ of π;
- for each $m \in \mathcal{M}$, a tuple $\vec{\mathsf{gen}}_m = (\mathsf{gen}_{m,1}, \ldots, \mathsf{gen}_{m,n}) \in (\widetilde{G})^n$ with $\pi(\vec{\mathsf{gen}}_m) \in X_m$.

Then output a public key pk consisting of $\widetilde{G}$, r_{ker}, $\vec{\mathsf{gen}}_m$ for all $m \in \mathcal{M}$, and the algorithms $\widetilde{A}_f$ for all $f \in \mathcal{F}$ appearing in the lift of the realization of $\mathcal{F}$; and output a secret key sk consisting of G, π, and X_m for all $m \in \mathcal{M}$.

$\mathsf{Enc}_{\mathsf{pk}}(m)$ for $m \in \mathcal{M}$: Sample n values $\vec{r}_{\mathrm{ker}} = (r_{\mathrm{ker},1}, \ldots, r_{\mathrm{ker},n})$ of the random variable r_{ker} independently, and then output $\vec{c} = (c_1, \ldots, c_n) \leftarrow \vec{\mathsf{gen}}_m \cdot \vec{r}_{\mathrm{ker}} \in (\widetilde{G})^n$.

$\mathsf{Dec}_{\mathsf{sk}}(c)$ for $\vec{c} \in (\widetilde{G})^n$: Compute $\pi(\vec{c}) \in G^n$, and if $\pi(\vec{c}) \in X_m$ for an $m \in \mathcal{M}$, then output the m. If no such m exists, then output $\perp$.

$\mathsf{Eval}_{\mathsf{pk}}(f; \vec{c}_1, \ldots, \vec{c}_{\ell_f})$ for $f \in \mathcal{F}$ and $\vec{c}_1, \ldots, \vec{c}_{\ell_f} \in (\widetilde{G})^n$: Output $\widetilde{A}_f(\vec{c}_1, \ldots, \vec{c}_{\ell_f}) \in (\widetilde{G})^n$.

The correctness of Enc is obvious; when $\vec{c} = \vec{\mathsf{gen}}_m \cdot \vec{r}_{\mathrm{ker}} \leftarrow \mathsf{Enc}_{\mathsf{pk}}(m)$, we have

$$\pi(\vec{c}) = \pi(\vec{\mathsf{gen}}_m) \cdot (\pi(r_{\mathrm{ker},1}), \ldots, \pi(r_{\mathrm{ker},n})) = \pi(\vec{\mathsf{gen}}_m) \cdot (1_G, \ldots, 1_G) = \pi(\vec{\mathsf{gen}}_m) \in X_m$$

as $r_{\mathrm{ker},i} \in \ker \pi$ for each i. The correctness of Eval is just a restatement of Lemma 1. On the other hand, for the security, we have the following result:

Theorem 1 *In the setting above, suppose that $\widetilde{G}$ is a finite group with polynomial-time computable group operators, and suppose either $n = 1$ or that the uniform* *distribution over $\widetilde{G}$ is polynomial-time samplable. Then, our proposed FHE scheme is CPA-secure if the subgroup membership problem for $\ker \pi \subseteq \widetilde{G}$ with respect to the random variable $r_{\ker}$ with auxiliary input* pk *is computationally hard, that is, for any PPT adversary $\mathcal{A}^{\dagger}$, the advantage $\mathsf{Adv}_{\mathcal{A}^{\dagger}}(\lambda) = |\Pr[b = b^{\dagger}] - 1/2|$ of $\mathcal{A}^{\dagger}$ in the following game is negligible:*

$$(\mathsf{pk}, \mathsf{sk}) \leftarrow \mathsf{Gen}(1^{\lambda}) \,;\; b^{\dagger} \leftarrow_R \{0, 1\} \,;\; \begin{cases} g^{\dagger} \leftarrow_R \widetilde{G} & (b^{\dagger} = 1) \\ g^{\dagger} \leftarrow r_{\ker} & (b^{\dagger} = 0) \end{cases} \,:\; b \leftarrow \mathcal{A}^{\dagger}(1^{\lambda}, \mathsf{pk}, g^{\dagger}) \,.$$

Proof Let $\mathcal{A}$ be any PPT CPA adversary for our scheme. Then we define an adversary $\mathcal{A}^{\dagger}$ for the subgroup membership problem specified in the statement as follows:

1. Given inputs 1^{λ}, pk, and $g^{\dagger}$ chosen according to the random bit $b^{\dagger}$, the adversary $\mathcal{A}^{\dagger}$ chooses $i \leftarrow_R \{1, \ldots, n\}$ and executes $\mathcal{A}(\mathsf{submit}, 1^{\lambda}, \mathsf{pk})$ to obtain a tuple (m_0, m_1, st).
2. The adversary $\mathcal{A}^{\dagger}$ chooses $b^* \leftarrow_R \{0, 1\}$ and executes $\mathcal{A}(\mathsf{guess}, 1^{\lambda}, \mathsf{pk}, \mathsf{st}, c^{b^*, b^{\dagger}, i})$ to obtain a bit b', where

$$c^{b^*, b^{\dagger}, i}$$
$$= (\mathsf{gen}_{m_{b^*}, 1} \rho_1, \ldots, \mathsf{gen}_{m_{b^*}, i-1} \rho_{i-1}, \mathsf{gen}_{m_{b^*}, i} g^{\dagger}, \mathsf{gen}_{m_{b^*}, i+1} u_{i+1}, \ldots, \mathsf{gen}_{m_{b^*}, n} u_n)$$

with independent random values $\rho_1, \ldots, \rho_{i-1}$ of $r_{\ker}$ and $u_{i+1}, \ldots, u_n \leftarrow_R \widetilde{G}$.
3. The adversary $\mathcal{A}^{\dagger}$ outputs $b = \mathsf{XOR}(b^*, b')$.

Note that this adversary $\mathcal{A}^{\dagger}$ is PPT as well as $\mathcal{A}$. Now we have

$$\mathsf{Adv}_{\mathcal{A}^{\dagger}}(\lambda) = |\Pr[b = b^{\dagger}] - 1/2| = \frac{1}{2} \left| \Pr[b = 0 \mid b^{\dagger} = 0] + \Pr[b = 1 \mid b^{\dagger} = 1] - 1 \right|$$

and

$$\Pr[b = 0 \mid b^{\dagger} = 0] = \Pr[b' = b^* \mid b^{\dagger} = 0]$$
$$= \sum_{i=1}^{n} \frac{1}{n} \Pr[b^* \leftarrow \mathcal{A}(\mathsf{guess}, 1^{\lambda}, \mathsf{pk}, \mathsf{st}, c^{b^*, 0, i})] \,,$$

while

$$\Pr[b = 1 \mid b^{\dagger} = 1] = 1 - \Pr[b' = b^* \mid b^{\dagger} = 1]$$
$$= 1 - \sum_{i=1}^{n} \frac{1}{n} \Pr[b^* \leftarrow \mathcal{A}(\mathsf{guess}, 1^{\lambda}, \mathsf{pk}, \mathsf{st}, c^{b^*, 1, i})] \,.$$

By the choice of $g^{\dagger}$, for each $i = 1, \ldots, n - 1$ and any choice of b^*, the two tuples $c^{b^*, 0, i}$ and $c^{b^*, 1, i+1}$ follow an identical probability distribution. Therefore, we have

$$\Pr[b=0 \mid b^{\dagger}=0] + \Pr[b=1 \mid b^{\dagger}=1] - 1$$
$$= \frac{1}{n} \Pr[b^* \leftarrow \mathcal{A}(\mathsf{guess}, 1^{\lambda}, \mathsf{pk}, \mathsf{st}, c^{b^*,0,n})] - \frac{1}{n} \Pr[b^* \leftarrow \mathcal{A}(\mathsf{guess}, 1^{\lambda}, \mathsf{pk}, \mathsf{st}, c^{b^*,1,1})] \ .$$

Now we have

$$c^{b^*,1,1} = (\mathsf{gen}_{m_{b^*},1} g^{\dagger}, \mathsf{gen}_{m_{b^*},2} u_2, \ldots, \mathsf{gen}_{m_{b^*},n} u_n)$$

and the element $g^{\dagger}$ when $b^{\dagger}=1$ is a uniformly random and independent element of $\widetilde{G}$ as well as $u_2, \ldots, u_n$. This implies that $c^{b^*,1,1}$ is uniformly random over $(\widetilde{G})^n$ regardless of the choice of b^*; therefore, we have

$$\Pr[b^* \leftarrow \mathcal{A}(\mathsf{guess}, 1^{\lambda}, \mathsf{pk}, \mathsf{st}, c^{b^*,1,1}) = \frac{1}{2}$$

and

$$\mathsf{Adv}_{\mathcal{A}^{\dagger}}(\lambda) = \frac{1}{2n} \left| \Pr[b^* \leftarrow \mathcal{A}(\mathsf{guess}, 1^{\lambda}, \mathsf{pk}, \mathsf{st}, c^{b^*,0,n})] - \frac{1}{2} \right| \ .$$

Moreover, we have

$$c^{b^*,0,n} = (\mathsf{gen}_{m_{b^*},1} \rho_1, \ldots, \mathsf{gen}_{m_{b^*},n-1} \rho_{n-1}, \mathsf{gen}_{m_{b^*},n} g^{\dagger})$$

and the element $g^{\dagger}$ when $b^{\dagger}=0$ is a random value of r_{ker} as well as $\rho_1, \ldots, \rho_{n-1}$. This implies that $c^{b^*,0,n}$ follows the same probability distribution as $\mathsf{Enc}_{\mathsf{pk}}(m_{b^*})$; therefore, we have

$$\mathsf{Adv}_{\mathcal{A}^{\dagger}}(\lambda) = \frac{1}{2n} \left| \Pr[b^* \leftarrow \mathcal{A}(\mathsf{guess}, 1^{\lambda}, \mathsf{pk}, \mathsf{st}, \mathsf{Enc}_{\mathsf{pk}}(m_{b^*}))] - \frac{1}{2} \right| = \frac{1}{2n} \mathsf{Adv}_{\mathcal{A}}(\lambda) \ .$$

As the adversary $\mathcal{A}^{\dagger}$ is PPT, the assumption in the statement implies that $\mathsf{Adv}_{\mathcal{A}^{\dagger}}(\lambda)$ is negligible; therefore, $\mathsf{Adv}_{\mathcal{A}}(\lambda)$ is also negligible as n is polynomially bounded. This completes the proof of Theorem 1. $\qquad\qquad\square$

4 Examples of Realizations of Functions in Groups

4.1 Deterministic Case: Known Result

The following result (which is restated according to our terminology here) was proved in the previous work (Khamsemanan et al. 2016; Ostrovsky and Skeith 2008) (see, e.g. Theorem 2.1 of Ostrovsky and Skeith 2008).

Proposition 1 (Khamsemanan et al. 2016; Ostrovsky and Skeith 2008) *Let G be any non-commutative finite simple group. Then there exists a deterministic, degree-1 group-theoretic realization of* NAND *in G.*

We note that its proof, utilizing the commutators $[g, h] = ghg^{-1}h^{-1}$ in a way analogous to Barrington's theorem (Barrington 1986), is in general not constructive. A concrete construction was given in Sect. 6 of Khamsemanan et al. (2016) only for the smallest case $G = A_5$, where the group word has a length 65.

4.2 Deterministic Case: Proposed Constructions

Here, we propose a completely different approach, which we call *approximate-then-adjust method*, to obtain deterministic realizations of operators in some small groups. An intuitive explanation is as follows. For example, the operations b_1 OR b_2 and $b_1 + b_2$ mod 3 have equal outputs for all but one input pairs $(b_1, b_2) \neq (1, 1)$ in $\{0, 1\}^2$, and $1 + 1$ mod $3 = 2$ (instead of 1 OR $1 = 1$) is "overflowed" from the correct output set $\{0, 1\}$. As the operation $b_1 + b_2$ mod 3 is easily realizable by using a cyclic subgroup of order 3, the problem has been reduced to realize the "adjusting function" $0 \mapsto 0$, $1 \mapsto 1, 2 \mapsto 1$ in a group.

In fact, by putting $\sigma_b = (1, 2, 3)^b \in S_5$ for $b \in \{0, 1, 2\}$ (where S_k denotes the symmetric group on k letters) and identifying each σ_b with b, the adjusting function mentioned above can be realized by a group word

$$w^{\text{out}}(g) = (1, 5)(2, 3, 4)g(2, 3, 4)g(3, 4)g^2(2, 3)(4, 5)g(2, 3, 4)g(3, 4)g^2(1, 4, 2, 5)$$

(formally, the left-hand side is an abbreviation of $w^{\text{out}}(g, \vec{y})$ where the variables in $\vec{y}$ take constant values over $G = S_5$ appearing in the right-hand side). This adjusting function defined by w^{out} is also applicable to other operations NAND, XOR, and EQ (= NOT ∘ XOR). Namely, by putting

$$w_{\text{OR}}^{\text{in}}(g_1, g_2) = g_1 g_2, \ w_{\text{NAND}}^{\text{in}}(g_1, g_2) = g_1^{-1} g_2^{-1} \sigma_1^2,$$
$$w_{\text{XOR}}^{\text{in}}(g_1, g_2) = g_1^{-1} g_2, \ w_{\text{EQ}}^{\text{in}}(g_1, g_2) = g_1 g_2 \sigma_1^{-1},$$

an output of each w_f^{in} for inputs in $\{\sigma_0, \sigma_1\}$ becomes either equal (via the identification $\sigma_b \leftrightarrow b$) to f, or σ_2 ($\leftrightarrow 2$) instead of σ_1 ($\leftrightarrow 1$). Hence, the composition $w^{\text{out}}(w_f^{\text{in}}(g_1, g_2))$ gives a correct group word to realize the operator f with $X_0 = \{\sigma_0 = 1_{S_5}\}$ and $X_1 = \{\sigma_1\}$. We also note that NOT is easily realized with the same X_0 and X_1 by $w^{\text{NOT}}(g) = g^{-1}\sigma_1$.

This method is also applicable to realizing arithmetic operations for $\mathbb{F}_3$. We put $\sigma_b = (1, 2, 3)^b \in S_5$ for $b \in \{0, 1, 2\}$ again, and set $X_b = \{\sigma_b\}$ for each b. Then the addition $+$ is easily realized by $w_+(g_1, g_2) = g_1 g_2$. For the multiplication $\times$, the following group word

$$w_\times^{\mathsf{in}}(g_1, g_2) = g_1((1, 4)(2, 3, 5))^{-1}g_2(1, 4)(2, 3, 5)$$

satisfies that $w_\times^{\mathsf{in}}(\sigma_{b_1}, \sigma_{b_2}) \in X'_{b_1 \times b_2 \bmod 3}$ for any $b_1, b_2 \in \{0, 1, 2\}$, where

$$
\begin{aligned}
X'_0 &= \{1_{S_5}, (2, 4, 5), (2, 5, 4), (1, 2, 3), (1, 3, 2)\} \ , \\
X'_1 &= \{(1, 2, 4, 5, 3), (1, 3, 2, 5, 4)\} \ , \\
X'_2 &= \{(1, 2, 5, 4, 3), (1, 3, 2, 4, 5)\} \ .
\end{aligned}
$$

On the other hand, by putting

$$
\begin{aligned}
w'_1(g) &= g^3, \ w'_2(g) = w'_3(g) = (2, 3, 4)^{-1}g^{-1}(3, 4, 5)g^2(3, 4, 5)^{-1}g(2, 3, 4) \ , \\
w'_4(g) &= g(1, 5, 3, 4, 2)g^{-1}(1, 5, 3, 4, 2)^{-1}g(1, 4, 2, 3, 5)g^{-1}(1, 4, 2, 3, 5)^{-1} \ ,
\end{aligned}
$$

the composed group word $w^{\mathsf{out}}(g) = w'_4(w'_3(w'_2(w'_1(g))))$ satisfies that $w^{\mathsf{out}}(g) = \sigma_b$ for any $b \in \{0, 1, 2\}$ and any $g \in X'_b$. Hence, the group word $w_\times(g_1, g_2) = w^{\mathsf{out}}(w_\times^{\mathsf{in}}(g_1, g_2))$ realizes the operator $\times$ for $\mathbb{F}_3$, as desired. We note that the group words in the arguments above are found by heuristic searches; a systematic method to find such group words is a future research topic.

4.3 Preliminaries: On Random Sampling of Group Elements

In the probabilistic constructions described below, the following result by Dixon (2008) on almost uniform sampling over any finite group G would be useful in implementation. We introduce a notation: for any $g_1, \ldots, g_L \in G$, let $\mathsf{Sample}[g_1, \ldots, g_L]$ denote the random variable that takes the value $g_1^{e_1} \cdots g_L^{e_L} \in G$ with $e_1, \ldots, e_L \leftarrow_R \{0, 1\}$.

Proposition 2 (Dixon 2008, Theorem 3) *Let G be a finite group, let $0 \leq \varepsilon < 1$, and let $\mathcal{U}$ be a random variable over G that is ε-close to the uniform random variable on G. Let L be a positive integer, and let $h, k \geq 0$. If*

$$L \geq \frac{\log_2 |G| + h + 2k - 2}{\log_2(2/(1 + \varepsilon))} \ ,$$

then we have $\Pr_{g_1, \ldots, g_L \leftarrow \mathcal{U}}[\,\mathsf{Sample}[g_1, \ldots, g_L]$ *is not* 2^{-k}*-close to uniform* $] < 2^{-h}$.

4.4 Probabilistic Case: "Commutator-Separable" Groups

We propose a degree-2 probabilistic realization of $\{\mathsf{NOT}, \mathsf{AND}\}$ in the following class of groups.

Definition 3 Let $\varepsilon > 0$. We say that a finite group G is ε-*commutator-separable*, if there exists a non-empty subset Y of $G \setminus \{1_G\}$ satisfying

$$\Pr_{u \leftarrow_R G} [[ugu^{-1}, g'] \notin Y] \leq \varepsilon \text{ for any } g, g' \in Y . \tag{1}$$

Moreover, we say that a family of finite groups $G = G_\lambda$ indexed by the security parameter λ is *commutator-separable*, if there exists a negligible function $\varepsilon = \varepsilon(\lambda)$ for which G is ε-commutator-separable for any λ.

Let G be an ε-commutator-separable group. We put

$$X_0 = \{(g_1, g_2) \in G^2 \mid g_1 \in Y, \ g_2 = 1_G\}, X_1 = \{(g_1, g_2) \in G^2 \mid g_1 \in Y, \ g_2 = g_1\} ,$$

where $Y \subseteq G \setminus \{1_G\}$ is as in Definition 3. Then NOT is easily realized by the group words (where $\vec{g} = (g_1, g_2)$)

$$\vec{w}_{\mathsf{NOT}}(\vec{g}) = (w_{\mathsf{NOT},1}(\vec{g}), w_{\mathsf{NOT},2}(\vec{g})) = (g_1, g_2^{-1} g_1) .$$

On the other hand, we define the (probabilistic) group words for AND by

$$\begin{aligned}
\vec{w}_{\mathsf{AND}}(\vec{g}, \vec{g}') &= (w_{\mathsf{AND},1}(\vec{g}, \vec{g}'), w_{\mathsf{AND},2}(\vec{g}, \vec{g}')) \\
&= ([ug_1 u^{-1}, g_1'], [ug_2 u^{-1}, g_2']) \text{ with } u \leftarrow_R G .
\end{aligned}$$

For any $\vec{g}, \vec{g}' \in X_0 \cup X_1$, the condition (1) implies that $\Pr[w_{\mathsf{AND},1}(\vec{g}, \vec{g}') \notin Y] \leq \varepsilon$ where the probability is taken over the random choice of u in $\vec{w}_{\mathsf{AND}}(\vec{g}, \vec{g}')$. Moreover, when $\vec{g} \in X_0$ or $\vec{g}' \in X_0$, we have $g_2 = 1_G$ or $g_2' = 1_G$; therefore, $w_{\mathsf{AND},2}(\vec{g}, \vec{g}') = 1_G$. On the other hand, when $\vec{g}, \vec{g}' \in X_1$, we have $g_2 = g_1$ and $g_2' = g_1'$; therefore, $w_{\mathsf{AND},2}(\vec{g}, \vec{g}') = w_{\mathsf{AND},1}(\vec{g}, \vec{g}')$. Summarizing, $\vec{w}_{\mathsf{AND}}(\vec{g}, \vec{g}')$ is a realization of AND with error probability $\leq \varepsilon$.

Remark 1 Although only the *existence* of such a subset Y is concerned in Definition 3, the efficient samplability of an element of Y is needed to be used as a part of our proposed framework for FHE. In general, this is at least probabilistically achievable if the ratio $|G \setminus Y|/|G|$ is negligible; now a uniformly random element of G is also an element of Y except for a negligible probability.

From now, we show that the groups $SL_2(\mathbb{F}_q)$ and $PSL_2(\mathbb{F}_q) = SL_2(\mathbb{F}_q)/\{\pm I\}$ are commutator-separable if the order q of the coefficient field $\mathbb{F}_q$ satisfies that $1/q$ is negligible. In the following, let $Z_H(g) = \{h \in H \mid gh = hg\}$ denote the centralizer of g in a group H. We note that $|Z_H(g)| = |H|/|g^H|$ for any $g \in H$, where $g^H = \{hgh^{-1} \mid h \in H\}$ denotes the conjugacy class of g in H.

Lemma 2 *Let H be a finite group, and let $X \subseteq H$. Then for any $x_1, x_2 \in H$, we have*

$$\Pr_{g \leftarrow_R H} [[gx_1 g^{-1}, x_2] \in X] \leq \frac{|X| \cdot |Z_H(x_1)| \cdot |Z_H(x_2)|}{|H|} .$$

Proof For $y \in X$, we have $[gx_1g^{-1}, x_2] = y$ if and only if $(gx_1g^{-1})x_2(gx_1g^{-1})^{-1} = yx_2$. As the mapping $h \mapsto hzh^{-1}$ is a $|Z_H(z)|$-to-1 mapping for any $z \in H$, there are at most $|Z_H(x_2)|$ possibilities of the value of gx_1g^{-1} to satisfy the condition $(gx_1g^{-1})x_2(gx_1g^{-1})^{-1} = yx_2$; and for each of them, there are at most $|Z_H(x_1)|$ possibilities of the value of g. This completes the proof. $\qquad\qquad\square$

Lemma 3 *Let $\varphi \colon H_1 \to H_2$ be a surjective group homomorphism between two finite groups, and let $x \in H_1$. Then we have $|Z_{H_2}(\varphi(x))| \leq |Z_{H_1}(x)|$.*

Proof As φ is a surjective homomorphism, it is a $(|H_1|/|H_2|)$-to-1 mapping and we have $\varphi(x^{H_1}) = \varphi(x)^{H_2}$. Therefore $|x^{H_1}| \leq (|H_1|/|H_2|) \cdot |\varphi(x)^{H_2}|$, or equivalently $|H_2|/|\varphi(x)^{H_2}| \leq |H_1|/|x^{H_1}|$. Hence the claim holds. $\qquad\qquad\square$

Lemma 4 *For any $A = \begin{pmatrix} a & b \\ c & d \end{pmatrix} \in \mathrm{SL}_2(\mathbb{F}_q)$ with $A \neq \pm I$, we have $|Z_{\mathrm{SL}_2(\mathbb{F}_q)}(A)| \leq 2q$ if $b \neq 0$ or $c \neq 0$, and $|Z_{\mathrm{SL}_2(\mathbb{F}_q)}(A)| = q - 1$ if $b = c = 0$.*

Proof Let $A = \begin{pmatrix} a & b \\ c & d \end{pmatrix} \in \mathrm{SL}_2(\mathbb{F}_q)$ with $A \neq \pm I$, and let $X = \begin{pmatrix} x & y \\ z & w \end{pmatrix} \in Z_{\mathrm{SL}_2(\mathbb{F}_q)}(A)$; therefore, $\det(X) = 1$ and $XA = AX$. Then we have

$$xw - yz = 1, \; cy = bz, \; bx + dy = ay + bw, \; az + cw = cx + dz \;.$$

First, suppose that $b \neq 0$. Then we have $z = b^{-1}cy$ and $w = x + b^{-1}(d - a)y$, therefore $x^2 + b^{-1}(d - a)xy - b^{-1}cy^2 = 1$. Now for each $y \in \mathbb{F}_q$, the quadratic equation in x has at most two solutions, and z and w are uniquely determined from x and y by the relations above. This implies that the number of the possible X is at most $2q$. The argument for the case $c \neq 0$ is similar; x and y are linear combinations of z and w, and w satisfies a quadratic equation when an element $z \in \mathbb{F}$ is fixed; therefore, the number of the possible X is at most $2q$.

On the other hand, suppose that $b = c = 0$. By the condition $\det(A) = 1$, we have $ad = 1$; therefore, $a \neq 0$ and $d \neq 0$. Now we have $dy = ay$ and $az = dz$, while the condition $A \neq \pm I$ implies that $a \neq d$. Therefore, we have $y = 0$ and $z = 0$. This implies that $xw = 1$; therefore, $w \neq 0$ and $x = w^{-1}$. Hence, the number of the possible X is $q - 1$. This completes the proof of Lemma 4. $\qquad\qquad\square$

Corollary 1 *We have $|Z_{\mathrm{PSL}_2(\mathbb{F}_q)}(A)| \leq 2q$ for any non-identity element $A \in \mathrm{PSL}_2(\mathbb{F}_q)$.*

Proof Apply Lemma 3 to the natural projection $\mathrm{SL}_2(\mathbb{F}_q) \to \mathrm{PSL}_2(\mathbb{F}_q)$ and use Lemma 4. $\qquad\qquad\square$

Theorem 2 *If $\dfrac{8q}{q^2 - 1} \leq \varepsilon$, or equivalently $q \geq \dfrac{4 + \sqrt{16 + \varepsilon^2}}{\varepsilon} \approx \dfrac{8}{\varepsilon}$, then $\mathrm{SL}_2(\mathbb{F}_q)$ and $\mathrm{PSL}_2(\mathbb{F}_q)$ are ε-commutator-separable with $Y = \mathrm{SL}_2(\mathbb{F}_q) \setminus \{\pm I\}$ and $Y = \mathrm{PSL}_2(\mathbb{F}_q) \setminus \{1_{\mathrm{PSL}_2(\mathbb{F}_q)}\}$, respectively.*

Proof Let $H \in \{\mathrm{SL}_2(\mathbb{F}_q), \mathrm{PSL}_2(\mathbb{F}_q)\}$. First, it is known that $|H| = q(q^2 - 1)/\eta$, where $\eta = 1$ if $H = \mathrm{SL}_2(\mathbb{F}_q)$ and $\eta = 2$ if $H = \mathrm{PSL}_2(\mathbb{F}_q)$. We also note that $|H \setminus Y| = 2/\eta$. Now for any $x_1, x_2 \in Y$, Lemma 4 and Corollary 1 imply that $|Z_H(x_1)|, |Z_H(x_2)| \le 2q$. Therefore, by Lemma 2, we have

$$\Pr_{g \leftarrow_R H}[[gx_1g^{-1}, x_2] \notin Y] \le \frac{(2/\eta) \cdot 2q \cdot 2q}{q(q^2 - 1)/\eta} = \frac{8q}{q^2 - 1} \le \varepsilon$$

by the condition for q in the statement. This completes the proof. $\qquad\square$

4.5 Probabilistic Case: Simple Groups

We also give a variant of the probabilistic realization described in Sect. 4.4. Although the correctness below relies on a heuristic assumption, the underlying group G for the realization can be taken as any sufficiently large non-commutative finite simple group.

The realization of **NOT** is similar to Sect. 4.4. Namely, we put

$$X_0 = \{(g_1, g_2) \in G^2 \mid g_1 \ne 1_G , g_2 = 1_G\}, X_1 = \{(g_1, g_2) \in G^2 \mid g_1 \ne 1_G , g_2 = g_1\}$$

and, for $\vec{g} = (g_1, g_2)$,

$$\vec{w}_{\mathsf{NOT}}(\vec{g}) = (w_{\mathsf{NOT},1}(\vec{g}), w_{\mathsf{NOT},2}(\vec{g})) = (g_1, g_2^{-1}g_1) .$$

From now, we consider the realization of **AND**. First we note that, for any $g \in G \setminus \{1_G\}$, the normal closure of $\{g\}$ in G is equal to the whole G as G is simple; hence, G is generated by the set g^G. Keeping this property in mind, we put the following heuristic assumption:

Assumption 1 Let $\varepsilon > 0$ be a negligible value, and let L be a sufficiently large parameter. We assume that, for any $g \in G \setminus \{1_G\}$, the probability distribution of the element $u_1gu_1^{-1} \cdots u_Lgu_L^{-1}$, where $u_1, \ldots, u_L \leftarrow_R G$, is ε-close to the uniform distribution over G.

Now we define $\vec{w}_{\mathsf{AND}}(\vec{g}, \vec{g}') = (w_{\mathsf{AND},1}(\vec{g}, \vec{g}'), w_{\mathsf{AND},2}(\vec{g}, \vec{g}'))$ by

$$w_{\mathsf{AND},i}(\vec{g}, \vec{g}') = [r_1g_ir_1^{-1} \cdots r_Lg_ir_L^{-1}, r_{L+1}g_i'r_{L+1}^{-1} \cdots r_{2L}g_i'r_{2L}^{-1}] \text{ for } i = 1, 2$$

where $r_1, \ldots, r_{2L} \leftarrow_R G$ are common to both $i = 1, 2$. Then an argument similar to Sect. 4.4 implies that, for $\vec{g} \in X_b$ and $\vec{g}' \in X_{b'}$, we have $\vec{w}_{\mathsf{AND}}(\vec{g}, \vec{g}') \in X_{b \text{ AND } b'}$ *provided* $w_{\mathsf{AND},1}(\vec{g}, \vec{g}') \ne 1_G$. To evaluate the latter probability, we use the following result by Guralnick and Robinson (Guralnick and Robinson 2006):

Proposition 3 (Guralnick and Robinson 2006, Theorem 9) *For any non-commutative finite simple group H, we have*

$$\Pr_{h_1,h_2 \leftarrow_R H}[\,[h_1,h_2] = 1_H\,] \leq |H|^{-1/2} \ .$$

Then we have the following result, implying that $\vec{w}_{\mathsf{AND}}$ realizes AND:

Theorem 3 *Assume that Assumption 1 holds. Then for any $\vec{g}, \vec{g}' \in X_0 \cup X_1$, we have*

$$\Pr_{r_1,\ldots,r_{2L} \leftarrow_R G}[w_{\mathsf{AND},1}(\vec{g}, \vec{g}'; r_1, \ldots, r_{2L}) = 1_G] \leq |G|^{-1/2} + 2\varepsilon \ ,$$

which is negligible when both $1/|G|$ and ε are negligible.

Proof First, if $h_1 = r_1 g_1 r_1^{-1} \cdots r_L g_1 r_L^{-1}$ and $h_2 = r_{L+1} g_1' r_{L+1}^{-1} \cdots r_{2L} g_1' r_{2L}^{-1}$ were uniformly random over G, then we would have $w_{\mathsf{AND},1}(\vec{g}, \vec{g}'; r_1, \ldots, r_{2L}) = [h_1, h_2] = 1_G$ with probability at most $|G|^{-1/2}$ by Proposition 3. Now note that $g_1, g_1' \neq 1_G$ as $\vec{g}, \vec{g}' \in X_0 \cup X_1$; therefore Assumption 1 implies that the probability distributions of h_1 and h_2 are independent and both ε-close to the uniform distribution over G. Hence, in fact, we have $w_{\mathsf{AND},1}(\vec{g}, \vec{g}'; r_1, \ldots, r_{2L}) = 1_G$ with probability at most $|G|^{-1/2} + 2\varepsilon$. This completes the proof. □

5 Towards Achieving Secure Lift of Realization

In this section, we give some observations towards constructing a lift of a realization of operators that will yield a secure FHE scheme based on our framework in Sect. 3; concrete candidates for the secure construction are not yet obtained and are an open problem.

5.1 A Remark on the Choice of Random Variables

Here, we give a remark on random variables $\tilde{r}_h$ involved in a lift of a realization of functions. First, for realizations of functions using a uniform random variable on a given target group G, such as those in Sects. 4.4 and 4.5, it may happen that sampling a uniformly random element of the source group $\tilde{G}$ is not easy even if uniformly random sampling on G is easy. In such a case, owing to Proposition 2, a uniform random variable on G may be approximated as follows: random elements $g_1, \ldots, g_L$ of G are chosen at the beginning, and each random sampling on G is done by taking $g_1^{e_1} \cdots g_L^{e_L}$ with $e_1, \ldots, e_L \leftarrow_R \{0, 1\}$. Provided L is sufficiently large, this approximation will work well except for a negligible probability in choosing $g_1, \ldots, g_L$. Then the corresponding random variable on $\tilde{G}$ is easily obtained by first taking elements $\widetilde{g}_1, \ldots, \widetilde{g}_L$ of $\widetilde{G}$ with $\pi(\widetilde{g}_i) = g_i$ for each i and then, for each sampling, generating $\widetilde{g}_1^{e_1} \cdots \widetilde{g}_L^{e_L}$ with $e_1, \ldots, e_L \leftarrow_R \{0, 1\}$.

On the other hand, for the random variable r_{ker} used by the algorithm **Gen**, it may also happen that uniformly random sampling over the subgroup $\ker \pi \subseteq \widetilde{G}$ seems not easy. In this case, we may choose a large number of elements $g_1', \ldots, g_{L'}'$ of $\ker \pi$ first and then sample an element of $\ker \pi$ by randomly multiplying elements from $g_1', \ldots, g_{L'}'$. It is naively expected that the probability distribution of the resulting element of $\ker \pi$ will be significantly random if L' is sufficiently large.

5.2 Insecurity of a Matrix-Based Naive Construction

In order to exhibit the difficult point in the problem, here we show an example of an *insecure* construction of a lift of a realization of functions and explain why the resulting FHE scheme based on this construction is not secure.

We start with the realization of **AND** and **NOT** in $G = SL_2(\mathbb{F}_q)$ proposed in Sect. 4.4. We define the corresponding group $\widetilde{G}$ by

$$\widetilde{G} = \left\{ T \begin{pmatrix} A & B \\ 0 & C \end{pmatrix} T^{-1} \mid A \in SL_2(\mathbb{F}_q), B \in M_{2,k}(\mathbb{F}_q), C \in GL_k(\mathbb{F}_q) \right\},$$

where k is a parameter and $T \in GL_{k+2}(\mathbb{F}_q)$ is a fixed, randomly chosen matrix that must be secret. Then the group homomorphism $\pi : \widetilde{G} \to G$ is defined as follows: for $g \in \widetilde{G}$, $\pi(g)$ is obtained by first computing the $(k + 2) \times (k + 2)$ matrix $T^{-1}gT$ and then extracting the upper left 2×2 block of $T^{-1}gT$ (i.e. A in the description of $\widetilde{G}$ above). The conjugation by the random T in the definition of $\widetilde{G}$ intends to hide the internal block upper triangular structure of elements of $\widetilde{G}$.

However, this construction is not secure by the following reason (this attack was pointed out by an anonymous reviewer in a previous submission of this work). First, any matrix of the form $\begin{pmatrix} A & B \\ 0 & C \end{pmatrix}$ with $A = I \in SL_2(\mathbb{F}_q)$ satisfies a constraint "the $(2, 1)$-component is zero", which is a *linear* constraint in terms of matrix components. By taking conjugation by T, this constraint is changed to another one, which is unknown but still a *linear* constraint in terms of matrix components. We denote the resulting constraint by "$F(g) = 0$", namely, any element g of $\ker \pi$ satisfies $F(g) = 0$.

Now we consider the linear subspace $\mathsf{span}(\ker \pi)$ generated by the set $\ker \pi$ in the matrix *ring* $M_{k+2,k+2}(\mathbb{F}_q)$. By the choice of the *linear* constraint F, $\mathsf{span}(\ker \pi)$ is a linear subspace of the space $V = \{g \in M_{k+2,k+2}(\mathbb{F}_q) \mid F(g) = 0\}$. Now by collecting sufficiently many elements $h_1, \ldots, h_L$ of $\ker \pi$, it is expected that $\mathsf{span}(\ker \pi)$ is generated by these $h_1, \ldots, h_L$. In this case, for a given element $g \in \widetilde{G}$, if $g \in \ker \pi$, then adding g to the subspace $\mathsf{span}(h_1, \ldots, h_L)$ (which is now equal to $\mathsf{span}(\ker \pi)$) does not increase the dimension of the subspace. On the other hand, if $g \notin \ker \pi$, then the constraint $F(g) = 0$ is not satisfied with high probability, and now the dimension is increased when g is added to $\mathsf{span}(h_1, \ldots, h_L)$, as $\mathsf{span}(h_1, \ldots, h_L) \subsetneq V$ and $g \notin V$. This yields a way for an adversary to decide whether a given $g \in \widetilde{G}$ belongs to ker π or not (hence to break the proposed FHE) by only comparing the dimensions of $\mathsf{span}(h_1, \ldots, h_L)$ and $\mathsf{span}(h_1, \ldots, h_L, g)$, even if the actual constraint F is not known to the adversary. This example suggests that the existence of a non-trivial *linear* constraint for the set ker π will yield a powerful tool for the adversary.

5.3 Observation for Avoiding Linear Constraints

In order to realize group homomorphisms in our framework without linear constraints for the kernel discussed in Sect. 5.2, our idea here is to utilize combinatorial group theory. Roughly speaking, we say that a group H has a *presentation* $\langle X \mid R \rangle$, if X is a generating set of H, R is a set of group words with variables in X, and H is (isomorphic to) the quotient group of the free group generated by X modulo the relations "$r(\vec{x}) = 1$" for all words $r(\vec{x}) \in R$. See, e.g. Johnson (1997) for basics in combinatorial group theory. For example, it is well known that the symmetric group S_n on n letters admits a presentation of the form $\langle s_1, \ldots, s_{n-1} \mid (s_i s_j)^{\Gamma(i,j)} (i, j = 1, \ldots, n-1) \rangle$ where each s_i is the adjacent transposition $(i, i+1)$ and Γ is a matrix given by $\Gamma(i, i) = 1$, $\Gamma(i, i+1) = \Gamma(i+1, i) = 3$, and $\Gamma(i, j) = 2$ when $|i - j| \geq 2$. (This is actually the Coxeter group of type A_{n-1}; see, e.g. Humphreys 1990 for basic theory of the Coxeter groups.) On the other hand, it is known that for any prime $p > 3$, the groups $SL_2(\mathbb{F}_p)$ and $PSL_2(\mathbb{F}_p)$ admit "compact" presentations with four generators and eight relations of lengths $O(\log p)$; see Theorem 3.6 and Remark 3.7 of Guralnick et al. (2008).

Our idea is based on the following fact implied by the fundamental theorem on homomorphisms for groups; if two groups H_1 and H_2 have presentations $\langle X \mid R_1 \rangle$ and $\langle X \mid R_2 \rangle$ with the same generating set X, and if every $r \in R_1$ is also equal to the unit element in H_2, then the identity map $X \to X$ induces a surjective group homomorphism $H_1 \to H_2$. As this kind of group homomorphism is obtained by a mechanism completely different from linear algebra, it is (naively) expected that such an approach would yield a desired group homomorphism without linear constraints.

Based on the argument above, we propose the following approach towards constructing a secure group homomorphism for our framework for FHE:

1. Take the group G associated to a realization of operations for plaintexts.
2. Take a semidirect product $H \rtimes G$ with a certain (possibly trivial) finite group H. Here, we require that a presentation of $H \rtimes G$ is efficiently computable. For example, when it is the direct product $H \times G$ and presentations for G and H are known, a presentation of $H \times G$ is obtained by introducing additional relations "generators of G and generators of H are mutually commutative" (see, e.g. Johnson 1997).

3. Let $\langle X \mid R_2 \rangle$ be the presentation of $H \rtimes G$. Then find a finite group $\widetilde{G}_0$ with presentation of the form $\langle X \mid R_1 \rangle$ and the associated surjective group homomorphism $\widetilde{G}_0 \to H \rtimes G$ as above.

4. Finally, randomly choose a group isomorphism from another group $\widetilde{G}$ to $\widetilde{G}_0$ in a certain way, subject to the condition that the group $\widetilde{G}$ admits a "compact" expression that yields efficient group operators for $\widetilde{G}$. Then the composition $\widetilde{G} \xrightarrow{\sim} \widetilde{G}_0 \to H \rtimes G \to G$ (where the last mapping is the natural projection) gives a candidate of the surjective homomorphism $\pi : \widetilde{G} \to G$.

In Step 4 of the approach described above, an easiest candidate of the "compact" expressions for the groups $\widetilde{G}_0$ and $\widetilde{G}$ is matrix expressions, i.e. embedding these groups into some matrix group. Now a candidate of the random isomorphism between them is taking the conjugation by a random secret matrix, just as in Sect. 5.2. In this case, due to the argument in Sect. 5.2, the kernel of the homomorphism $\widetilde{G}_0 \to H \rtimes G$ must avoid a linear constraint. Here we note that, even though the homomorphism from $\widetilde{G}_0 = \langle X \mid R_1 \rangle$ to $H \rtimes G = \langle X \mid R_2 \rangle$ is based on the mechanism of combinatorial group theory, this does not always guarantee that the resulting homomorphism is free from linear constraints.

For example, let $\widetilde{G}_0$ be the Coxeter group of type B_n, with presentation

$$\langle s_1, \ldots, s_n \mid (s_i s_j)^{\Gamma'(i,j)} \ (i, j = 1, \ldots, n) \rangle,$$

where $\Gamma'(i, j) = \Gamma(i, j)$ for $i, j \in \{1, \ldots, n-1\}$, $\Gamma'(n, n) = 1$, $\Gamma'(n, n-1) = \Gamma'(n-1, n) = 4$, and $\Gamma'(n, i) = \Gamma'(i, n) = 2$ for $1 \leq i \leq n-2$. If the value of $\Gamma'(n, n-1) = \Gamma'(n-1, n)$ is changed from 4 to 2, then it results in the direct product $S_n \times H$ with $H = \langle s_n \rangle$ being the cyclic group of order two. This implies that there is a natural surjective homomorphism $\widetilde{G}_0 \to S_n \times H$; hence, we obtain a surjective homomorphism $\widetilde{G}_0 \to S_n \times H \to S_n = G$. Now by using the expression of $\widetilde{G}_0$ as a "signed" permutation group (see, e.g. Humphreys 1990), it can be proved that the kernel of $\widetilde{G}_0 \to G$ is an elementary abelian 2-group generated by the elements $s_j s_{j+1} \cdots s_{n-1} s_n s_{n-1} \cdots s_{j+1} s_j$ with $j = 1, \ldots, n$. Moreover, in the standard matrix representation for the Coxeter groups (see, e.g. Humphreys 1990), these elements $s_j s_{j+1} \cdots s_{n-1} s_n s_{n-1} \cdots s_{j+1} s_j$ are all expressed as lower triangular matrices. Hence, the kernel of the homomorphism above has a linear constraint "upper triangular components are 0", which is not desirable. We also note that, owing to the classification result on finite Coxeter groups (see, e.g. Humphreys 1990), the group of type B_n mentioned above is essentially (i.e. without using direct products) the unique choice for a surjective, but not bijective, homomorphism from a finite Coxeter group onto the group S_n with $n \geq 5$. Consequently, the candidates for the group $\widetilde{G}_0$ in the case $G = S_n$ should be searched from outside the class of the Coxeter groups. Finding a concrete candidate for $\widetilde{G}_0$ in this case is left as an open problem.

5.4 Another Trial Using Tietze Transformations

Another trial for realizing the approach in Sect. 5.3 is as follows. Recall that, we are supposing that the group $H \rtimes G$ has a presentation of the form $\langle X \mid R_2 \rangle$. When the presentation is constructed naively, it might happen that the natural projection $H \rtimes G \to G$ is easy to compute by using the presentation of the group. Now the idea is choosing $\widetilde{G}_0 = H \rtimes G$ and constructing the isomorphic group $\widetilde{G}$ by randomly rewriting the original presentation $\langle X \mid R_2 \rangle$ while keeping the isomorphic class of groups. By letting the rewriting process be a part of the secret key, it is expected to be difficult to compute the map $\widetilde{G} \xrightarrow{\sim} H \rtimes G \to G$ without the secret key, while the secret key enables to compute the map by reversing the rewriting process above.

Such a rewriting of presentations that keeps the group isomorphic can be performed by using *Tietze transformation*. Namely, the following fact is known:

Lemma 5 (see, e.g. Johnson 1997) *Given a presentation $\langle X \mid R \rangle$ of a group, let w be a group word with variables in X and let y be a symbol not belonging to X. Then, the group $\langle X \cup \{y\} \mid R \cup \{wy^{-1}\} \rangle$ is isomorphic to $\langle X \mid R \rangle$ where each element of X in the group $\langle X \mid R \rangle$ corresponds to the same element in the group $\langle X \cup \{y\} \mid R \cup \{wy^{-1}\} \rangle$.*

We also have the following result, which utilizes presentations of the trivial group:

Lemma 6 *Given a presentation $\langle X \mid R \rangle$ of a group, let $\langle Y \mid T \rangle$ be a presentation of the trivial group (i.e. the group with a single element), and for each $y \in Y$, choose an element r_y of R. Let $T(r_y \mid y \in Y)$ denote the set of words of the form $t(r_y \mid y \in Y)$ with $t(\vec{y}) \in T$, where $t(r_y \mid y \in Y)$ denotes the group word with variables in X obtained by substituting the word r_y into the variable y in the word $t(\vec{y})$ for each $y \in Y$. Then the subsets R and $R' = (R \setminus \{r_y \mid y \in Y\}) \cup T(r_y \mid y \in Y)$ have the same normal closure in the free group $\mathsf{Free}(X)$ generated by X; therefore, $\langle X \mid R' \rangle$ is isomorphic to $\langle X \mid R \rangle$.*

Proof The definition of the words $t(r_y \mid y \in Y)$ implies that R' is a subset of the normal closure $\langle R \rangle_{\mathrm{normal}}$ of R. To prove the opposite relation $R \subseteq \langle R' \rangle_{\mathrm{normal}}$, it suffices to show that $r_y \in \langle R' \rangle_{\mathrm{normal}}$ for each $y \in Y$. Now as $\langle Y \mid T \rangle$ is a trivial group, y is the product of words of the form $u(\vec{y})t(\vec{y})u(\vec{y})^{-1}$ with $u(\vec{y}) \in \mathsf{Free}(Y)$ and $t(\vec{y}) \in T$. By substituting the word $r_{y'}$ into the variable y' for each $y' \in Y$, it follows that r_y is the product of words of the form $u(r_{y'} \mid y' \in Y)t(r_{y'} \mid y' \in Y)u(r_{y'} \mid y' \in Y)^{-1}$ with $u(r_{y'} \mid y' \in Y) \in \mathsf{Free}(X)$ and $t(r_{y'} \mid y' \in Y) \in T(r_{y'} \mid y' \in Y)$. This implies that $r_y \in \langle R' \rangle_{\mathrm{normal}}$, as desired. This completes the proof. $\square$

We note that the current idea of randomly rewriting the presentation of the group $H \rtimes G$ has (at least) one unsolved problem from the viewpoint of efficiency and two from the viewpoint of security. For the efficiency, we recall that the expression of the resulting group $\widetilde{G}$ should enable efficient computation for group operators. However, with a randomly chosen presentation $\langle X \mid R \rangle$ of $\widetilde{G}$, in general, it seems not easy to compute the product of two elements. More precisely, each element of $\widetilde{G}$ is now expressed as a group word on X, and the product corresponds to the concatenation of the two words. This concatenation of words increases the length of the word; therefore, the word has to be replaced with a shorter equivalent word by using relations in R before the word length becomes too long. However, this process of reducing the word length by using the relations in R is not efficient in general. It is an open problem to develop rewriting methods for group presentations while keeping efficiency of group operations.

From the viewpoint of security, first, it has not been evaluated how many random rewriting steps for the presentation of the group are sufficient to securely conceal the structure of the group. On the other hand, even if the sufficient number of the rewriting steps has been estimated, it may still happen that the resulting FHE scheme is not secure when the component H in $H \rtimes G$ is not appropriately chosen.

Namely, let $E = E(g)$ be a (deterministic) group word, which we call an "equation" over groups. We suppose that both of the probabilities $\mathrm{Pr}_{u \leftarrow_R H}[E(u) = 1]$ and $\mathrm{Pr}_{u \leftarrow_R H \rtimes G}[E(u) \neq 1]$ are non-negligible and at least one of them is noticeable. Then an adversary can distinguish a random element of $\ker \pi \simeq H$ (where $\pi : \widetilde{G} \to G$) from a random element of $\widetilde{G} \simeq H \rtimes G$ by checking whether a given random element u satisfies $E(u) = 1$ or not. Hence, it should be difficult to find a non-trivial equation E for which $\mathrm{Pr}_{u \leftarrow_R H}[E(u) = 1]$ is non-negligible.

For example, when the underlying group is the direct product $H \times G$, it should not be feasible to find a non-identity element w of the group for which its H-component is an identity element. Indeed, for any such "target" element w, it commutes with every element of $H \subseteq H \times G$, while it is likely not commutative with a random element of $H \times G$. Hence, the equation $E(g) = [w, g]$ will satisfy the attacking condition above. In particular, H should satisfy $|H| \geq 2^{2\lambda}$ for security parameter λ due to Birthday Paradox, as a collision in the H-components of two elements yields a target element. Moreover, the center of H should not be large, as otherwise the commutator $[w_1, w_2]$ for random elements w_1, w_2 will yield a target element with high probability.

For a general case of the semidirect product $H \rtimes G$, a candidate of such an equation E is $E(g) = g^k$ for some fixed value k; therefore, it is important to study the distribution of the orders of elements in H. For example, suppose that $H = A_\ell$ with $\ell \geq 4$. Let p be the largest odd prime with $p \leq \ell$. Then the number of elements of A_ℓ that are cyclic permutations on p letters is $\binom{\ell}{p}(p-1)! = \dfrac{2}{p \cdot (\ell - p)!} \cdot |A_\ell|$. This implies that $\displaystyle \Pr_{u \leftarrow_R H}[u^p = 1] = \dfrac{2}{p \cdot (\ell - p)!} + \dfrac{1}{|H|!}$. As $\ell - p$ is small for reasonable choices of ℓ (e.g. $\ell - p \leq 6$ for $\ell \leq 80$), the probability above is significantly high, which is not desirable to avoid the attack above.

On the other hand, we consider the choice $H = \mathrm{SL}_2(\mathbb{F}_q)$ for an odd prime q for which $1/q$ is negligible, and study the element orders in the group. Following the argument in Sect. 5.2 of Fulton and Harris (1991), we choose a generator ζ of the cyclic group $(\mathbb{F}_q)^\times$. Put $A_i = \begin{pmatrix} \zeta^i & 0 \\ 0 & \zeta^{-i} \end{pmatrix}$ for $i = 0, 1, \ldots, q - 2$. On the other hand, by considering the quadratic extension field $\mathbb{F}_{q^2}$ of $\mathbb{F}_q$, ζ has a square root $\sqrt{\zeta}$

Table 1 The conjugacy classes in $SL_2(\mathbb{F}_q)$ for odd prime $q > 3$ (see the text for notations)

Type	Representative x in the class	Cardinality	Order of x
1	$\begin{pmatrix} 1 & 0 \\ 0 & 1 \end{pmatrix}$	1	1
2	$\begin{pmatrix} -1 & 0 \\ 0 & -1 \end{pmatrix}$	1	2
3	$\begin{pmatrix} 1 & 1 \\ 0 & 1 \end{pmatrix}$	$\dfrac{q^2 - 1}{2}$	q
4	$\begin{pmatrix} 1 & \zeta \\ 0 & 1 \end{pmatrix}$	$\dfrac{q^2 - 1}{2}$	q
5	$\begin{pmatrix} -1 & 1 \\ 0 & -1 \end{pmatrix}$	$\dfrac{q^2 - 1}{2}$	$2q$
6	$\begin{pmatrix} -1 & \zeta \\ 0 & -1 \end{pmatrix}$	$\dfrac{q^2 - 1}{2}$	$2q$
7-i	$A_i \left(1 \le i < \dfrac{q-1}{2} \right)$	$q^2 + q$	$\dfrac{q-1}{\gcd(q-1, i)}$
8-i	$B_{(q-1)i}$ $\left(1 \le i < \dfrac{q+1}{2} \right)$	$q^2 - q$	$\dfrac{q+1}{\gcd(q+1, i)}$

in $(\mathbb{F}_{q^2})^\times \setminus (\mathbb{F}_q)^\times$ (as q is odd). This yields a bijection $\mathbb{F}_q \times \mathbb{F}_q \to \mathbb{F}_{q^2}$, $(a, b) \mapsto a + b\sqrt{\zeta}$. Choose a generator υ of the cyclic group $(\mathbb{F}_{q^2})^\times$. For $i = 0, 1, \ldots, q^2 - 2$, put $B_i = \begin{pmatrix} a & b \\ b\zeta & a \end{pmatrix}$ where a, b satisfy $\upsilon^i = a + b\sqrt{\zeta}$. By using these notations, the list of conjugacy classes in $SL_2(\mathbb{F}_q)$ is obtained as in Table 1, where the second and the third columns are quoted (with slightly different notations) from Sect. 5.2 of Fulton and Harris (1991).

In Table 1, the ratio to $|H|$ of the cardinality of each conjugacy class of type 1 to 6 is at most a negligible value $\dfrac{(q^2 - 1)/2}{q(q^2 - 1)} = \dfrac{1}{2q}$; therefore, these conjugacy classes can be ignored. On the other hand, for each divisor k of $q - 1$, an element x of the conjugacy class of type 7-i satisfies $x^k = 1$ if and only if i is a multiple of $(q - 1)/k$. Therefore, the number of such elements x is at most $\dfrac{(q - 1)/2}{(q - 1)/k}(q^2 + q) = \dfrac{k}{2}(q^2 + q)$, whose ratio to $|H| = q(q^2 - 1)$ is $\dfrac{k}{2(q - 1)}$. To make the ratio non-negligible, one must find a divisor k of $q - 1$ which is almost as large as $q - 1$; this is expected to be difficult *provided the size q of the coefficient field $\mathbb{F}_q$ is not known*. The same also holds for conjugacy classes of type 8. Summarizing, the attack using the equations of the form $E(g) = g^k$ will be not effective for the group $H = SL_2(\mathbb{F}_q)$ provided the size of the coefficient field $\mathbb{F}_q$ is appropriately concealed by the random rewriting of the presentation of the group. A further analysis of attacks using other kind of equations will be a future research topic.

Acknowledgements The author thanks members of Shin-Akarui-Angou-Benkyou-Kai for their helpful comments. In particular, the author thanks Shota Yamada for inspiring the author with motivation to this work, and Takashi Yamakawa, Takahiro Matsuda, Keita Emura, Yoshikazu Hanatani, Jacob C. N. Schuldt, and Goichiro Hanaoka for giving many precious comments on the work. The author also thanks the anonymous reviewers of previous submissions of the paper for their careful reviews and valuable comments. This work was supported by JST PRESTO Grant Number JPMJPR14E8, JST CREST Grant Number JPMJCR14D6, and JSPS KAKENHI Grant Number JP19H01804.

References

D.A. Barrington, Bounded-width polynomial-size branching programs recognize exactly those languages in NC^1, in *Proceedings of STOC 1986*, pp. 1–5 (1986)

Z. Brakerski, V. Vaikuntanathan, Efficient fully homomorphic encryption from (Standard) LWE, in *Proceedings of FOCS 2011*, pp. 97–106 (2011)

J.H. Cheon, D. Stehlé, Fully homomophic encryption over the integers revisited, in *Proceedings of EUROCRYPT 2015* (1), LNCS 9056, pp. 513–536 (2015)

I. Chillotti, N. Gama, M. Georgieva, M. Izabachène, Faster fully homomorphic encryption: bootstrapping in less than 0.1 seconds, in *Proceedings of ASIACRYPT 2016* (1), LNCS 10031, pp. 3–33 (2016)

M. van Dijk, C. Gentry, S. Halevi, V. Vaikuntanathan, Fully homomorphic encryption over the integers, in *Proceedings of EUROCRYPT 2010*, LNCS 6110, pp. 24–43 (2010)

J.D. Dixon, Generating random elements in finite groups. Electron. J. Comb. **15** (2008), no. R94

L. Ducas, D. Micciancio, FHEW: bootstrapping homomorphic encryption in less than a second, in *Proceedings of EUROCRYPT 2015* (1), LNCS 9056, pp. 617–640 (2015)

W. Fulton, J. Harris, *Representation Theory*, vol. 129 (Springer, Berlin, 1991)

C. Gentry, Fully homomorphic encryption using ideal lattices, in *Proceedings of STOC 2009*, pp. 169–178 (2009)

C. Gentry, S. Halevi, Fully homomorphic encryption without squashing using depth-3 arithmetic circuits, in *Proceedings of FOCS 2011*, pp. 107–109 (2011)

C. Gentry, S. Halevi, N.P. Smart, Better bootstrapping in fully homomorphic encryption, in *Proceedings of PKC 2012*, LNCS 7293, pp. 1–16 (2012)

R.M. Guralnick, W.M. Kantor, M. Kassabov, A. Lubotzky, Presentations of finite simple groups: a quantitative approach. J. Am. Math. Soc. **21**, 711–774 (2008)

R.M. Guralnick, G.R. Robinson, On the commuting probability in finite groups. J. Algebr. **300**, 509–528 (2006)

J.E. Humphreys, *Reflection Groups and Coxeter Groups* (Cambridge University Press, Cambridge, 1990)

D.L. Johnson, *Presentations of Groups*, vol. 15, 2nd edn., London Mathematical Society Student Texts (Cambridge University Press, Cambridge, 1997)

N. Khamsemanan, R. Ostrovsky, W.E. Skeith III, On the black-box use of somewhat homomorphic encryption in noninteractive two-party protocols. SIAM J. Discret. Math. **30**(1), 266–295 (2016)

K. Nuida, K. Kurosawa, (Batch) Fully homomorphic encryption over integers for non-binary message spaces, in *Proceedings of EUROCRYPT 2015* (1), LNCS 9056, pp. 537–555 (2015)

R. Ostrovsky, W.E. Skeith III, Communication complexity in algebraic two-party protocols, in *Proceedings of CRYPTO 2008*, LNCS 5157, pp. 379–396 (2008)

D.J.S. Robinson, *A Course in the Theory of Groups*, vol. 80, 2nd edn. (Springer, Berlin, 1996)

A. Silverberg, Fully homomorphic encryption for mathematicians. IACR Cryptology ePrint Archive 2013/250 (2013). http://eprint.iacr.org/2013/250

D. Stehlé, R. Steinfeld, Faster fully homomorphic encryption, in *Proceedings of ASIACRYPT 2010*, LNCS 6477, pp. 377–394 (2010)

From the Bloch Sphere to Phase-Space Representations with the Gottesman–Kitaev–Preskill Encoding

L. García-Álvarez, A. Ferraro, and G. Ferrini

Abstract In this work, we study the Wigner phase-space representation of qubit states encoded in continuous variables (CV) by using the Gottesman–Kitaev–Preskill (GKP) mapping. We explore a possible connection between resources for universal quantum computation in discrete-variable (DV) systems, i.e. non-stabilizer states, and negativity of the Wigner function in CV architectures, which is a necessary requirement for quantum advantage. In particular, we show that the lowest Wigner logarithmic negativity corresponds to encoded stabilizer states, while the maximum negativity is associated with the most non-stabilizer states, H-type and T-type quantum states.

Keywords Continuous variables quantum computation · Quantum advantage · Wigner function · Wigner logarithmic negativity · Gottesman–Kitaev–Preskill code

1 Introduction

Quantum computers, i.e. quantum devices in which information can be encoded, processed, and read out, are predicted to solve certain computational problems faster than classical computers Shor (1999). Specifically, a problem is said to be hard to solve if its solution requires a number of steps exponential in the size of the input, while polynomial time solutions are called efficient. An example of a problem believed to be hard to solve classically that can be efficiently solved by a quantum computer is factorization. While known classical algorithms factorize integer numbers in a time

L. García-Álvarez (✉) · G. Ferrini
Department of Microtechnology and Nanoscience (MC2), Chalmers University of Technology, 412 96 Göteborg, Sweden
e-mail: lauraga@chalmers.se

A. Ferraro
Centre for Theoretical Atomic, Molecular and Optical Physics, Queen's University Belfast, Belfast BT7 1NN, UK

© The Author(s) 2021   79
T. Takagi et al. (eds.), *International Symposium on Mathematics, Quantum Theory, and Cryptography*, Mathematics for Industry 33, https://doi.org/10.1007/978-981-15-5191-8_9

which scales exponentially with the size of the integer to factor, a quantum algorithm exists that only requires a polynomial time.

This technologically appealing property is referred to as *quantum advantage*, and has recently motivated the undertaking of a global effort toward building a quantum computer. However, a conclusive experimental evidence of quantum advantage for computation is still lacking, since it has not yet been possible to build a quantum computer with enough elementary components to practically beat classical machines. Furthermore, the ultimate origin of quantum advantage is still unclear.

The traditional approach to encode information in quantum systems, based on two-level quantum systems with finite-dimensional Hilbert spaces, i.e. qubits, is an example of the discrete-variable (DV) approach. An alternative approach for information encoding uses continuous variables (CVs), i.e. quantized variables with a continuous spectrum, such as the amplitude (q) and phase (p) quadratures of the quantized electromagnetic field, defined in an infinite-dimensional Hilbert space. Within this approach, one million optical modes have been entangled Yoshikawa et al. (2016), Chen et al. (2014). Beyond the optical realm, new CV implementations are studied in opto-mechanics Aspelmeyer et al. (2014) and with microwaves coupled to superconducting devices Ofek et al. (2016), Wilson et al. (2011), where high-order nonlinearities can be engineered.

A fundamental tool for studying a classical dynamical system is the probability distribution on a phase space in which all possible states of the system are represented. Similarly, quantum systems can be conveniently and unambiguously described with quasi-probability distributions defined on the classical phase space Wigner (1932), Hillery et al. (1984), Gibbons et al. (2004). Although these useful mathematical constructs, such as the Wigner function, retain some properties of classical probability distributions, they can take negative values for quantum states.

A series of theorems has progressively narrowed down the characteristics that both DV and CV quantum computing architectures must possess in order to display quantum advantage. In DV quantum information processors, the Gottesman–Knill theorem states that the so-called Clifford circuits, which are composed, for example, of Hadamard, $\pi/2$-phase, and CNOT gates, when acting on stabilizer states, i.e. those generated with Clifford gates acting on the initial n-qubit register $|0\rangle_1 \otimes |0\rangle_2 \otimes \cdots \otimes |0\rangle_n$, and followed by a Pauli measurement, can be efficiently simulated on a classical computer Gottesman (1999), Aaronson and Gottesman (2004). Non-stabilizer pure states are called magic, and are hence necessary to yield quantum advantage when acted on by Clifford circuits with Pauli measurements Bravyi et al. (2005). In CV quantum computation, it has been shown firstly that circuits with input, evolution, and measurements solely described by Gaussian Wigner functions are efficiently simulatable by classical computers Bartlett et al. (2002). Later it was shown that negativity of the Wigner function is a necessary requirement for quantum advantage, since quantum states and operations with positive Wigner functions (strictly including Gaussian circuits) can be classically efficiently simulated Mari and Eisert (2012). Minimal extensions of positive Wigner function circuits that exhibit quantum advantage, where either the input, or the evolution, or the measurement are described by negative Wigner functions, have been studied Chabaud et al. (2017), Douce et al. (2017), Hamilton et al. (2017), Chakhmakhchyan and Cerf (2017), Douce et al. (2019). Finally, the criteria for efficient classical simulatability have been extended by using other phase-space representations, namely Husimi and Glauber–Sudarshan Rahimi et al. (2016).

A bridge between the DV and the CV worlds is provided by CV-codes, i.e. by sets of CV states that allow for encoding DV states such that orthogonal wavefunctions represent different DV states. One such example is the Gottesman–Kitaev–Preskill (GKP) code, where the qubit logical states are encoded in trains of delta functions at different locations Gottesman et al. (2001). The encoding of discrete quantum information into infinite-dimensional quantum systems is used to get a high-quality qubit protected from environmental noise Menicucci (2014). The GKP code is particularly suitable for our analysis since Clifford gates on the qubit encoded states are given by Gaussian operations, which in principle lead us to an analogy between DV and CV requirements for classical efficient simulatability of quantum operations.

In this manuscript, we analyze the negativity of the Wigner function for any single-qubit state mapped in CV architectures with the GKP code, with the aim of establishing a relation between DV and CV criteria for quantum advantage. In Sect. 2, we review in detail the GKP code that we use in our work. In Sect. 3, we compute the Wigner function of any single-qubit GKP encoded state, and we compare the results for encoded stabilizer and non-stabilizer states. In Sect. 4, we quantify the negativity of the Wigner function for both cases, and we observe that stabilizer encoded states saturate the lower bound of negativity, while the most non-stabilizer states, also known as magic states, show the maximum amount of negativity. We conclude in Sect. 5 with our final remarks.

2   GKP Encoding of Qubit States

The formal GKP encoding maps a qubit into an oscillator using non-normalizable superpositions of infinitely squeezed states in the position q and momentum p quadratures of the oscillator Gottesman et al. (2001). We review the GKP qubit states used in this work, which are defined as

$$|0\rangle = \sum_{s=-\infty}^{\infty} |q = 2\sqrt{\pi}s\rangle$$

$$|1\rangle = \sum_{s=-\infty}^{\infty} |q = \sqrt{\pi}(1+2s)\rangle, \tag{1}$$

for which the wavefunction $\Psi(q) = \langle q|\Psi\rangle$ is a sum of delta functions, since $\langle q|q = x\rangle = \delta(x)$.

In practice, the qubit states must be normalizable, and thus are defined approximating the previous expression with finitely squeezed states, and weighting the infinite sum of squeezed states by a Gaussian envelope. The approximated states are quasi-orthogonal states given by

$$
|\bar{0}\rangle \propto \sum_{s=-\infty}^{\infty} \int_{-\infty}^{\infty} e^{-2\pi\kappa^2 s^2} e^{-\frac{(q-2\sqrt{\pi}s)^2}{2\sigma^2}} |q\rangle dq
$$

$$
|\bar{1}\rangle \propto \sum_{s=-\infty}^{\infty} \int_{-\infty}^{\infty} e^{-2\pi\kappa^2 s^2} e^{-\frac{(q-(2s+1)\sqrt{\pi})^2}{2\sigma^2}} |q\rangle dq, \tag{2}
$$

with κ^{-1}, the width of the Gaussian envelope, and σ, the width of the Gaussian peaks substituting the delta functions. These imperfect GKP states are suitable for numerical computations but introduce a probability of error in the identification of $|\bar{0}\rangle$ and $|\bar{1}\rangle$. In our calculations, we use the perfect GKP states given in Eq. (1) for obtaining analytical results, and imperfect GKP states in Eq. (2) for numerical results.

3 Phase-Space Wigner Representation of GKP Encoded States

The Wigner function of a pure state $|\Psi\rangle$ is defined as

$$
W(q, p) \equiv \frac{1}{2\pi} \int_{-\infty}^{\infty} dx e^{ipx} \Psi\left(q + \tfrac{x}{2}\right)^* \Psi\left(q - \tfrac{x}{2}\right), \tag{3}
$$

with $\Psi(x) = \langle x|\Psi\rangle$ the wavefunction of the quantum system.

We consider infinitely squeezed GKP states, that is, the ideal logical qubit GKP states $|j\rangle$ with $j = 0, 1$ given in Eq. (1). The corresponding Wigner function reads Gottesman et al. (2001)

$$
W_j(q, p) = \frac{1}{4\sqrt{\pi}} \sum_{st} (-1)^{st} \delta\left(p - \tfrac{\sqrt{\pi}}{2}s\right) \delta\left(q - \sqrt{\pi}j - \sqrt{\pi}t\right). \tag{4}
$$

We now take into account arbitrary pure qubit states given by superpositions of GKP states as $|\Psi\rangle = \cos\frac{\theta}{2}|0\rangle + e^{i\phi}\sin\frac{\theta}{2}|1\rangle$, which can be represented in the surface of the Bloch sphere as shown in Fig. 1. The Wigner function for a qubit state depends consequently on the the angles θ, ϕ of its Bloch sphere representation. It reads

$$
\begin{aligned}
W(\theta, \phi; q, p) = \frac{1}{2\pi} \int_{-\infty}^{\infty} dx e^{ipx} \Big[&\cos^2\tfrac{\theta}{2}\Psi_0\left(q + \tfrac{x}{2}\right)^* \Psi_0\left(q - \tfrac{x}{2}\right) \\
&+ \sin^2\tfrac{\theta}{2}\Psi_1\left(q + \tfrac{x}{2}\right)^* \Psi_1\left(q - \tfrac{x}{2}\right) \\
&+ \cos\tfrac{\theta}{2}\sin\tfrac{\theta}{2}e^{i\phi}\Psi_0\left(q + \tfrac{x}{2}\right)^* \Psi_1\left(q - \tfrac{x}{2}\right) \\
&+ \cos\tfrac{\theta}{2}\sin\tfrac{\theta}{2}e^{-i\phi}\Psi_1\left(q + \tfrac{x}{2}\right)^* \Psi_0\left(q - \tfrac{x}{2}\right) \Big],
\end{aligned} \tag{5}
$$

Fig. 1 Geometrical representation of pure qubit states in the Bloch sphere

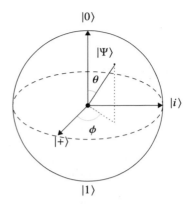

with Ψ_i, $i = 0, 1$, the wavefunctions corresponding to the GKP states $|i\rangle$, $i = 0, 1$. A detailed derivation can be found in Appendix 1. Explicitly, we have

$$W(\theta, \phi; q, p) = \cos^2 \tfrac{\theta}{2} W_0(q, p) + \sin^2 \tfrac{\theta}{2} W_1(q, p)$$
$$+ \frac{\sin\theta}{4\sqrt{\pi}} \sum_{st} (-1)^{st} \cos\left(\phi + s\tfrac{\pi}{2}\right) \delta\left(q - \tfrac{\sqrt{\pi}}{2}(1+2t)\right) \delta\left(p - \tfrac{s\sqrt{\pi}}{2}\right), \quad (6)$$

which can be pictured in a grid of square cells of $\Delta q = \Delta p = \frac{\sqrt{\pi}}{2}$. By analyzing Eqs. (4) and (6), we thus observe that the Wigner function consists of a sum of delta functions positioned at all the sites of the lattice in phase space with coordinates $(l, m) \equiv (q = l\frac{\sqrt{\pi}}{2}, p = m\frac{\sqrt{\pi}}{2})$ for l and m integer numbers. The coefficients for each site are given by

$$w_{lm}(\theta, \phi) = \begin{cases} \frac{1}{4\sqrt{\pi}}\left(\cos^2\tfrac{\theta}{2} + \sin^2\tfrac{\theta}{2}\right) & \text{for } l \text{ even, } m \text{ even} \\[4pt] \frac{1}{4\sqrt{\pi}}\left(\cos^2\tfrac{\theta}{2} - \sin^2\tfrac{\theta}{2}\right) & \text{for } l = 4u, \, m \text{ odd} \\[4pt] \frac{1}{4\sqrt{\pi}}\left(\sin^2\tfrac{\theta}{2} - \cos^2\tfrac{\theta}{2}\right) & \text{for } l = 4u+2, \, m \text{ odd} \\[4pt] \frac{1}{4\sqrt{\pi}} \sin\theta\cos\phi & \text{for } \begin{cases} l = 4u+3, \, m = 4v \\ l = 4u+1, \, m = 4v \end{cases} \\[10pt] \frac{-1}{4\sqrt{\pi}} \sin\theta\cos\phi & \text{for } \begin{cases} l = 4u+3, \, m = 4v+2 \\ l = 4u+1, \, m = 4v+2 \end{cases} \\[10pt] \frac{-1}{4\sqrt{\pi}} \sin\theta\sin\phi & \text{for } \begin{cases} l = 4u+3, \, m = 4v+3 \\ l = 4u+1, \, m = 4v+1 \end{cases} \\[10pt] \frac{1}{4\sqrt{\pi}} \sin\theta\sin\phi & \text{for } \begin{cases} l = 4u+3, \, m = 4v+1 \\ l = 4u+1, \, m = 4v+3 \end{cases} \end{cases} \quad (7)$$

with u and v integer numbers.

In particular, we consider the six single-qubit stabilizer pure states, corresponding to the eigenvectors of the Pauli matrices σ_x, σ_y, and σ_z,

$$\sigma_x : \quad |+\rangle = \frac{1}{\sqrt{2}}(|0\rangle + |1\rangle) \quad |-\rangle = \frac{1}{\sqrt{2}}(|0\rangle - |1\rangle),$$

$$\sigma_y : \quad |i\rangle = \frac{1}{\sqrt{2}}(|0\rangle + i|1\rangle) \quad |-i\rangle = \frac{1}{\sqrt{2}}(|0\rangle - i|1\rangle),$$

$$\sigma_z : \quad |0\rangle \qquad\qquad\qquad |1\rangle. \tag{8}$$

The Wigner functions of single-qubit stabilizer states mapped in CV via the GKP code are shown in Fig. 2. We observe a similar pattern repeated periodically and isotropically in the whole phase space, with one quarter of negative delta functions with respect to the total amount of peaks. It is possible to obtain from the initial state $|0\rangle$ all stabilizer states with Clifford operations, which for a single qubit are generated in DV by the Hadamard H, and $\frac{\pi}{2}$-phase gates $R_{\frac{\pi}{2}}$,

$$H : \quad |0\rangle \rightarrow |+\rangle, \quad |1\rangle \rightarrow |-\rangle,$$

$$R_{\frac{\pi}{2}} : \quad |0\rangle \rightarrow |0\rangle, \quad |1\rangle \rightarrow e^{i\frac{\pi}{2}}|1\rangle. \tag{9}$$

With the GKP encoding, these gates in CV correspond to the Fourier transform F, and the $\pi/2$-phase gate P, which are the symplectic transformations

$$F : \quad q \rightarrow p, \quad p \rightarrow -q,$$

$$P : \quad q \rightarrow q, \quad p \rightarrow p - q. \tag{10}$$

Let us consider now the single-qubit magic states $|T\rangle$ and $|H\rangle$,

$$|T\rangle = \cos\frac{\theta}{2}|0\rangle + \sin\frac{\theta}{2}e^{i\frac{\pi}{4}}|1\rangle \quad \text{with} \quad \theta = \arccos\left(\frac{1}{\sqrt{3}}\right)$$

$$|H\rangle = \frac{1}{\sqrt{2}}\left(|0\rangle + e^{i\frac{\pi}{4}}|1\rangle\right), \tag{11}$$

which are the maximal non-stabilizer states in the Bloch sphere and in the equatorial plane of the Bloch sphere, respectively Bravyi et al. (2005). There are 8 T-type magic states and 12 H-type magic states, which can be obtained from the states in Eq. (11) with Clifford transformations (see Fig. 4).

The Wigner function of the quantum states $|T\rangle$ and $|H\rangle$ mapped in CV via the GKP code are shown in Fig. 3. Both the numerical computations and the analytical expression indicate that the number of negative peaks increases with respect to the Wigner function of stabilizer states, although the proportion remains as before: one quarter of negative delta functions and three quarters of positive ones. As one can observe comparing Figs. 2 and 3, it is not possible to obtain a non-stabilizer Wigner function pattern from a stabilizer one with single-qubit Clifford GKP encoded operations as those given in Eq. (10).

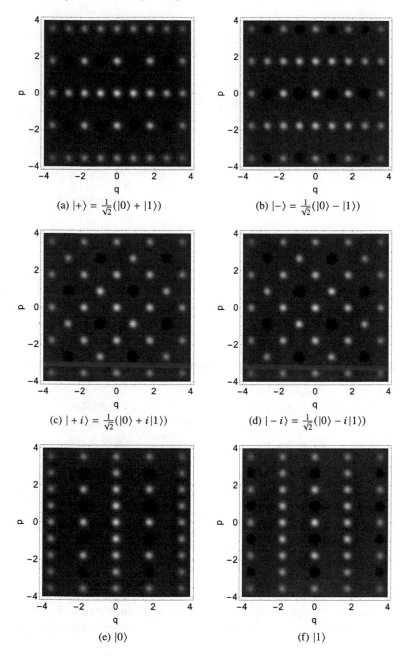

(a) $|+\rangle = \frac{1}{\sqrt{2}}(|0\rangle + |1\rangle)$

(b) $|-\rangle = \frac{1}{\sqrt{2}}(|0\rangle - |1\rangle)$

(c) $|+i\rangle = \frac{1}{\sqrt{2}}(|0\rangle + i|1\rangle)$

(d) $|-i\rangle = \frac{1}{\sqrt{2}}(|0\rangle - i|1\rangle)$

(e) $|0\rangle$

(f) $|1\rangle$

Fig. 2 Wigner function of qubit GKP encoded stabilizer states. The function acquires nonzero values on the dark and white peaks, where it has a negative value (dark) and positive value (white), respectively. We consider finitely squeezed states as in Eq. (2), with $\sigma = \kappa = 0.2$

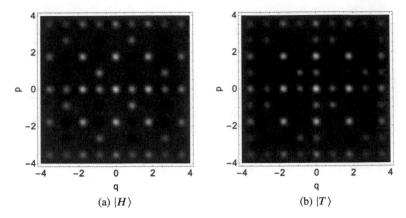

(a) $|H\rangle$ (b) $|T\rangle$

Fig. 3 Wigner function of qubit GKP encoded magic states. The function acquires nonzero values on the dark and white peaks, where it has a negative value (dark) and positive value (white), respectively. We consider finitely squeezed states as in Eq. (2), with $\sigma = \kappa = 0.2$. **a** $|H\rangle$ state, and **b** $|T\rangle$ state, both given in Eq. (11)

4 Quantification of Negativity of the Wigner Function for GKP Encoded States

We now aim at quantifying the volume of the negative part of the Wigner function for the different types of states that we have introduced. The quantification of the volume of the negative part of the Wigner function in CV is related to the monotone *Wigner logarithmic negativity* (WLN) Kenfack et al. (2004), Albarelli et al. (2018), defined as

$$\mathcal{W}(\rho) = \log_2 \left(\int dq dp |W(q, p)| \right), \tag{12}$$

with $W(q, p)$ the Wigner function of the state or operator ρ. The WLN has allowed for the derivation of a bound in the number of necessary copies of an input state for the conversion to a target state Albarelli et al. (2018).

As we have already mentioned, the proportion of negative delta functions compared to positive ones in the Wigner function of both stabilizer and magic encoded states is one quarter. However, we observe in Figs. 2 and 3 that the Wigner function of non-stabilizer states is composed of more peaks in the phase space, resulting in a higher number of negative delta peaks. We now use the WLN for analyzing the differences in both kinds of states, since it tracks the amount of negativity instead of the proportion.

We consider the Wigner function of perfect GKP states in Eq. (6). The negativity takes an infinite value since the Wigner function has support in the whole phase space $\mathbb{R}^2$, but the delta functions are periodically arranged following symmetric patterns that are repeated along the two axes in a similar way for each qubit superposition state. Therefore, we may consider the same square unit cell of dimension $(\Delta q, \Delta p) =$

$(2\sqrt{\pi}, 2\sqrt{\pi})$ for all cases, and compare the negativity within the same finite area in phase space. We choose the unit cell corresponding to $s = t = 0$ in Eq. (7), which contains sixteen delta functions given by l and m with values in the set $\{0, 1, 2, 3\}$.

Explicitly, the Wigner function in the unit cell domain $q \in [0, 2\sqrt{\pi})$ and $p \in [0, 2\sqrt{\pi})$ is given by

$$W_{\text{cell}}(\theta, \phi; q, p) = \sum_{l,m=0}^{3} w_{lm}(\theta, \phi)\delta\left(q - l\tfrac{\sqrt{\pi}}{2}\right)\delta\left(p - m\tfrac{\sqrt{\pi}}{2}\right), \qquad (13)$$

where the coefficients correspond to those defined in Eq. (7). The absolute value of the Wigner function for the unit cell can be taken as the absolute value of the summands, since for any coordinate (q_i, p_i) in the domain only one of the terms is different from zero due to the properties of the delta functions. Thus,

$$|W_{\text{cell}}(\theta, \phi; q, p)| = \sum_{l,m=0}^{3} |w_{lm}(\theta, \phi)|\delta\left(q - l\tfrac{\sqrt{\pi}}{2}\right)\delta\left(p - m\tfrac{\sqrt{\pi}}{2}\right). \qquad (14)$$

As a result, the WLN corresponding to a unit cell in the phase space for any pure qubit GKP encoded state $|\Psi\rangle = \cos\tfrac{\theta}{2}|0\rangle + e^{i\phi}\sin\tfrac{\theta}{2}|1\rangle$ characterized in the Bloch sphere by angles (θ, ϕ) is given by

$$\mathcal{W}_{\text{cell}}(\theta, \phi) = \log_2\left(\int dq\,dp|W_{\text{cell}}(\theta, \phi; q, p)|\right)$$

$$= \log_2 \sum_{l,m=0}^{3} |w_{lm}(\theta, \phi)|\left(\int dq\,dp\,\delta\left(q - \tfrac{l\sqrt{\pi}}{2}\right)\delta\left(p - \tfrac{m\sqrt{\pi}}{2}\right)\right)$$

$$= \log_2 \sum_{l,m=0}^{3} |w_{lm}(\theta, \phi)|. \qquad (15)$$

Explicitly, the WLN per cell of a qubit state is then given by

$$\mathcal{W}_{\text{cell}}(\theta, \phi) = \log_2\left[\frac{1}{\sqrt{\pi}}\left[1 + \left|\cos^2\tfrac{\theta}{2} - \sin^2\tfrac{\theta}{2}\right| + |\sin\theta\cos\phi| + |\sin\theta\sin\phi|\right]\right]. \qquad (16)$$

Now, we compare the finite WLN per cell, $\mathcal{W}_{\text{cell}}$, for different magic and stabilizer states by analyzing for simplicity the integral over a unit cell of the absolute value of the Wigner function $\int dq\,dp|W_{\text{cell}}|$, i.e. the argument of the logarithm in Eq. (15). The corresponding values are provided in Table 1. We observe that the WLN per cell for GKP encoded qubit stabilizer states is lower than for non-stabilizer states. Since all GKP encoded qubit states have a proportion of one quarter of negative delta functions, the WLN is different from zero for all of them. This Wigner negativity is intrinsic to the use of the GKP encoding, that is, it is only attributed to the fact that we are using an encoding where even the stabilizer states are represented by non-Gaussian

Table 1 Integral over a unit cell of the absolute value of the Wigner function for stabilizer states and magic states

	θ	ϕ	$\sqrt{\pi} \int |W_{\text{cell}}|$
$|0\rangle$	0	0	2
$|+\rangle$	$\pi/2$	0	2
$|i\rangle$	$\pi/2$	$\pi/2$	2
$|H\rangle$	$\pi/2$	$\pi/4$	$1 + \sqrt{2} \approx 2.41$
$|T\rangle$	$\arccos\left(1/\sqrt{3}\right)$	$\pi/4$	$1 + \sqrt{3} \approx 2.73$

wavefunctions exhibiting Wigner negativity. This intrinsic Wigner negativity in GKP states might be sufficient to promote Gaussian quantum circuits to universal quantum computation Baragiola et al. (2019).

We now compute the lower bound of this intrinsic negativity by considering

$$\int dq\,dp |W_{\text{cell}}(\theta, \phi; q, p)| \geq \left| \int dq\,dp\, W_{\text{cell}}(\theta, \phi; q, p) \right| = \frac{2}{\sqrt{\pi}}. \quad (17)$$

We observe that stabilizer states saturate the lower bound of the integral over a unit cell of the absolute value of the Wigner function, $\int |W_{\text{cell}}|$, and therefore they are the least negative qubit GKP encoded states.

We show in Fig. 4 the function $\sqrt{\pi} \int |W_{\text{cell}}(\theta, \phi; q, p)| dq\,dp$, which is proportional to the argument of the logarithm in the WLN. It is computed for all qubit states, characterized in the Bloch sphere with (θ, ϕ), with $\theta \in [0, \pi)$ and $\phi \in [0, 2\pi)$. We observe that the stabilizer states are the least negative, whereas the maxima appears for $|T\rangle$ qubit states, which are the most non-stabilizer single-qubit states. On the equatorial plane of the Bloch sphere (see Fig. 1), $\theta = \frac{\pi}{2}$, the maxima appears for $|H\rangle$ states, which are the most non-stabilizer states on that plane.

5 Conclusions

In this work, we use CV tools as the Wigner phase-space representation for studying DV single-qubit states encoded in infinite Hilbert spaces with the GKP mapping. We give an analytical expression for the Wigner function of any GKP encoded qubit state, and quantify the amount of negativity with the WLN. All qubit states have nonzero WLN, and therefore we cannot distinguish which states and processes are classically efficiently simulatable with current criteria for quantum advantage in CV systems. On the other hand, our quantitative analysis of the WLN for GKP encoded states shows differences for stabilizer and non-stabilizer states, since the first ones are the least negative, saturating the lower bound of negativity. The most non-stabilizer states, H-type and T-type quantum states, reach the maximum negativity. Our results suggest a possible connection between a DV characterization of resources for universal quantum computation and CV necessary criteria for quantum advantage.

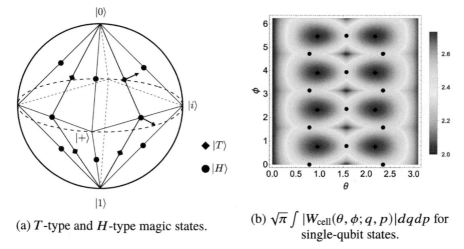

(a) T-type and H-type magic states.

(b) $\sqrt{\pi}\int|W_{\text{cell}}(\theta,\phi;q,p)|dqdp$ for single-qubit states.

Fig. 4 **a** Representation of single-qubit states on the Bloch sphere. Stabilizer states correspond to the vertices of an octahedron embedded in the sphere. The most non-stabilizer states are those projected on the surface of the sphere from the middle points of the edges of the octahedron, H-type magic states (circle), and perpendicularly from the center of the faces, T-type magic states (diamond), as indicated by the arrows (Bravyi et al. 2005). **b** Quantification of negativity of the Wigner function of qubit GKP encoded states with $\sqrt{\pi}\int|W_{\text{cell}}|$. We consider all qubit states, described by the angles (θ,ϕ), with $\theta\in[0,\pi)$ and $\phi\in[0,2\pi)$

A natural perspective stemming from this work is to explore the relation between different states with nonzero WLN and the computational complexity of quantum circuits including these states.

Acknowledgements We thank P. Milman and A. Ketterer for sharing with us a Mathematica code that was useful in the explorative stage of this project. L. G.-Á. and G. F. acknowledge support from the Wallenberg Center for Quantum Technology (WACQT), and G. F. acknowledges financial support from the Swedish Research Council through the VR project QUACVA.

Appendix 1

A detailed derivation of Eq. (6) is provided here. Firstly, we can conveniently rewrite the Wigner function in Eq. (5) as follows:

$$W(\theta,\phi;q,p) = \frac{1}{2\pi}\int_{-\infty}^{\infty} dx\,e^{ipx}\left[\cos^2\frac{\theta}{2}\Psi_0\left(q+\frac{x}{2}\right)^*\Psi_0\left(q-\frac{x}{2}\right)\right.$$
$$+\sin^2\frac{\theta}{2}\Psi_1\left(q+\frac{x}{2}\right)^*\Psi_1\left(q-\frac{x}{2}\right)$$
$$\left.+\cos\frac{\theta}{2}\sin\frac{\theta}{2}e^{i\phi}\Psi_0\left(q+\frac{x}{2}\right)^*\Psi_1\left(q-\frac{x}{2}\right)\right.$$

$$+ \cos \tfrac{\theta}{2} \sin \tfrac{\theta}{2} e^{-i\phi} \Psi_1 \left(q + \tfrac{x}{2}\right)^* \Psi_0 \left(q - \tfrac{x}{2}\right) \bigg]$$

$$= \cos^2 \tfrac{\theta}{2} W_0(q, p) + \sin^2 \tfrac{\theta}{2} W_1(q, p) + \frac{1}{2\pi} \cos \tfrac{\theta}{2} \sin \tfrac{\theta}{2} e^{i\phi} \widetilde{W}_{01}(q, p)$$

$$+ \frac{1}{2\pi} \cos \tfrac{\theta}{2} \sin \tfrac{\theta}{2} e^{-i\phi} \widetilde{W}_{10}(q, p), \tag{18}$$

where we have defined the cross terms as follows:

$$\widetilde{W}_{jk}(q, p) \equiv \int_{-\infty}^{\infty} dx e^{ipx} \Psi_j \left(q + \tfrac{x}{2}\right)^* \Psi_k \left(q - \tfrac{x}{2}\right). \tag{19}$$

We simplify the cross terms as follows:

$$\widetilde{W}_{jk}(q, p) = \int dx e^{ipx} \left[\sum_s \delta \left(q - \sqrt{\pi}(j + 2s) + \tfrac{x}{2}\right) \right] \left[\sum_t \delta \left(q - \sqrt{\pi}(k + 2t) - \tfrac{x}{2}\right) \right]$$

$$= \sum_{st} e^{i2p[q - \sqrt{\pi}(k+2t)]} \delta \left(q - \tfrac{\sqrt{\pi}}{2}(j + k + 2s + 2t)\right)$$

$$= \sum_{st} e^{i2p[q - \sqrt{\pi}(k+2t-2s)]} \delta \left(q - \tfrac{\sqrt{\pi}}{2}(j + k + 2t)\right)$$

$$= \sum_{st} e^{i2p\sqrt{\pi} 2s} e^{ip\sqrt{\pi}(j-k-2t)} \delta \left(q - \tfrac{\sqrt{\pi}}{2}(j + k + 2t)\right)$$

$$= \frac{\sqrt{\pi}}{2} \sum_{st} e^{ip\sqrt{\pi}(j-k-2t)} \delta \left(p - s\tfrac{\sqrt{\pi}}{2}\right) \delta \left(q - \tfrac{\sqrt{\pi}}{2}(j + k + 2t)\right)$$

$$= \frac{\sqrt{\pi}}{2} \sum_{st} (-1)^{\frac{s}{2}(j-k-2t)} \delta \left(p - s\tfrac{\sqrt{\pi}}{2}\right) \delta \left(q - \tfrac{\sqrt{\pi}}{2}(j + k + 2t)\right). \tag{20}$$

Now, combining Eqs. (18) and (20), we have

$$W(\theta, \phi; q, p) = \cos^2 \tfrac{\theta}{2} W_0(q, p) + \sin^2 \tfrac{\theta}{2} W_1(q, p) + \frac{1}{4\sqrt{\pi}} \cos \tfrac{\theta}{2} \sin \tfrac{\theta}{2}$$

$$\times \Bigg[e^{i\phi} \sum_{st} (-1)^{\frac{s}{2}(-1-2t)} \delta \left(p - s\tfrac{\sqrt{\pi}}{2}\right) \delta \left(q - \tfrac{\sqrt{\pi}}{2}(1 + 2t)\right)$$

$$+ e^{-i\phi} \sum_{st} (-1)^{\frac{s}{2}(1-2t)} \delta \left(p - s\tfrac{\sqrt{\pi}}{2}\right) \delta \left(q - \tfrac{\sqrt{\pi}}{2}(1 + 2t)\right) \Bigg]$$

$$= \cos^2 \tfrac{\theta}{2} W_0(q, p) + \sin^2 \tfrac{\theta}{2} W_1(q, p)$$

$$+ \frac{1}{8\sqrt{\pi}} \sin \theta \sum_{st} (-1)^{st} \left(e^{i\phi} (-1)^{\frac{s}{2}} + e^{-i\phi} (-1)^{-\frac{s}{2}} \right)$$

$$\times \delta \left(q - \tfrac{\sqrt{\pi}}{2}(1 + 2t)\right) \delta \left(p - s\tfrac{\sqrt{\pi}}{2}\right). \tag{21}$$

Then, it follows that the Wigner function for arbitrary superpositions of GKP states is given by Eq. (6) in the main text.

Appendix 2

The table below summarizes the estimated climate footprint of this work, including air travel for collaboration purposes. Estimations have been calculated using the examples of ScientificCO$_2$nduct https://scientific-conduct.github.io/.

Transport	
Total CO$_2$-Emission For Transport (kg)	6645
Were The Emissions Offset?	**No**
Total CO$_2$-Emission (kg)	6645

References

https://scientific-conduct.github.io/

S. Aaronson, D. Gottesman, Improved simulation of stabilizer circuits. Phys. Rev. A **70**, 052328 (2004). https://doi.org/10.1103/PhysRevA.70.052328, https://link.aps.org/doi/10.1103/PhysRevA.70.052328

F. Albarelli, M.G. Genoni, M.G.A. Paris, A. Ferraro, Resource theory of quantum non-Gaussianity and Wigner negativity. Phys. Rev. A **98**(5), 052350 (2018). https://doi.org/10.1103/PhysRevA.98.052350, https://link.aps.org/doi/10.1103/PhysRevA.98.052350

M. Aspelmeyer, T.J. Kippenberg, F. Marquardt, Cavity optomechanics. Rev. Mod. Phys. **86**, 1391 (2014)

B.Q. Baragiola, G. Pantaleoni, R.N. Alexander, A. Karanjai, N.C. Menicucci, All-Gaussian universality and fault tolerance with the Gottesman-Kitaev-Preskill code (2019). arXiv:1903.00012 [quant-ph], http://arxiv.org/abs/1903.00012

S.D. Bartlett, B.C. Sanders, S.L. Braunstein, K. Nemoto, Efficient classical simulation of continuous variable quantum information processes. Phys. Rev. Lett. **88**, 9 (2002)

S. Bravyi, A. Kitaev, Universal quantum computation with ideal Clifford gates and noisy ancillas. Phys. Rev. A **71**(2), 022316 (2005). https://doi.org/10.1103/PhysRevA.71.022316, https://link.aps.org/doi/10.1103/PhysRevA.71.022316

U. Chabaud, T. Douce, D. Markham, P. van Loock, E. Kashefi, G. Ferrini, Continuous-variable sampling from photon-added or photon-subtracted squeezed states. Phys. Rev. A **96**, 062307 (2017)

L. Chakhmakhchyan, N.J. Cerf, Boson sampling with Gaussian measurements. Phys. Rev. A **96**, 032326 (2017)

M. Chen, N.C. Menicucci, O. Pfister, Experimental realization of multipartite entanglement of 60 modes of a quantum optical frequency comb. Phys. Rev. Lett. **112**, 120505 (2014)

T. Douce, D. Markham, E. Kashefi, E. Diamanti, T. Coudreau, P. Milman, P. van Loock, G. Ferrini, Continuous-variable instantaneous quantum computing is hard to sample. Phys. Rev. Lett. **118**, 070503 (2017)

T. Douce, D. Markham, E. Kashefi, P. van Loock, G. Ferrini, Probabilistic fault-tolerant universal quantum computation and sampling problems in continuous variables. Phys. Rev. A **99**, 012344 (2019)

K.S. Gibbons, M.J. Hoffman, W.K. Wootters, Discrete phase space based on finite fields. Phys. Rev. A **70**, 062101 (2004)

D. Gottesman, The Heisenberg representation of quantum computers, in *Group22: Proceedings of the XXII International Colloquium on Group Theoretical Methods in Physics, Cambridge, MA* (International Press, 1999), arXiv:quant-ph/9807006

D. Gottesman, A. Kitaev, J. Preskill, Encoding a qubit in an oscillator. Phys. Rev. A **64**, 012310 (2001)

C.S. Hamilton, R. Kruse, L. Sansoni, S. Barkhofen, C. Silberhorn, I. Jex, Gaussian Boson sampling. Phys. Rev. Lett. **119**, 170501 (2017)

M. Hillery, R. O'Connell, M. Scully, E. Wigner, Distribution functions in physics: fundamentals. Phys. Rep. **106**(3), 121–167 (1984)

A. Kenfack, K. Zyczkowski, Negativity of the Wigner function as an indicator of non-classicality. J. Opt. B: Quantum Semiclassical Opt. **6**(10), 396–404 (2004)

A. Mari, J. Eisert, Positive wigner functions render classical simulation of quantum computation efficient. Phys. Rev. Lett. **109**, 230503 (2012)

N.C. Menicucci, Fault-tolerant measurement-based quantum computing with continuous-variable cluster states. Phys. Rev. Lett. **112**(12), 120504 (2014)

N. Ofek, A. Petrenko, R. Heeres, P. Reinhold, Z. Leghtas, B. Vlastakis, Y. Liu, L. Frunzio, S.M. Girvin, L. Jiang, M. Mirrahimi, M.H. Devoret, R.J. Schoelkopf, Demonstrating quantum error correction that extends the lifetime of quantum information. Nature **536**, 441 (2016)

S. Rahimi-Keshari, T.C. Ralph, C.M. Caves, Sufficient conditions for efficient classical simulation of quantum optics. Phys. Rev. X **6**, 021039 (2016)

P.W. Shor, Polynomial-time algorithms for prime factorization and discrete logarithms on a quantum computer. SIAM Rev. **41**, 303 (1999)

E. Wigner, On the quantum correction for thermodynamic equilibrium. Phys. Rev. **40**, 749–759 (1932)

C.M. Wilson, G. Johansson, A. Pourkabirian, M. Simoen, J.R. Johansson, F. Nori, P. Delsing, Observation of the dynamical casimir effect in a superconducting circuit. Nature **479**, 376 (2011)

J. Yoshikawa, S. Yokoyama, T. Kaji, C. Sornphiphatphong, Y. Shiozawa, K. Makino, A. Furusawa, Invited article: generation of one-million-mode continuous-variable cluster state by unlimited time-domain multiplexing. APL Photonics **1**(6), 060801 (2016)

Quantum Interactions

Number Theoretic Study in Quantum Interactions

Masato Wakayama

Abstract The quantum interaction models, with the quantum Rabi model as a distinguished representative, are recently appearing ubiquitously in various quantum systems including cavity and circuit quantum electrodynamics, quantum dots and artificial atoms, with potential applications in quantum information technologies including quantum cryptography and quantum computing (Haroche and Raimond 2008; Yoshihara et al. 2018). In this extended abstract, based on the contents of the talk at the conference, we describe shortly certain number theoretical aspects arising from the *non-commutative harmonic oscillators* (NCHO: see Parmeggiani and Wakayama 2001; Parmeggiani 2010) and *quantum Rabi model* (QRM: see Braak 2011 for the integrability) through their respective spectral zeta functions.

The quantum interaction models, with the quantum Rabi model as a distinguished representative, are recently appearing ubiquitously in various quantum systems including cavity and circuit quantum electrodynamics, quantum dots and artificial atoms, with potential applications in quantum information technologies including quantum cryptography and quantum computing (Haroche and Raimond 2008; Yoshihara et al. 2018). In this extended abstract, based on the contents of the talk at the conference, we describe shortly certain number theoretical aspects arising from the *non-commutative harmonic oscillators* (NCHO: see Parmeggiani and Wakayama 2001; Parmeggiani 2010) and *quantum Rabi model* (QRM: see Braak 2011 for the integrability) through their respective spectral zeta functions.

In physics, given a quantum interaction model, one of the main interests is to know the heat kernel (or equivalently the evolution operator) since, among other reasons, the heat kernel gives the partition function by taking the trace. With partition function of the model, we may also get the analytic properties of the spectral zeta function

M. Wakayama (✉)
Department of Mathematics, Tokyo University of Science, 1-3 Kagura-zaka, Shinjyuku-ku, Tokyo 162-8601, Japan
e-mail: wakayama@rs.tus.ac.jp; wakayama@imi.kyushu-u.ac.jp

Institute of Mathematics for Industry, Kyushu University, 744 Motooka, Nishi-ku Fukuoka 819-0395, Japan

© The Author(s) 2021
T. Takagi et al. (eds.), *International Symposium on Mathematics, Quantum Theory, and Cryptography*, Mathematics for Industry 33, https://doi.org/10.1007/978-981-15-5191-8_10

95

by means of the Mellin transform. A spectral zeta function is defined, in general, as the Dirichlet series formed by the spectrum (eigenvalues) of the corresponding Hamiltonian (Ichinose and Wakayama 2005; Sugiyama 2018). Notice that knowing the spectral zeta function is essentially equivalent to knowing the partition function in any quantum system.

In the case of the NCHO, the Hamiltonian is given by

$$
Q = \begin{pmatrix} \alpha & 0 \\ 0 & \beta \end{pmatrix} \left(-\frac{1}{2}\frac{d^2}{dx^2} + \frac{1}{2}x^2 \right) + \begin{pmatrix} 0 & -1 \\ 1 & 0 \end{pmatrix} \left(x\frac{d}{dx} + \frac{1}{2} \right),
$$

with $\alpha, \beta > 0$ and $\alpha\beta > 1$ (the condition for having only a discrete spectrum with positive eigenvalues), and the spectral zeta function by

$$
\zeta_Q(s) := \sum_{n=1}^{\infty} \lambda_n^{-s} \quad (\Re(s) > 1),
$$

where $(0 <)\lambda_1 < \lambda_2 \leq \lambda_3 \leq \ldots (\nearrow \infty)$ are the eigenvalues of NCHO. Note that the lowest eigenstate is multiplicity free (Hiroshima and Sasaki 2014) and the multiplicity of general eigenstate is less than or equal to two (Wakayama 2016). The function $\zeta_Q(s)$ is meromorphically continued to the whole complex plane with a unique simple pole at $s = 1$ and has trivial zeros at the even non-positive integers (Ichinose and Wakayama 2005). Although our study is very much influenced by the classical algebro-geometric work on Apéry numbers for the Riemann zeta function in Beukers (1987) and its subsequent developments, since the family of generating functions for *Apéry-like numbers* (Kimoto and Wakayama 2006) arising via the NCHO possesses a remarkable hierarchical structure, there is a decisive difference between these two (Ichinose and Wakayama 2005; Kimoto and Wakayama 2019).

For instance, there are congruence properties of the (normalized) Apéry-like numbers that have arisen naturally from the special values $\zeta_Q(2)$ at $s = 2$. This can be seen by the same idea that guided the studies for the Apéry numbers for $\zeta(2)(= \pi^2/6)$ in Beukers (1985). These congruence properties led us further to observe that the generating function w_2 of the Apéry-like numbers for $\zeta_Q(2)$ is interpreted as a $\Gamma(2)$-modular form of weight 1 (Kimoto and Wakayama 2007) in the same way as in a pioneering study by Beukers (1983, 1987) for the Apéry numbers. It is worth mentioning that the recurrence equation of these Apéry-like numbers defined in Kimoto and Wakayama (2006) provides one of the particular examples listed in Zagier (2009) (it gives #19 in the list).[1] Also, recently, certain congruence relations among these Apéry-like numbers conjectured in Kimoto and Wakayama (2006) resembling Rodriguez–Villegas type congruences (Mortenson 2003) were proved in Long et al. (2016). It is, however, hard in general to obtain precise information, in the same level of $\zeta_Q(2)$, of the higher special values of $\zeta_Q(n)$ $(n > 2)$. Thus, we introduce the Apéry-like numbers $J_k(n)$ $(k = 0, 1, 2, \ldots)$ for each n defined through

---

[1] Although the terminology "*Apéry-like*" is the identical, the usage/definition of the name in the current paper is different from the one in the title of Zagier (2009).

the *first anomaly* of $\zeta_Q(n)$ $(n > 2)$ (Kimoto and Wakayama 2019) (see also Kimoto (2016)). These Apéry-like numbers share the properties of the one for $\zeta_Q(2)$, e.g. satisfy a similar recurrence relation as in the case of $\zeta_Q(2)$ and hence the ordinary differential equation satisfied by the generating function follows from the recurrence relation. Remarkably, the homogeneous part of each of the differential equations is identified with a (n dependent) power of the homogeneous part of the one corresponding to $\zeta_Q(2)$. Further, we observe that the meta-generating functions of Apéry-like numbers $J_k(n)$ are described explicitly by the modular Mahler measures studied by Rodriguez–Villegas in Rodriguez (1999). Through this relation, we may find an interesting aspect of a discrete dynamical system behind NCHO defined by a certain limit of finite abelian group via *(weighted) Cayley graphs* studied in Dasbach and Lalin (2009). Moreover, we note here (Kimoto and Wakayama 2012, 2019) that the generating function w_{2n} of Apéry-like numbers corresponding to the first anomaly in $\zeta_Q(2n)$ when $n = 2$ is given by an automorphic integral with a rational period function in the sense of Knopp (1978). This is obviously a generalization of our earlier result (Kimoto and Wakayama 2007) showing that w_2 is interpreted as a $\Gamma(2)$-modular form of weight 1.

Furthermore, we show certain congruence relations among these normalized Apéry-like numbers which are the generalization of the results in Kimoto and Wakayama (2006). A possible generalization of the results in Liu (2018) seems very interesting. We also conjecture much stronger results based on numerical experiments in Kimoto and Wakayama (2019).

The Hamiltonian H_{Rabi} of the QRM is precisely given by

$$H_{\text{Rabi}} := \omega a^\dagger a + \Delta \sigma_z + g(a + a^\dagger)\sigma_x.$$

Here, $a^\dagger$ and a are the creation and annihilation operators of the single bosonic mode $([a, a^\dagger] = 1)$, σ_x, σ_z are the Pauli matrices (sometimes written as σ_1 and σ_3, but since there is no risk of confusion with the variable x to appear below in the heat kernel, we use the usual notations), 2Δ is the energy difference between the two levels, and g denotes the coupling strength between the two-level system and the bosonic mode with frequency ω (subsequently, we set $\omega = 1$ without loss of generality). The integrability of the QRM was established in Braak (2011) using the well-known $\mathbb{Z}_2$-symmetry of the Hamiltonian H_{Rabi}, usually called parity.

In the case of QRM, we recently obtained the (analytic formula of) heat kernel (Reyes and Wakayama 2019) using the Trotter–Kato product formula by extensive discussions of combinatorics and graph theory including quantum Fourier transform. Concretely, the heat kernel $K_{\text{Rabi}}(t, x, y)$ of the QRM is given by

$$K_{\text{Rabi}}(t, x, y) = \widetilde{K}_0(x, y, g, t) \sum_{\lambda=0}^{\infty} (t\Delta)^\lambda \Phi_\lambda(x, y, g, t).$$

Here the 2×2 matrix-valued function $\Phi_\lambda(g, t)$ for $\lambda \geq 0$ is given by

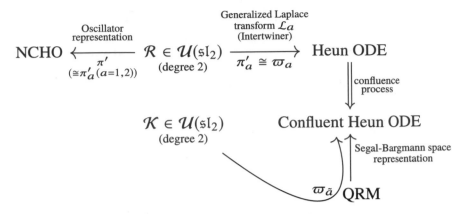

Fig. 1 From the NCHO to QRM (Heun's Pictures)

$$\Phi_\lambda(x, y, g, t) = \int \cdots \int_{0 \le \mu_1 \le \cdots \le \mu_\lambda \le 1} e^{\phi(\mu_\lambda, t) + \xi_\lambda(\mu_\lambda, t)} \begin{bmatrix} (-1)^\lambda \cosh & (-1)^{\lambda+1} \sinh \\ -\sinh & \cosh \end{bmatrix}$$
$$\times (\theta_\lambda(x, y, \mu_\lambda, t)) \, d\mu_\lambda,$$

where $\mu_\lambda = (\mu_1, \mu_2, \cdots, \mu_\lambda)$ and $d\mu_\lambda = d\mu_1 d\mu_2 \cdots d\mu_\lambda$ with $\mu_0 = 0$ and $d\mu_0 = 1$. For the definition of the functions $\phi, \xi_\lambda, \theta_\lambda$ and $\widetilde{K}_0$, (Mehler's kernel) the reader is directed to Reyes and Wakayama (2019).

This is the first time an explicit determination of the heat kernel is obtained for an interacting system (though certain partial results have been discussed, e.g. in Legget 1987 for the Spin-Boson model and Anderson et al. 1970; Chakravarty 1995 for the Kondo effect using the Feynman–Kac formula.) The heat kernel formula allows us to have the contour integral representation of the spectral zeta function of the QRM (Sugiyama 2018) and open the study of the special values of negative integral points using it (Reyes and Wakayama 2019).

Further, although NCHO is not confirmed as a practical physical model, it may be considered as a "*covering*" model of QRM through the respective Heun ODE pictures (Wakayama 2016) (Fig. 1). Thus, in addition to the study of the respective number theoretical aspects of the models independently, the comparison of the number theoretic objects appearing from each model is an interesting and significant problem.

In addition to the number theoretic structure described above, we remark here that there appear certain algebraic curves, including elliptic and super elliptic curves, in the description of degenerations of the eigenstates for the asymmetric QRM with an integral perturbation parameter (Wakayama 2017; Kimoto et al. 2020; Reyes and Wakayama 2017). This shows another mathematical structure behind the asymmetric and symmetric QRM.

The following figure (Fig. 2) illustrates the position of this extended abstract from our whole interest. Particularly, the talk focused on the special values of such zeta

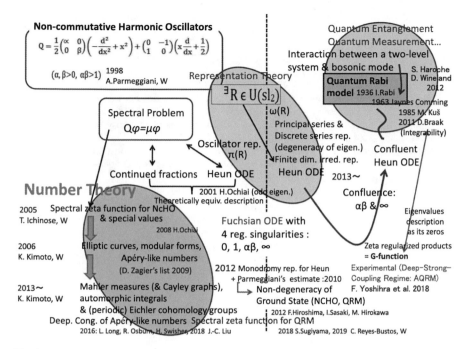

Fig. 2 Non-commutative harmonic oscillator and (asymmetric and symmetric) quantum Rabi models

functions (Ichinose and Wakayama 2005; Ochiai 2008; Kimoto and Wakayama 2006, 2007, 2012; Long et al. 2016; Liu 2018; Kimoto and Wakayama 2019). We note that special values of zetas may be considered as the moments of the partition function of the corresponding model.

References

P.W. Anderson, G. Yuval, D.R. Hamann, Exact results in the kondo problem. II. Scaling theory, qualitatively correct solution, and some new results on one-dimensional classical statistical models. Phys. Rev. B **1**, 4464 (1970)

F. Beukers, Irrationality of π^2, periods of an elliptic curve and $\Gamma_1(5)$, Diophantine approximations and transcendental numbers (Luminy, 1982). Progr. Math. **31**, 47–66, Birkhäuser, Boston (1983)

F. Beukers, Some congruences for the Apéry numbers. J. Number Theo. **21**, 141–155 (1985)

F. Beukers, Irrationality proofs using modular forms. Soc. Math. France, Astérisque **21**, 271–283 (1987)

D. Braak, Integrability of the Rabi model. Phys. Rev. Lett. **107**, 100401 (2011)

S. Chakravarty, J. Rudnick, Dissipative dynamics of a two-state system, the Kondo problem, and the inverse-square Ising model. Phys. Rev. Lett. **75**, 501 (1995)

O. Dasbach, M. Lalin, Mahler measure under variations of the base group. Forum Math. **21**, 621–637 (2009)

S. Haroche, J.M. Raimond, *Exploring the Quantum - Atoms Cavities and Photons* (Oxford University Press, Oxford, 2008)

F. Hiroshima, I. Sasaki, Spectral analysis of non-commutative harmonic oscillators: the lowest eigenvalue and no crossing. J. Math. Anal. Appl. **105**, 595–609 (2014)

T. Ichinose, M. Wakayama, Zeta functions for the spectrum of the non-commutative harmonic oscillators. Commun. Math. Phys. **258**, 697–739 (2005)

T. Ichinose, M. Wakayama, Special values of the spectral zeta function of the non-commutative harmonic oscillator and confluent Heun equations. Kyushu J. Math. **59**, 39–100 (2005)

K. Kimoto, Generalized Apéry numbers arising from the non-commutative harmonic oscillator. Ryukyu Math. J. **29**, 1–31 (2016)

K. Kimoto, C. Reyes-Bustos, M. Wakayama, Determinant expressions of constraint polynomials and degeneracies of the asymmetric quantum Rabi model. Int. Math. Res. Notices (2020), Published online 20, April, arXiv:1712.04152

K. Kimoto, M. Wakayama, Apéry-like numbers arising from special values of spectral zeta functions for non-commutative harmonic oscillators. Kyushu J. Math. **60**, 383–404 (2006)

K. Kimoto, M. Wakayama, Elliptic curves arising from the spectral zeta function for non-commutative harmonic oscillators and $\Gamma_0(4)$- modular forms, in *Proceedings Conference on L-Functions*, eds. by L. Weng, M. Kaneko, pp. 201–218. World Scientific (2007)

K. Kimoto, M. Wakayama, Spectrum of non-commutative harmonic oscillators and residual modular forms, in *Noncommutative Geometry and Physics*, eds. by G. Dito, et al (World Scientific, Singapore, 2012), pp. 237–267

K. Kimoto, M. Wakayama, Apéry-like numbers for non-commutative harmonic oscillators and automorphic integrals, http://arxiv.org/abs/1905.01775

I. Knopp, Rational period functions of the modular group. With an appendix by Georges Grinstein. Duke Math. J. **45**, 47–62 (1978)

A.J. Legget et al., Dynamics of the dissipative two-state system. Rev. Mod. Phys. **59**, 1–85 (1987)

J.-C. Liu, A generalized supercongruence of Kimoto and Wakayama. J. Math. Anal. Appl. **467**, 15–25 (2018)

L. Long, R. Osburn, H. Swisher, On a conjecture of Kimoto and Wakayama. Proc. Amer. Math. Soc. **144**, 4319–4327 (2016)

E. Mortenson, A supercongruence conjecture of Rodriguez-Villegas for a certain truncated hypergeometric function. J. Number Theo. **99**, 139–147 (2003)

H. Ochiai, A special value of the spectral zeta function of the non-commutative harmonic oscillators. Ramanujan J. **15**, 31–36 (2008)

A. Parmeggiani, *Spectral Theory of Non-Commutative Harmonic Oscillators: An Introduction*, Lecture Notes in Mathematics, vol. 1992 (Springer, Berlin, 2010)

A. Parmeggiani, M. Wakayama, Oscillator representations and systems of ordinary differential equations. Proc. Nat'l. Acad. Sci. USA **98**, 26–30 (2001)

C. Reyes-Bustos, M. Wakayama, Spectral degeneracies in the asymmetric quantum Rabi model, in *Mathematical Modelling for Next-Generation Cryptography*, eds. by T. Takagi et al., Mathematics for Industry, vol. 29 (Springer, Berlin, 2017), pp. 117–137

C. Reyes-Bustos, M. Wakayama, The heat kernel and spectral zeta function for the quantum Rabi model, arXiv:1906.09597

F. Rodriguez-Villegas, *Modular Mahler Measures I*, Topics in Number Theory (Kluwer, Berlin, 1999), pp. 17–48

S. Sugiyama, Spectral zeta functions for the quantum Rabi models. Nagoya Math. J. **229**, 52–98 (2018)

M. Wakayama, Equivalence between the eigenvalue problem of non-commutative harmonic oscillators and existence of holomorphic solutions of Heun differential equations, eigenstates degeneration and the Rabi model. Int. Math. Res. Notices **2016**(3), 759–794 (2016)

M. Wakayama, Symmetry of asymmetric quantum Rabi models. J. Phys. A: Math. Theor. **50**, 174001 (2017)

F. Yoshihara et al., Inversion of Qubit Energy Levels in Qubit-Oscillator Circuits in the Deep-Strong-Coupling Regime. Phys. Rev. Lett. **120**, 183601 (2018)

D. Zagier, Integral solutions of Apéry-like recurrence equations, in *Groups and Symmetries*, CRM Proceedings & Lecture Notes, vol. 47 (American Mathematical Society, Providence, 2009), pp. 349–366

A Data Concealing Technique with Random Noise Disturbance and a Restoring Technique for the Concealed Data by Stochastic Process Estimation

Tomohiro Fujii and Masao Hirokawa

Abstract We propose a technique to conceal data on a physical layer by disturbing them with some random noises, and moreover, a technique to restore the concealed data to the original ones by using the stochastic process estimation. Our concealing-restoring system manages the data on the physical layer from the data link layer. In addition to these proposals, we show the simulation result and some applications of our concealing-restoring technique.

Keywords Concealing-restoring system · OSI · Physical layer · Data link layer · Noise-disturbance · Stochastic process estimation · Noise-filtering · Kalman filter · Particle filter

1 Introduction

Micro-device technology in the near future realizes the remote control of microprocessor chips in several things such as household electric appliances, information-processing equipment, and even brain–computer/brain–machine interfaces from the outside through wireless communications or the so-called IoT (i.e., Internet of Things). Moreover, it enables the automatic operation of such things with the remote control. They are going to infiltrate society and play several important roles in every area of society. We then have to establish the data security for them (Youm 2017; Román-Castro et al. 2018; Lin et al. 2018; Clausen et al. 2017). In particular, we have to stem the hacking of the remote control and the wiretapping of the data of communication. We are interested in a data concealing technique with disturbance on a physical layer and a restoring technique for those concealed data. Here, the

T. Fujii (✉) · M. Hirokawa (✉)
Graduate School of Engineering, Hiroshima University, Hiroshima, Japan
e-mail: fujii@amath.hiroshima-u.ac.jp

M. Hirokawa
Graduate School of ISEE, Kyushu University, Fukuoka, Japan
e-mail: hirokawa@inf.kyushu-u.ac.jp

© The Author(s) 2021
T. Takagi et al. (eds.), *International Symposium on Mathematics,
Quantum Theory, and Cryptography*, Mathematics for Industry 33,
https://doi.org/10.1007/978-981-15-5191-8_11

physical layer is the lowest layer of the open systems interconnection (OSI) (Kain and Agrawala 1992) (see Fig. 1). OSI is a reference model to grasp and analyze how data are sent and received over a computation or communication network. Some methods using disturbance have been presented to conceal data for storage and communication. For instance, chaotic cryptology (Cuomo and Oppenheim 1993; Grassi and Mascolo 1999; Lenug and Lam 1997; Wu and Chua 1993) uses chaos to make the disturbance. The method using cryptographic hash functions for the disturbance has lately been gaining a practical position (Merkle 1979, 1989; Damgård 1989; Schneier 2015). There have been some endeavors for the concealing technique on physical layers: the chaos multiple-input multiple-output (Okamoto and Iwanami 2006; Zheng 2009; Okamoto 2011; Okamoto and Inaba 2015; Ito et al. 2019). Meanwhile, it is noteworthy that the secured telecommunication using noises has been actively studied (Wyner 1975; Hero 2003; Goel and Negi 2008; Swindlehurst 2009; Mukherjee and Swindlehurst 2011). In that technique, we send some noises from interference antennas to the signal on a carrier wave sent from an antenna; we have the signal interfering with the noises and make it an interference wave. There, however, may be a way to remove the noises from the interference wave and to wiretap the original signal (Ohno et al. 2012).

We take interest in how to conceal data on a physical layer using some random noise disturbances and how to restore those concealed data applying a stochastic filtering theory to maintain the safety of data over a proper period of time, which is different from the interference wave method. Thus, our concealing-restoring system should be installed on a data link layer above the physical layer (see Fig. 1). Although we employ the disturbance by random noises instead of the chaotic one, we can design our concealing-restoring system so that it includes the chaotic disturbance (Fujii and Hirokawa 2020). The idea of the concealing-restoring system was primarily originated in keeping security for the data processed on the physical layer of our developing quantum-sensing equipment over a necessary period. This equipment detects and handles some ultimate personal information. Since we must remove several noises on the physical layer in any case, we make our concealing-restoring system coexist with the denoising system of the equipment. We then consider the information concealing method for qubits (i.e., quantum bits) using the random noises in classical physics. The qubits $|0\rangle$ and $|1\rangle$ are represented by spin states $|\uparrow\rangle$ and $|\downarrow\rangle$, namely, $|0\rangle = |\uparrow\rangle = (1, 0)$ and $|1\rangle = |\downarrow\rangle = (0, 1)$. A general qubit $|q\rangle$ can be described with the superposition of the qubits $|0\rangle$ and $|1\rangle$: $|q\rangle = \alpha|0\rangle + \beta|1\rangle$ for some complex numbers α and β with $|\alpha|^2 + |\beta|^2 = 1$. Thus, the qubit can have the representation, $|q\rangle = (\Re\alpha, \Im\alpha, \Re\beta, \Im\beta)$, and an information sequence of qubits, $|q_1\rangle, |q_2\rangle, \ldots, |q_v\rangle$, is expressed with a finite sequence,

$$\Re\alpha_1 \; \Im\alpha_1 \; \Re\beta_1 \; \Im\beta_1 \; \Re\alpha_2 \; \Im\alpha_2 \; \Re\beta_2 \; \Im\beta_2 \; \ldots \; \Re\alpha_v \; \Im\alpha_v \; \Re\beta_v \; \Im\beta_v.$$

We transform it into an electrical signal X_t, $0 \leq t \leq 4v$, using linear interpolation. We process the electrical signal in a microprocessor, made by some semiconductors, of our quantum-sensing equipment. Since the microprocessor is for the conventional computation (i.e., not quantum computation), we need to transport the electrical

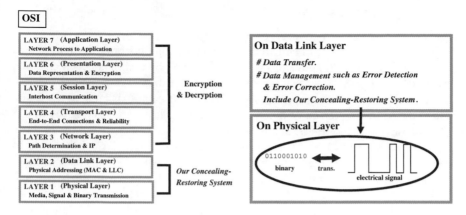

Fig. 1 The left picture shows that the OSI consists of 7 layers. The encryption and decryption are usually done on one out of layers between Layer 3 and Layer 7, typically on the presentation layer. The right picture shows what we aim our concealing-restoring system at

signal to memory or register according to a microarchitecture. To keep the security for the electric signal X_t while processing, storing, and saving it, we employ a mathematical idea to conceal it using the noise disturbance. In this paper, we introduce that mathematical idea for more general signals on the physical layer and more broad applications.

As some applications derive therefrom, we first establish a mathematical technique for concealing data by the disturbance with randomness of the noises, and moreover, a mathematical technique for restoring the concealed data by the stochastic process estimation. In addition to these establishments, we show the simulation result and some applications for the two techniques. The idea of our method to conceal data comes from an image of the scene when we conceal a treasure map, and it is so simple as follows:

(c1) we plaster over the treasure map at random and make it messy;
(c2) we repeat c1 and plaster it over repeatedly.

In this paper, we mathematically realize c1 and c2, and make their implementation on conventional computers. In addition to c1 and c2, we can consider that

(c3) we tear the muddled map by c1 and c2, and split it into several pieces, though we do not make its implementation in this paper.

We are planning that we use the concealed data for saving them in memory or for sending them for telecommunication. We expect to use our methods in the situation where the physical layer is under restrictions in the implementation space due to a small consumed electric power, a small arithmetic capacity, a small line capacity, and a bad access environment. Concretely, we hope to apply the implementation of our techniques to the remote control of drones and devices on them, and to the security of some data sent from those devices. Moreover, we suppose the situation where it is too harsh to make a remote maintenance of the physical layer, for example, in outerspace development or seafloor development.

2 Mathematical Setups

We first explain the outline of how to make our concealing-restoring system for data
X_t, $t \in \mathbb{R}$. The concealing-restoring system is given by a simultaneous equation
system (SES). This SES consists of some stochastic differential equations (SDEs),
linear equations, and a nonlinear equation (NLE). The data X_t is input as the initial
data of the SES. We prepare N functionals F_i, $i = 1, 2, \ldots, N$, making the SDEs.
We suppose that each form of the individual functional F_i is known only by those
who conceal the original data X_t and restore the concealed data. We use the forms of
the functionals as well as the composition of the SES for secret keys or common keys.
We prepare $2N$ random noises $W_t^{j,i}$, $j = 1, 2$; $i = 1, 2, \ldots, N$, for the SDEs, and
a nonlinear bijection f for the NLE. The SDEs for processes X_t^i, $i = 1, 2, \ldots, N$,
and the NLE for the process X_t^{N+1} are used to introduce the noise disturbance in
our concealing-restoring system. We also use the means, variances, and distributions
of the random noises as well as the nonlinear bijection as secret keys. As shown
below, we obtain $N + 1$ concealed data, U_t^i, $i = 1, 2, \ldots, N, N + 1$, using the
SDEs and the NLE. We use them as the data for saving in a digital memory such as
a semiconductor memory or an analog memory such as a magnetic tape. We may
also put the concealed data on a carrier wave and send them. This is the outline of
the data concealing. Meanwhile, the data restoration is done in the following. Using
the stochastic filtering theory and the inverse function f^{-1}, we remove the random
noises from every concealed data U_t^i, and we estimate the process X_t^i. We denote the
estimate by $\widehat{X}_t^i$, and call it *estimated data* for the process X_t^i. We regard the estimate
$\widehat{X}_t^1$ as the *restoration* of the original data X_t. We denote it by $\widehat{X}_t$.

We here explain how to make the data X_t from binary data. We use the low/high-
signal for the binary data in this paper though there are many other ways. Thus,
we represent 'low' by 0 and 'high' by 1. For $n + 1$ bits, $a_0, a_1, \ldots, a_n \in \{0, 1\}$,
we concatenate them and make a word $a_0 a_1 \ldots a_n$. We employ the following linear
interpolation as a simple digital–analog (D/A) transformation. We first define X_i by

$$X_i = \begin{cases} +1 & \text{if } a_i = 1, \\ -1 & \text{if } a_i = 0, \end{cases} \qquad i = 0, 1, \ldots, n.$$

We connect X_i and X_{i+1} with a straight line for each $i = 0, 1, \ldots, n-1$, and we have
a polygonal line X_t, $0 \le t \le n$. When the data X_t are made from the binary word
$a_0 a_1 \ldots a_n$, we call X_t a *binary pulse* for the word $a_0 a_1 \ldots a_n$. As for the restoration
of the word, we use the simple analog–digital (A/D) transformation to seek the
character $\widehat{a}_i \in \{0, 1\}$ for each $i = 0, 1, \ldots, n$, and make a word $\widehat{a}_0 \widehat{a}_1 \ldots \widehat{a}_n$ for
the original word $a_0 a_1 \ldots a_n$ in the following. We determine a threshold in advance
between those who conceal the binary pulse and restore its concealed data to it. The
threshold is basically determined taking into account the mean and variance of the
random noises when used for concealing data. For each $i = 0, 1, \ldots, n$, we define
the character $\widehat{a}_i$ by

$$\widehat{a_i} = \begin{cases} 1 & \text{if } \widehat{X}_i > \text{threshold,} \\ 0 & \text{if } \widehat{X}_i \leq \text{threshold.} \end{cases}$$

We call the word $\widehat{a}_0 \widehat{a}_1 \ldots \widehat{a}_n$ *restored word* from $\widehat{X}_t$. We note that the mean and the variance play important roles to define a threshold between 'low' and 'high' of signals, in particular, when we use ν-adic numbers such as octal numbers and hexadecimal numbers instead of binary numbers.

From now on, we explain mathematical details for our data concealing technique and restoring technique. We give our secret SES by

$$F_i(X_t^i, \dot{X}_t^i, U_t^i, W_t^{1,i}) = 0, \qquad i = 1, 2, \ldots, N, \tag{1}$$
$$X_t^{i+1} = c^i X_t^i + W_t^{2,i}, \qquad i = 1, 2, \ldots, N, \tag{2}$$
$$U_t^{N+1} = f\left(X_t^{N+1}\right). \tag{3}$$

In the above system, $\dot{X}_t^i$ stands for the time derivative dX_t^i/dt of the process X_t^i, and c^i is a constant. The initial data X_t^1 is given by $X_t^1 = X_t$. The concealed data $U_t^i, i = 1, 2, \ldots, N, N + 1$, are directly defined by Eqs. (1) and (3), not Eq. (2). That is, we can hide the linear part of our system because we do not have to make an interference wave. This is the point of our method that is different from that of telecommunication using noises (Wyner 1975; Hero 2003; Goel and Negi 2008; Swindlehurst 2009; Mukherjee and Swindlehurst 2011). Introducing functionals, G_i, $i = 1, 2, \ldots, N$, and using them for Eq. (2), we can introduce the chaotic disturbance in our concealing-restoring system (Fujii and Hirokawa 2020).

Equations (1) and (3) are the mathematical realization of c1. The repetition of Eq. (1) from $i = 1$ to $i = N$ with the help of Eq. (2) is for the realization of c2. We can mathematically realize c3 as follows: Take numbers $r_\ell, \ell = 1, 2, \ldots, M$, with $\sum_{\ell=1}^{M} r_\ell = 0$, and define

$$U_t^\ell - \frac{1}{M}\left(U_t^i + r_\ell U_t^j\right), \quad \ell = 1, 2, \ldots, M,$$

where $i \neq j$. Then, we can split the data U_t^i into the data $U_t^\ell, \ell = 1, 2, \ldots, M$. In the case $M = 2$, for instance, we generate a random number r with $r \neq 0$, and set r_1 and r_2 as $r_1 = r$ and $r_2 = -r$. From the split data, $U_t^\ell, \ell = 1, 2, \ldots, M$, we can restore the data U_t^ℓ to the data U_t^i and U_t^j by

$$U_t^i = \sum_{\ell=1}^{M} U_t^\ell \quad \text{and} \quad U_t^j = r_\ell^{-1}\left(M U_t^\ell - U_t^i\right)$$

for an ℓ satisfying $r_\ell \neq 0$. We can also use the sequence, $r_1, r_2, \ldots, r_M$, as a secret or common key.

We note that the last stochastic process appearing in Eq. (3) has the form,

$$X_t^{N+1} = c^1 \cdots c^N X_t + \sum_{i=1}^{N-1}\left(\prod_{j=i+1}^{N} c^j\right) W_t^{2,i} + W_t^{2,N}. \tag{4}$$

2.1   How to Conceal Data

We take the original data X_t as initial data,

$$X_t^1 = X_t.$$

Inputting it into Eq. (1) with the noise $W_t^{1,1}$, we conceal it by the SDE,

$$F_1(X_t^1, \dot{X}_t^1, U_t^1, W_t^{1,1}) = 0.$$

We seek U_t^1 in the above and obtain a concealed data U_t^1. By Eq. (2),

$$X_t^2 = c^1 X_t^1 + W_t^{2,1},$$

we have data X_t^2 for the next step. These data X_t^2 consist of the superposition (i.e., linear combination) of X_t^1 and $W_t^{2,1}$, and thus, there is a possibility that a wiretapper removes the noise $W_t^{2,1}$ and wiretap X_t^1. Thus, to improve the security with another noise-disturbance, we have the same procedure again. We input the data X_t^2 into Eq. (1) with the noise $W_t^{1,2}$,

$$F_2(X_t^2, \dot{X}_t^2, U_t^2, W_t^{1,2}) = 0.$$

We then obtain the concealed data U_t^2. Repeating the same procedures, we obtain the concealed data, $U_t^1, U_t^2, \ldots, U_t^N$, and hide the data, $X_t^1, X_t^2, \ldots, X_t^N$.

At last, input the concealed data X_t^N into Eq. (2) and get the data X_t^{N+1}. We input this into Eq. (3) and hide it. We then obtain the last concealed data U_t^{N+1}. In this way, the sequence of the concealed data, $U_t^1, U_t^2, \ldots, U_t^N, U_t^{N+1}$, is created.

In the case where the original data are digital, and they give the binary pulse X_t, the concealed data, U_t^i, $i = 1, 2, \ldots, N, N+1$, merely become analog data. So, a wiretapper has to know A/D transformation to obtain the original digital data as getting the concealed data. Therefore, the D/A and A/D transformations play an important role for the concealing-restoring system for some digital data. We can also use them as secret or common keys.

2.2   How to Restore Data

Since the nonlinear function f is bijective, we can restore the concealed data U_t^{N+1} to the data X_t^{N+1} by

$$X_t^{N+1} = f^{-1}\left(U_t^{N+1}\right).$$

In the light of the stochastic filtering theory, Eqs. (1) and (2) are the state equation and the observation equation, respectively, and they make the system of the noise-filtering. Inputting the above X_t^{N+1} into Eq. (2), and the concealed data U_t^N into Eq. (1), we have simultaneous equations to seek the data X_t^N,

$$F_N(X_t^N, \dot{X}_t^N, U_t^N, W_t^{1,N}) = 0,$$
$$X_t^{N+1} = c^N X_t^N + W_t^{2,N}.$$

Since we cannot completely restore the noises to the original ones, $W_t^{1,N}$ and $W_t^{2,N}$, we cannot completely seek the stochastic process X_t^N. Thus, we estimate it with the help of a proper stochastic filtering theory to remove the random noises. We then obtain the estimated data $\widehat{X}_t^N$.

Inputting the estimated data $\widehat{X}_t^N$ into the slot of X_t^N of Eq. (2), and the concealed data U_t^{N-1} into Eq. (1), we reach simultaneous equations to seek the data X_t^{N-1},

$$F_{N-1}(X_t^{N-1}, \dot{X}_t^{N-1}, U_t^{N-1}, W_t^{1,N-1}) = 0,$$
$$\widehat{X}_t^N = c^{N-1} X_t^{N-1} + W_t^{2,N-1}.$$

In the same way as in the above, the stochastic filtering theory gives us the next estimated data $\widehat{X}_t^{N-1}$. We repeat this procedure, and obtain the estimated data, $\widehat{X}_t^N, \widehat{X}_t^{N-1}, \ldots, \widehat{X}_t^2, \widehat{X}_t^1$, by turns, and we pick up the last estimate $\widehat{X}_t^1$. This is the restoration $\widehat{X}_t$ of the original data X_t.

3 Example of Functionals and Simulation

As for how to determine each functional, F_i, $i = 1, 2, \ldots, N$, any definition of it is fine so long as a noise-filtering theory is established for the system with F_i. To restore the concealed data, $U_t^1, U_t^2, \ldots, U_t^N, U_t^{N+1}$, generally speaking, we have to know the concrete forms of the functionals, and the noise-filtering theory. Therefore, we must hide both for securing the original data. In this paper, however, we disclose one of examples of the concrete definition of the functionals and one of examples of the noise-filterings, which should actually be supposed to be in secret. We point out that the example of concealing-restoring system introduced in this section is not valid for other functionals. In particular, it is not tolerant of nonlinearity. See Sect. 5.

3.1 An Example of the Set of Functionals

We release an example of functionals in this section. We determine functions $A^i(t)$, $v^i(t)$, and non-zero constants b_u^i, b^i in secret. Here $v^i(t)$ can be a random noise. For instance, we often make $v^i(t)$ by the linear interpolation based on normal random numbers. Namely, we first assign a normal random number with $N(0, \sigma_v^2)$ to $v^i(k)$ for each i and k, and then, connect them by linear interpolation. Here, $N(0, \sigma_v^2)$

means the normal distribution whose mean and standard deviation are, respectively, 0 and σ_v. We give each functional F_i such that it makes a SDE,

$$dX_t^i = \left(A^i(t) - 1\right) X_t^i dt + b_u^i U_t^i dt + b^i v^i(t) dt - b_u^i dB_t^i, \tag{5}$$

for $i = 1, 2, \ldots, N$. That is,

$$\dot{X}_t^i = \left(A^i(t) - 1\right) X_t^i + b_u^i U_t^i + b^i v^i(t) - b_u^i W_t^{1,i}. \tag{6}$$

Here, $W_t^{1,i}$ and $W_t^{2,i}$ are Gaussian white noises whose mean $m^{j,i}$ and variance $V^{j,i}$ are, respectively, 0 and $(\sigma_j^i)^2$. B_t^i is the Brownian motion given by $W_t^{1,i} = dB_t^i/dt$, $i = 1, 2, \ldots, N$. We assume that the noises $W_t^{1,i}$ and $W_t^{2,i}$ are independent for each $i = 1, 2, \ldots, N$, but the noises $W_t^{2,i}$, $i = 1, 2, \ldots, N$, are not always independent. Thus, in the case where they are not independent, the linear combination of white noises appearing in Eq. (4) is not always white noise.

We regard the functions $A^i(t)$, the constants b_u^i, b^i, and the mean $m^{j,i}$ and variance $V^{j,i} = (\sigma_j^i)^2$ of the white noises as secret keys which are known only by the administrator of our concealing-restoring system. We use functions $v^i(t)$ as common keys. Since Eqs. (5) and (2), respectively, play the individual roles of the state equation and observation equation in the stochastic filtering theory, we employ the linear Kalman filtering theory (Kalman 1960; Kallianpur 1980; Bain and Crisan 2009; Grewal and Andrews 2015) to obtain the restoration $\widehat{X}_t$.

Using Eq. (6) we give the concealed data U_t^i, $i = 1, 2, \ldots, N$, by

$$U_t^i = \frac{1}{b_u^i} \left\{ dX_t^i + \left(1 - A^i(t)\right) X_t^i - b^i v^i(t) \right\} + dB_t^i. \tag{7}$$

In addition to these concealed data, we give the last concealed data U_t^{N+1} by Eq. (3). Conversely, since we obtain the data X_t^{N+1} by $X_t^{N+1} = f^{-1}(U_t^{N+1})$, we can estimate the data, $X_t^N, X_t^{N-1}, \ldots, X_t^1$, from the concealed data, $U_t^N, U_t^{N-1}, \ldots, U_t^1$, using the linear Kalman filtering theory.

3.2 Simulation of Concealing and Restoring Data on Physical Layer

In our simulation of concealing and restoring data on the physical layer, we employ the message digest (Rivest 1991, 1992a, b; Suhaili and Watanabe 2017; MessageDigest 2020) to check the coincidence of the original word $a_0 a_1 \ldots a_n$ and its restored word $\widehat{a}_0 \widehat{a}_1 \ldots \widehat{a}_n$ though the message digest works on upper layers. Moreover, we can use the message digest to detect any falsification of the concealed data. We take the original word $a_0 a_1 \ldots a_n$ as a message, and then, produce its digest. We also produce the digest for the restored word $\widehat{a}_0 \widehat{a}_1 \ldots \widehat{a}_n$. Comparing hash values of the two digests, we can make the check of the coincidence and the detection of the falsification at the same time. The check and detection should be performed on a layer out of layers between Layer 3 and Layer 7. In our simulation, we employ SHA-256 to make the hash values (Secure Hash Standard 2015).

To make the estimation in the simulation, we employ the linear Kalman filtering theory under the following conditions. We make Eqs. (1)–(3) for $N = 2$ with $A^i(t) = 0.1$ (constant function), $b^i = 1$, $b^i_u = 1$, and $c^i = 1$ for each $i = 1, 2$. We define the common key $v^i(t)$ by the linear interpolation based on a normal random number with $N(0, 1^2)$. We assume that the means of white noises are all 0. The standard deviation of the white noise $W_t^{j,1}$ is $\sigma_j^1 = 0.1$, and that of the white noise $W_t^{j,2}$ is $\sigma_j^2 = 1$. The length of the word $a_0 a_1 \ldots a_n$ is 100, and therefore, $n = 99$.

Our original word $a_1 a_2 \ldots a_{99}$ is given by Eq. (8). We here note that we remove the character a_0 because we cannot estimate the first bit in our concealing-restoring system.

$$00001100100111001000100000101110111111111001000110$$
$$10100111101111011001010101000101101111001101111001. \tag{8}$$

Then, we get its binary pulse X_t as in Fig. 2. The hash value of the digest made from the original word (8) is

$$979bca61579e002c9097c78088740e9fdaf21535d6a5c5876bd8623a86185292. \tag{9}$$

We make the concealed data, U_t^1 and U_t^2, by Eq. (7) with the help of the linear equation given in Eq. (2). We finally make the concealed data U_t^3 using the nonlinear equation given in Eq. (3) with $f(\xi) = \xi^3$. Their graphs are in Figs. 3 and 4. Following the Kalman filtering theory, we remove the white noises, and estimate the binary pulse X_t. Then, we obtain the restoration $\widehat{X}_t$ as in Fig. 5. The concrete algorithm to seek the restoration $\widehat{X}_t$ comes out in Ref. Fujii and Hirokawa (2020). Let us take 0 as the

Fig. 2 The binary pulse X_t transformed from the original word (8)

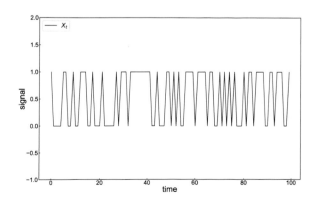

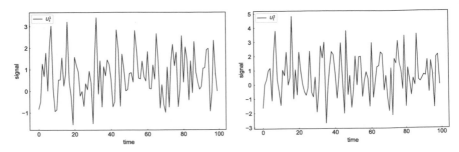

Fig. 3 The concealed data, U_t^1 (left) and U_t^2 (right), for the binary pulse X_t in Fig. 2

Fig. 4 The concealed data U_t^3 for the binary pulse X_t in Fig. 2

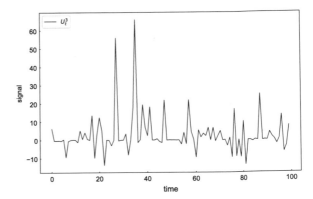

Fig. 5 The restoration $\widehat{X}_t$ for the binary pulse X_t in Fig. 2

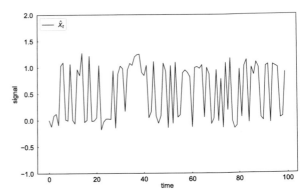

threshold. Then, we obtain the restored word $\widehat{a}_1\widehat{a}_2\ldots\widehat{a}_{99}$ and the hash value of its digest made from the restoration $\widehat{X}_t$. We can achieve positive results that they are the same as Eqs. (8) and (9), respectively.

We note that the graphs in Figs. 3 and 4 say that the concealed data, U_t^1, U_t^2, and U_t^3, are merely analog data. If a wiretapper becomes aware that the concealed data are for digital ones and knows our A/D transformation in some way, then the wiretapper gets a binary word from the concealed data as follows:

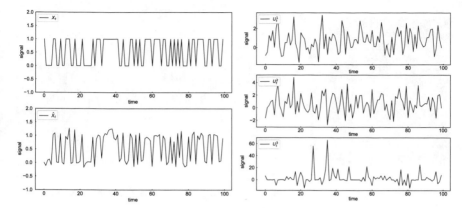

Fig. 6 X_t (Fig. 2) and $\widehat{X}_t$ (Fig. 5) from the above of the left 2 graphs. U_t^1 (Fig. 3), U_t^2 (Fig. 3), and U_t^3 (Fig. 4) from the above of the right 3 graphs. Here $t \in [0, 99]$

$$00111011000111011000111000001001101011111001101100$$
$$11011111101001111000010111100101101011000111100110$$

for U_t^1,

$$00011011000111011010110000100100111001111011001010$$
$$01011001001001111010010101111010100001000111011010110$$

for U_t^2, and

$$10000000000010110101110000010001001100111100100100$$
$$00000101100111110101100010100010000001000111011001$$

for U_t^3. Here, since the wiretapper does not know that we removed the first bit, every concealed data U_t^i makes the word consisting of 100 characters.

In Fig. 6 we show the comparison of the original binary pulse X_t, its restoration $\widehat{X}_t$, and the concealed data U_t^i, $i = 1, 2, 3$.

4 Application to Data on Physical Layer and Presentation Layer

4.1 Binary Data of Pictorial Image

We now apply the technology of our mathematical method to the binary data of a pictorial image. We use digital data of a pictorial image in the ORL Database of

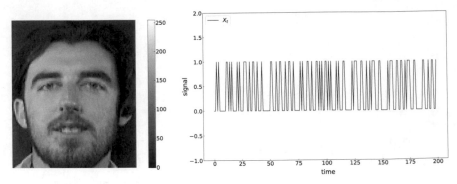

Fig. 7 The original pictorial image (left) with the digital data, and its binary pulse X_t (right) only for $t \in [0, 200]$

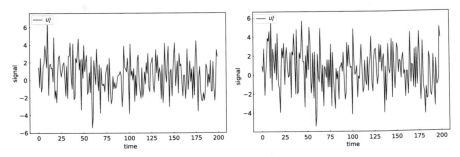

Fig. 8 The concealed data, U_t^1 (left) and U_t^2 (right), for the binary pulse X_t in Fig. 7. Here $t \in [0, 200]$ only

Faces, an archive of AT&T Laboratories Cambridge (The ORL Database of Faces 2020). The data have the grayscale value of 256 gradations (8bit/pixel). We set our parameters as $A = A^i = 0.1, b = b^i = 1, b_u = b_u^i = 1, c = c^i = 1, \sigma_1 = \sigma_1^i = 0.1$, and $\sigma_2 = \sigma_2^i = 1$. We determine the common key $v^i(t)$ in the same way as in Sect. 3.2 with $\sigma_v = 2$. The original pictorial image and its binary pulse X_t are obtained as in Fig. 7. Here, the upper bound of t is $92 \times 112 = 10304$ and t runs over $[0, 10304]$. We obtain the concealed data, U_t^1 and U_t^2, by Eq. (7) as in Fig. 8, and the concealed data U_t^3 by Eq. (3) as in Fig. 9. The restoration $\widehat{X}_t$ and the restored pictorial image from it are in Fig. 10.

If a wiretapper tries to get the original pictorial image from the concealed data $U_t^i, i = 1, 2, 3$, since the concealed data are analog as in Figs. 8 and 9, the wiretapper has to know our A/D transformation, and our transformation from the digital data to a pictorial image as well as some keys used in SES. The latter transformation should be done on upper layers. We now assume that the wiretapper can know the transformations. Then, each pictorial image of the concealed data, $U_t^i, i = 1, 2, 3$, is in Fig. 11. The format of the pictorial image of Fig. 7 is PGM (i.e., portable gray map). In fact, we cannot restore the PGM header from the concealed data, that is, the header of the PGM is completely broken. Thus, the wiretapper has to realize that

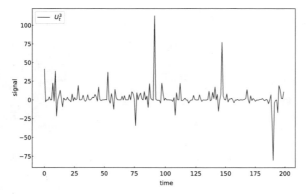

Fig. 9 The concealed data U_t^3 for the binary pulse X_t in Fig. 7. Here $t \in [0, 200]$ only

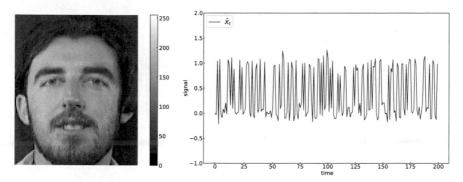

Fig. 10 The restoration $\widehat{X}_t$ for the binary pulse X_t in Fig. 7 only for $t \in [0, 200]$ (right) and the restored pictorial image (left) of $\widehat{X}_t$

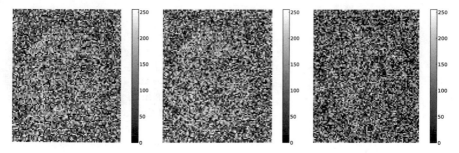

Fig. 11 From the left, pictorial images of the concealed data, U_t^1, U_t^2 in Fig. 8, and U_t^3 in Fig. 9, for the binary pulse X_t in Fig. 7. Here $(\sigma_v)^2 = 4$

the concealed data are for PGM in some way, and he/she has to write the header by himself/herself to restore the pictorial image.

As for the role of the common key $v^i(t)$, comparing Fig. 12 with Fig. 11, we can realize the effect of the variance of the common key $v^i(t)$ and the nonlinear function

Fig. 12 From the left, pictorial images of the concealed data, U_t^1, U_t^2 in Fig. 8, and U_t^3 in Fig. 9, for the binary pulse X_t in Fig. 7. Here $(\sigma_v)^2 = 1$

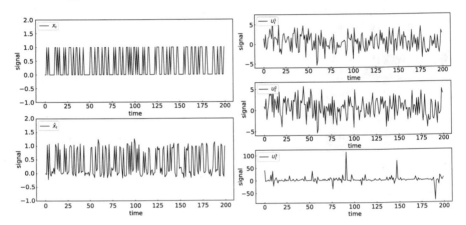

Fig. 13 X_t (Fig. 7) and $\widehat{X}_t$ (Fig. 10) from the above of the left 2 graphs. U_t^1 (Fig. 8), U_t^2 (Fig. 8), and U_t^3 (Fig. 9) from the above of the right 3 graphs. Here $t \in [0, 200]$ only

$f(\xi)$. The variance of the common key $v^i(t)$ is smaller in Fig. 12 than it is in Fig. 11, that is, $(\sigma_v)^2 = 4$ for Fig. 11 and $(\sigma_v)^2 = 1$ for Fig. 12, though other parameters for Fig. 12 are the same as for Fig. 11. The contour of the face in the pictorial image of U_t^1 in Fig. 12 stands out more clearly than in Fig. 11. Meanwhile, the nonlinearity conceals the contour as in the pictorial image of U_t^3 in Fig. 12.

In Fig. 13 we show the comparison of the original binary pulse X_t, its restoration $\widehat{X}_t$, and the concealed data U_t^i, $i = 1, 2, 3$.

4.2 Analog Data of Pictorial Image

We use analog data of a pictorial image in the Olivetti faces database (The Olivetti Faces Database 2020), where the data of pictorial images are transformed to analog data from the original ones in the ORL Database of Faces, an archive of AT&T

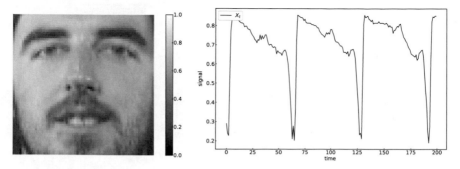

Fig. 14 The original pictorial image (left) with the analog data, and the analog data X_t only for $t \in [0, 200]$ (right)

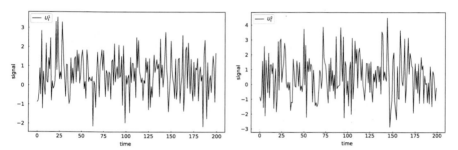

Fig. 15 The concealed data, U_t^1 (left) and U_t^2 (right), for the analog data X_t in Fig. 14. Here, $t \in [0, 200]$ only

Laboratories Cambridge (The ORL Database of Faces 2020). The data have the grayscale value of 256 gradations (8bit/pixel). Our parameters are $A = A^i = 0.1$, $b = b^i = 1$, $b_u = b_u^i = 1$, $c = c^i = 1$, $\sigma_1 = \sigma_1^i = 0.1$, and $\sigma_2 = \sigma_2^i = 1$ again. We also use the common key $v^i(t)$ in the same way as in Sect. 3.2 with $\sigma_v = 2$. The original analog data X_t and their pictorial image are in Fig. 14. Here, the upper bound of t is $64 \times 64 = 4096$ and t runs over $[0, 4096]$. The concealed data, U_t^1 and U_t^2, defined by Eq. (7) are in Fig. 15, and the concealed data U_t^3 defined by Eq. (3) are in Fig. 16. We can restore the pictorial image with the restoration $\widehat{X}_t$ as in Fig. 17. If a wiretapper becomes aware of our method to make a pictorial image from analog data, then the wiretapper gets pictorial images from the concealed data U_t^i, $i = 1, 2, 3$, as in Fig. 18.

In Fig. 19 we show the comparison of the original binary pulse X_t, its restoration $\widehat{X}_t$, and the concealed data U_t^i, $i = 1, 2, 3$.

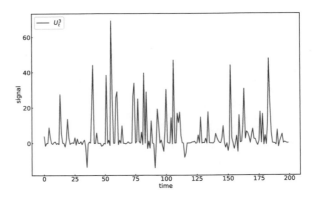

Fig. 16 The concealed data U_t^3 for the analog data X_t, $t \in [0, 200] \subset [0, 4096]$, in Fig. 14

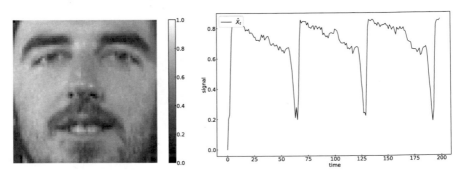

Fig. 17 The restoration $\widehat{X}_t$ (right) for the analog data X_t in Fig. 14 only for $t \in [0, 200]$, and the pictorial image (left) of $\widehat{X}_t$

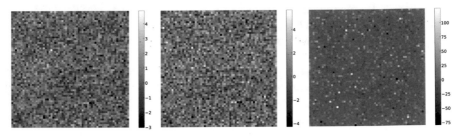

Fig. 18 From the left, pictorial images of the concealed data, U_t^1 (Fig. 15), U_t^2 (Fig. 15), and U_t^3 (Fig. 16)

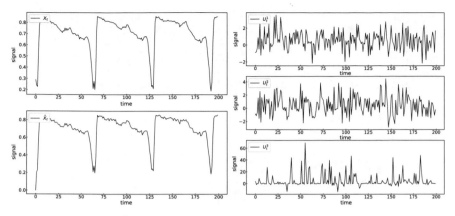

Fig. 19 X_t (Fig. 14) and $\widehat{X}_t$ (Fig. 17) from the above of the left 2 graphs. U_t^1 (Fig. 15), U_t^2 (Fig. 15), and U_t^3 (Fig. 16) from the above of the right 3 graphs. Here $t \in [0, 200]$ only

5 Conclusion and Future Work

We have proposed a mathematical technique for concealing data on the physical layer of the OSI reference model by using random noise disturbance, and moreover, a mathematical technique for restoring the concealed data by using the stochastic process estimation. In this concealing-restoring system, the functionals determining SDEs play a role of secret or common keys. Then, the proper noise-filtering theory forms a nucleus to restore the concealed data. In addition, we have showed the simulation result for the data on physical layer and some applications of the two techniques to the pictorial images. We have opened one of examples of the functionals. Then, we have showed how to conceal the data by using the noise-disturbance, and have demonstrated how to restore the data by removing the noises. Here, the significant point to be emphasized is that any composition of the SES and any form of the individual functional will do so long as a proper noise-filtering method is established for them. We make briefly some comments about it at the tail end of this section.

We have used the scalar-valued processes, and thus, prepared just one common key for one SDE. We can prepare some common keys for one SDE by using the vector-valued processes.

Although we have employed the message digest to make the check of the coincidence of the binary word and the detection of the falsification at the same time, we are now developing a method with low complexity so that we can make them for data on the physical layer.

Fig. 20 From the left, the original pictorial image, the individual pictorial images of the concealed data U_t^1 and U_t^2, and the pictorial image of the restored data. The original pictorial image is a bitmap image, and the parameter t of the original data X_t runs over [0, 90123byte]

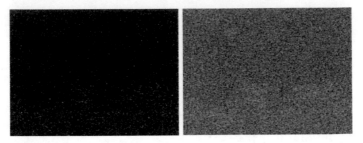

Fig. 21 Comparison between the pictorial images of U_t^2 with nonlinearity (left) and $X_t^2 = f^{-1}(U_t^2)$ without nonlinearity (right)

According to our several experiments including the concrete examples in Sect. 4, we think that the nonlinearity enhances the noise-disturbance. For instance, the pictorial images in Fig. 20 are the case $N = 1$. Comparing the pictorial images of U_t^2 and $X_t^2 = f^{-1}(U_t^2)$ in Fig. 21, we can say that the enhancement of noise-disturbance appears with the black color. We will study the roles of several parameters including the nonlinearity. We here introduce the effect coming from the nonlinearity beforehand. The state space determined by Eq. (5) is constructed by the linear Gaussian model, and thus, we used the linear Kalman filtering theory in Sects. 3 and 4. We can make it more general: nonlinear, non-Gaussian state space. Then, we should employ another noise-filtering theory such as the particle filtering theory (Bain and Crisan 2009). In fact, putting a concrete nonlinearity N_A or another nonlinearity N_B in the functional F_i of Eq. (1), we have concealed data $U_t^{A,i}$ or $U_t^{B,i}$, $i = 1, 2, 3$, different from those in this paper. Then, the linear Kalman filtering theory is not useful any longer. For instance, we respectively conceal the data in Figs. 7 and 14 using such functionals with the nonlinearity N_A or N_B. Then, we cannot estimate the data from the concealed ones by the linear Kalman filter to our satisfaction. See Figs. 22, 23, 24, and 25. The difference between the restorations in Figs. 22 and 23 or between those in Figs. 24 and 25 depends on the degree of nonlinearity. We show the restoring system using the particle filter in Ref. Fujii and Hirokawa (2020).

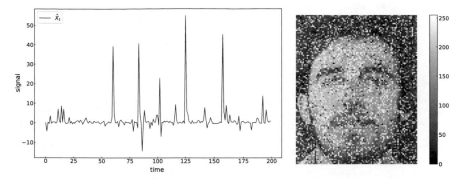

Fig. 22 The left graph is restoration $\widehat{X}_t$, $0 \le t \le 200$, from the concealed data, $U_t^{A,i}$, $i = 1, 2, 3$, with the nonlinearity N_A using the Kalman filtering. The right picture is the pictorial image restored from such a restoration $\widehat{X}_t$

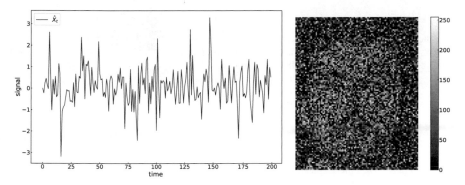

Fig. 23 The left graph is restoration $\widehat{X}_t$, $0 \le t \le 200$, from the concealed data, $U_t^{B,i}$, $i = 1, 2, 3$, with the nonlinearity N_B using the linear Kalman filtering. The right picture is the pictorial image restored from such a restoration $\widehat{X}_t$

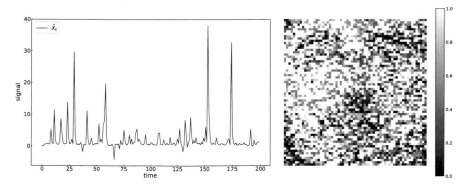

Fig. 24 The left graph is restoration $\widehat{X}_t$, $0 \le t \le 200$, from the concealed data, $U_t^{A,i}$, $i = 1, 2, 3$, with the nonlinearity N_A using the linear Kalman filtering. The right picture is the pictorial image restored from such a restoration $\widehat{X}_t$

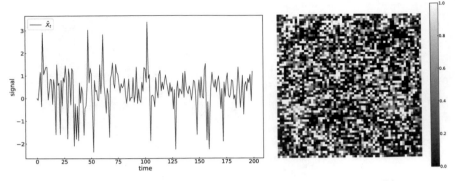

Fig. 25 The left graph is restoration $\widehat{X}_t$, $0 \leq t \leq 200$, from the concealed data, $U_t^{B,i}$, $i = 1, 2, 3$, with the nonlinearity N_B using the Kalman filtering. The right picture is the pictorial image restored from such a restoration $\widehat{X}_t$

Acknowledgements This work is partially based on Fujii's bachelor thesis at Hiroshima University in March, 2019. For useful comments and discussion, the authors thank the following: Kirill Morozov (University of North Texas), Shuichi Ohno (Hiroshima University), Kouichi Sakurai (Kyushu University), Takeshi Takagi (Hiroshima University), and Tatsuya Tomaru (Hitachi, Ltd.).

References

A. Bain, D. Crisan, *Fundamentals of Stochastic Filtering* (Springer, Berlin, 2009)

J. Clausen, E. Fetz, J. Donoghue, J. Ushiba, U. Spörhase, J. Chandler, N. Birbaumer, S.R. Soekadar, Help, hope, and hype: ethical dimensions of neuroprosthetics. Science **356**, 1338–1339 (2017)

K.M. Cuomo, A.V. Oppenheim, Synchronization of Lorenz-based chaotic circuits with applications to communications. IEEE Trans. Circuit Syst. **40**, 626–633 (1993)

I. Damgård, A design principle for hash functions, in *Advances in Cryptology – CRYPTO'89 Proceedings*. Lecture Notes in Computer Science, vol. 435 (Springer, 1989), pp. 416–427

T. Fujii, M. Hirokawa, Nonlinear concealing-restoring system with random noise disturbance for data on physical layer arXiv:1910.03214

S. Goel, R. Negi, Guaranteeing secrecy using artificial noise. IEEE Trans. Wirel. Commun. **7**, 2180–2189 (2008)

G. Grassi, S. Mascolo, A system theory approach for designing cryptosystems based on hyperchaos. IEEE Trans. Circuit Syst. -I: Fundam. Theory Appl. **46**, 1135–1138 (1999)

M.S. Grewal, A.P. Andrews, *Kalman Filtering. Theory and Practice Using MATLAB* (Wiley, New York, 2015)

A. Hero, Secure space-time communication. IEEE Trans. Inf. Theory **49**, 3235–3249 (2003)

K. Ito, Y. Masuda, E. Okamoto, A chaos MIMO-based polar concatenation code for secure channel coding, in *2019 International Conference on Information Networking* (2019), pp. 262–267

B.N. Kain, A.K. Agrawala, *Open Systems Interconnection: Its Architecture and Protocols* (McGraw-Hill, London, 1992)

G. Kallianpur, *Stochastic Filtering Theory* (Springer, Berlin, 1980)

R.E. Kalman, A new approach to linear filtering and prediction problems. Trans. ASME - J. Basic Eng. (Ser. D) **82**, 35–45 (1960)

H. Lenug, J. Lam, Design of demodulator for the chaotic modulation communication system. Trans. Circuit Syst. **44**, 262–267 (1997)

C. Lin, D. He, N. Kumar, K.-K.R. Choo, A. Vinel, X. Huang, Security and privacy for the internet of drones: challenges and solutions. IEEE Commun. Mag. **56**, 64–69 (2018)

R.C. Merkle, Secrecy, authentication, and public key systems, Ph.D. thesis (Stanford University, 1979)

R.C. Merkle, A certified digital signature, in *Advances in Cryptology - CRYPTO'89 Proceedings*. Lecture Notes in Computer Science, vol. 435 (Springer, 1989), pp. 218–238

MessageDigest, Android developers, https://developer.android.com/reference/java/security/MessageDigest

A. Mukherjee, A.L. Swindlehurst, Robust beam-forming for security in MIMO wiretap channels with imperfect CSI. IEEE Trans. Signal Process. **59**, 351–361 (2011)

S. Ohno, H. Kaida, T. Kodani, Secret communication with multiple antennas may not be secure against eavesdropping using blind equalization (in Japanese). IEICE Trans. Commun. **J95-B**, 751–759 (2012)

E. Okamoto, A chaos MIMO transmission scheme for secure communications on physical layer, in *Proceedings of the 2011 IEEE 73rd Vehicular Technology Conference (VTC Spring)* (2011), pp. 1–5

E. Okamoto, Y. Inaba, A chaos MIMO transmission scheme using turbo principle for secure channel-coded transmission. IEICE Trans. Commun. **E98.B**, 1482–1491 (2015)

E. Okamoto, Y. Iwanami, A trellis-coded chaotic modulation scheme, in *2006 IEEE International Conference on Communications*, vol. 11 (2006), pp. 5010–5015

R. Rivest, The MD4 message digest algorithm, in *Advances in Cryptology - CRYPTO'90 Proceedings*. Lecture Notes in Computer Science, vol. 37 (Springer, 1991), pp. 303–311

R. Rivest, The MD4 message digest algorithm. RFC **1320** (1992a) (MIT and RSA Data Security, Inc.)

R. Rivest, The MD5 message digest algorithm. RFC **1321** (1992b) (MIT and RSA Data Security, Inc.)

R. Román-Castro, J. López, S. Gritzalis, Evolution and trends in IoT security. Computer **51**, 16–25 (2018)

B. Schneier, *Applied Cryptography: Protocols, Algorithms and Source Code in C* (Wiley, New York, 2015)

Secure hash standard (SHS), FIPS PUB 180-4, Federal Information Processing Standards Publication (2015)

S.b. Suhaili, T. Watanabe, High-throughput message digest (MD5) design and simulation-based power estimation using unfolding transformation. J. Signal Process. **21**, 233–238 (2017)

A.L. Swindlehurst, Fixed SINR solution for the MIMO wiretap channel, in *Proceedings of the IEEE International Conference on Acoustics, Speech and Signal Processing* (2009), pp. 2437–2440

The Olivetti faces database, https://scikit-learn.org/0.19/datasets/olivetti_faces.html

The ORL database of faces, https://www.cl.cam.ac.uk/research/dtg/attarchive/facedatabase.html

C.W. Wu, L.O. Chua, A simple way to synchronize chaotic systems with applications to secure communication systems. Int. J. Bifurc. Chaos **3**, 1619–1627 (1993)

A.D. Wyner, The wire tap channel. Bell Syst. Tech. J. **54**, 1355–1387 (1975)

H.Y. Youm, An overview of security and privacy issues for internet of things. IEICE Trans. Inf. Syst. **E100–D**, 1649–1662 (2017)

G. Zheng, Secure communication based on multi-input multi-output chaotic system with large message amplitude. Chaos Solitons Fractals **41**, 1510–1517 (2009)

Quantum Optics with Giant Atoms—the First Five Years

Anton Frisk Kockum

Abstract In quantum optics, it is common to assume that atoms can be approximated as point-like compared to the wavelength of the light they interact with. However, recent advances in experiments with artificial atoms built from superconducting circuits have shown that this assumption can be violated. Instead, these artificial atoms can couple to an electromagnetic field at multiple points, which are spaced wavelength distances apart. In this chapter, we present a survey of such systems, which we call *giant atoms*. The main novelty of giant atoms is that the multiple coupling points give rise to interference effects that are not present in quantum optics with ordinary, small atoms. We discuss both theoretical and experimental results for single and multiple giant atoms, and show how the interference effects can be used for interesting applications. We also give an outlook for this emerging field of quantum optics.

Keywords Quantum optics · Giant atoms · Waveguide QED · Relaxation rate · Lamb shift · Superconducting qubits · Surface acoustic waves · Cold atoms

1 Introduction

Natural atoms are so small (radius $r \approx 10^{-10}$ m) that they can be considered point-like when they interact with light at optical frequencies (wavelength $\lambda \approx 10^{-6}$–10^{-7} m) (Leibfried et al. 2003). If the atoms are excited to high Rydberg states, they can reach larger sizes ($r \approx 10^{-8}$–10^{-7} m), but quantum-optics experiments with such atoms have them interact with microwave radiation, which has much longer wavelength ($\lambda \approx 10^{-2}$–10^{-1} m) (Haroche 2013). It has thus been well justified in theoretical treatments of quantum optics to assume $r \ll \lambda$, called the *dipole approximation*,

A. Frisk Kockum (✉)
Wallenberg Centre for Quantum Technology, Chalmers University of Technology, 412 96 Gothenburg, Sweden
e-mail: anton.frisk.kockum@chalmers.se

© The Author(s) 2021
T. Takagi et al. (eds.), *International Symposium on Mathematics, Quantum Theory, and Cryptography*, Mathematics for Industry 33, https://doi.org/10.1007/978-981-15-5191-8_12

125

which simplifies the description of the interaction between light and matter (Walls and Milburn 2008).

In recent years, experimental investigations of quantum optics have expanded to systems with *artificial atoms*, i.e., engineered quantum systems such as quantum dots (Hanson et al. 2007) and superconducting quantum bits (qubits) (You and Nori 2011; Xiang et al. 2013; Gu et al. 2017; Kockum and Nori 2019), which emulate essential aspects of natural atoms. The circuits making up superconducting qubits can be large, reaching sizes up to $r \approx 10^{-4}$–10^{-3} m, but this is still small when compared with the wavelength of the microwave fields they interact with.

In 2014, one experiment (Gustafsson et al. 2014) forced quantum opticians to reconsider the dipole approximation. In that experiment, a superconducting transmon qubit (Koch et al. 2007) was coupled to surface acoustic waves (SAWs) (Datta 1986; Morgan 2007). Due to the low propagation velocity of SAWs, their wavelength was $\lambda \approx 10^{-6}$ m, and the qubit, due to its layout with an interdigitated capacitance, coupled to the SAWs at multiple points, which were spaced $\lambda/4$ apart.

Motivated by this experiment, theoretical investigations on *giant atoms* were initiated (Kockum et al. 2014). The main finding was that the multiple coupling points lead to interference effects, e.g., the coupling of the giant atom to its environment becomes frequency-dependent (Kockum et al. 2014).

These initial experimental and theoretical works on giant atoms were published 5 years ago, at the time of writing for this book chapter. In this chapter, we give a brief survey of the developments in the field of quantum optics with giant atoms that have followed since. We begin in Sect. 2 with theory for giant atoms, looking first at the properties of a single giant atom (Sect. 2.1), including what happens when the coupling points are extremely far apart (Sect. 2.2), and then at multiple giant atoms (Sect. 2.3). In Sect. 3, we survey the different experimental systems where giant atoms have been implemented or proposed. We conclude with an outlook (Sect. 4) for future work on giant atoms, pointing to several areas where interesting results can be expected.

2   Theory for Giant Atoms

The experimental setup where giant atoms were first implemented (Gustafsson et al. 2014) falls into the category of waveguide quantum electrodynamics (QED). In waveguide QED (Gu et al. 2017; Roy et al. 2017), a continuum of bosonic modes can propagate in a one-dimensional (1D) waveguide and interact with atoms coupled to this waveguide. As reviewed in Gu et al. (2017), Roy et al. (2017), there is an abundance of theoretical papers dealing with one, two, or more atoms coupled to a 1D waveguide, but they almost all assume that the dipole approximation is valid, or, in other words, that the atoms are "small".

The difference between small and giant atoms is illustrated in Fig. 1. While a small atom, because of its diminutive extent, can be described as being connected to the waveguide at a single point, a giant atom couples to the waveguide at multiple

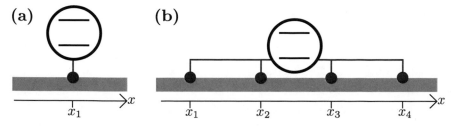

Fig. 1 The difference between a small atom and a giant atom. **a** A small atom (two levels) couples to the 1D waveguide (grey) at a single point (red, coordinate x_1). **b** A giant atom couples to the waveguide at multiple points (labelled k, coordinates x_k). The distance between two coupling points k and n, $|x_k - x_n|$, is *not* negligible compared to the wavelength of the modes in the waveguide that the atom interacts with

points, and the distance between these points *cannot* be neglected in comparison to the wavelength of the modes in the waveguide that couple to the atom. The relevant wavelength λ to compare with is set by the (angular) transition frequency ω_a of the atom and the propagation velocity v in the waveguide: $\lambda = 2\pi v/\omega_a$.

2.1 One Giant Atom

Quantum optics with a single giant atom was first studied theoretically in Kockum et al. (2014), prompted by the experiment in Gustafsson et al. (2014) (discussed in Sect. 3.1). For a small atom coupled to a continuum of modes, like in Fig. 1a, standard quantum-optics procedure is to derive a master equation by assuming that the coupling to the modes is relatively weak and tracing out the modes (Carmichael 1999; Gardiner and Zoller 2004; Walls and Milburn 2008). When considering whether the same procedure can be applied to a giant atom, there is a new timescale to take into account: the time it takes to travel in the waveguide between coupling points. In Kockum et al. (2014), this time was assumed small compared to the time it takes for an excitation in the atom to relax into the waveguide. With this assumption, the system is *Markovian*, i.e., the time evolution of the atom only depends on the present state of the system, not on the past (for the non-Markovian case, see Sect. 2.2). Thus, the standard master-equation derivation from quantum optics with small atoms can be applied here as well.

2.1.1 Master Equation for a Giant Atom

The derivation of a master equation for a giant atom starts from the total system Hamiltonian (we use units where $\hbar = 1$ throughout this chapter),

$$H = H_a + H_{wg} + H_I, \tag{1}$$

with the bare atomic Hamiltonian

$$H_a = \sum_m \omega_m |m\rangle\langle m|, \tag{2}$$

the bare waveguide Hamiltonian

$$H_{wg} = \sum_j \omega_j \left(a_{Rj}^\dagger a_{Rj} + a_{Lj}^\dagger a_{Lj}\right), \tag{3}$$

and the interaction Hamiltonian

$$H_I = \sum_{j,k,m} g_{jkm} \left(\sigma_-^{(m)} + \sigma_+^{(m)}\right)$$
$$\times \left(a_{Rj} e^{-i\omega_j x_k/v} + a_{Lj} e^{i\omega_j x_k/v} + a_{Rj}^\dagger e^{i\omega_j x_k/v} + a_{Lj}^\dagger e^{-i\omega_j x_k/v}\right). \tag{4}$$

Here, the atomic levels are labelled $m = 0, 1, 2, \ldots$, have energies ω_m, and are connected through lowering and raising operators $\sigma_-^{(m)} = |m\rangle\langle m+1|$ and $\sigma_+^{(m)} = |m+1\rangle\langle m|$. The bosonic modes in the waveguide are labelled with indices j and with an index R (L) for right-moving (left-moving) modes. The corresponding annihilation and creation operators are a and $a^\dagger$, respectively. The difference to the case of a small atom is the sum over coupling points labelled by k in Eq. (4). The phase factors $e^{\pm i\omega_j x_k/v}$ are not present for a small atom. These phase factors give rise to interference effects. Note that the coupling strengths g_{jkm} can depend on both j, k, and m.

Following the standard master-equation derivation using the Born-Markov approximation, the resulting master equation becomes

$$\dot{\rho} = -i \left[\sum_m (\omega_m + \Delta_m) |m\rangle\langle m|, \rho\right] + \sum_m \Gamma_{m+1,m} \mathcal{D}\left[\sigma_-^{(m)}\right]\rho, \tag{5}$$

where ρ is the density matrix for the atom, $\mathcal{D}[X]\rho = X\rho X^\dagger - \frac{1}{2}X^\dagger X\rho - \frac{1}{2}\rho X^\dagger X$ is the Lindblad superoperator describing relaxation (Lindblad 1976), and we have assumed negligible temperature T, i.e., $\omega_m \gg k_B T$. The relaxation rates for the atomic transitions $|m+1\rangle \rightarrow |m\rangle$ are

$$\Gamma_{m+1,m} = 4\pi J\left(\omega_{m+1,m}\right)\left|A_m\left(\omega_{m+1,m}\right)\right|^2, \tag{6}$$

where $\omega_{a,b} = \omega_a - \omega_b$, $J(\omega)$ is the density of states at frequency ω in the waveguide, and we have defined

$$A_m\left(\omega_j\right) = \sum_k g_{jkm} e^{i\omega_j x_k/v}. \tag{7}$$

The frequency shifts Δ_m of the atomic energy levels are Lamb shifts (Lamb and Retherford 1947; Bethe 1947) given by

$$\Delta_m = 2\mathcal{P} \int_0^\infty d\omega \frac{J(\omega)}{\omega} \left(\frac{|A_m(\omega)|^2 \omega_{m+1,m}}{\omega + \omega_{m+1,m}} - \frac{|A_{m-1}(\omega)|^2 \omega_{m,m-1}}{\omega - \omega_{m,m-1}} \right). \quad (8)$$

Both the relaxation rates and the Lamb shifts acquire a strong dependence on the atomic transition frequencies, encoded in the factor $A_m(\omega_j)$. For the case of a small atom, $A_m(\omega_j) = g_{jm}$, which is a constant provided that g_{jm} does not depend strongly on j. The effect of this frequency dependence for giant atoms can be seen clearly if one considers the simple case of an atom with two coupling points x_1 and x_2 [compare Fig. 1b] having equally strong coupling to the waveguide. If the two points are half a wavelength apart, i.e., $|x_1 - x_2| = \pi v/\omega_{m+1,m}$, there will be destructive interference between emission from the two points, and the relaxation for the corresponding atomic transition is completely suppressed: $\Gamma_{m+1,m} = 0$. If the two points are one wavelength apart, there is instead constructive interference and the relaxation rate is enhanced.

2.1.2 Frequency-Dependent Relaxation Rate

To further understand the frequency-dependence of the relaxation rates and the Lamb shifts, consider the case of a two-level atom coupled to the waveguide at N equidistant points with equal coupling strength at each point. In this case, introducing the notation $\varphi = \omega_{1,0}(x_2 - x_1)/v$, we obtain (Kockum et al. 2014)

$$\Gamma_{1,0} = \gamma \frac{\sin^2\left(\frac{N}{2}\varphi\right)}{\sin\left(\frac{1}{2}\varphi\right)} = \gamma \frac{1 - \cos(N\varphi)}{1 - \cos(\varphi)}, \quad (9)$$

$$\Delta_1 = \gamma \frac{N \sin(\varphi) - \sin(N\varphi)}{2\left[1 - \cos(\varphi)\right]}, \quad (10)$$

where γ is the relaxation rate that the atom would have had if it was coupled to the waveguide only at a single point. To obtain the Lamb shift, we have also made the simplifying assumption that $J(\omega)$ is constant, that the lower limit of the integral in Eq. (8) can be extended down to $-\infty$, and that only the dominating second term in that integral contributes. Since $\Delta_0 = 0$ with these assumptions, Eq. (10) gives the full frequency shift for the two-level atom. In fact, the relaxation rate and the Lamb shift are related through a Hilbert transform due to Kramers–Kronig relations (Cohen-Tannoudji et al. 1998).

The relaxation rates and Lamb shifts in Eqs. (9)–(10) are plotted for two values of N in Fig. 2. The central peak corresponds to the distance between neighbouring coupling points being one wavelength. Note that the frequency dependence becomes sharper when more coupling points are added; in frequency units, the width of the central peak is approximately $\omega_{1,0}/2\pi N$. This sharpness can be used to determine when the Markovian approximation underlying the master-equation derivation breaks down, which happens roughly when the relaxation rate changes noticeably within the linewidth of the atom, i.e., when $\Gamma_{1,0} \approx \omega_{1,0}/2\pi N$. Interestingly, this is

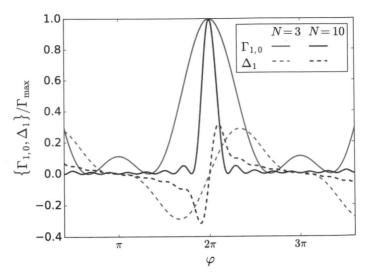

Fig. 2 Relaxation rates and Lamb shifts for a giant two-level atom with symmetrically spaced coupling points all having the same coupling strength. Red curves: $N = 3$ coupling points. Blue curves: $N = 10$ coupling points. Solid curves: Relaxation rates $\Gamma_{1,0}$. Dashed curves: Lamb shifts Δ_1. The relaxation rates and Lamb shifts are scaled to the maximum relaxation rate Γ_{max} for each N. Figure adapted from Kockum et al. (2014) with permission

approximately the same condition as when the travelling time between the outermost coupling points, $2\pi(N-1)/\omega_{1,0}$, becomes comparable to the relaxation time $1/\Gamma_{1,0}$.

An attractive feature of giant atoms is that the frequency-dependence of their relaxation rates (and Lamb shifts) can be *designed* (Kockum et al. 2014). The frequency dependence is directly determined by Eq. (7), which simply is a discrete Fourier transform of the coupling-point coordinates, weighted by the coupling strength in each point. With N coupling points, an experimentalist thus has $2N - 1$ knobs to turn (the translational invariance of the setup removes one degree of freedom). With enough coupling points, the curves in Fig. 2 can be moulded into any shape. Note that although the coupling-point coordinates and coupling strengths will be fixed in an experiment, superconducting qubits offer the possibility to tune the atomic frequency widely in situ (Gu et al. 2017; Kockum and Nori 2019), making it possible to move between regions with high and low relaxation rates during an experiment.

If we consider more than two atomic levels, other interesting applications of the frequency-dependent relaxation rate open up. As illustrated in Fig. 3, if the atomic transition frequencies $\omega_{1,0} \neq \omega_{2,1}$, it is possible to engineer the relaxation rates such that $\Gamma_{2,1}$ is at a maximum when $\Gamma_{1,0}$ is at a minimum. At that point, one can then create population inversion, and thus lasing, by driving the transition from $|0\rangle$ to $|2\rangle$ (Kockum et al. 2014). Recent experiments have been making use of this possibility to control the ratio of relaxation rates to enable electromagnetically induced transparency (EIT) (Andersson et al. 2020; Vadiraj et al. 2020).

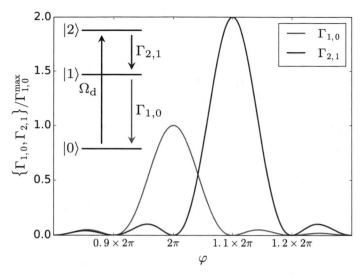

Fig. 3 Engineering population inversion in a giant atom. The blue curve and the red curve are the relaxation rates $\Gamma_{1,0}$ and $\Gamma_{2,1}$, respectively, as a function of transition frequency $\omega_{1,0}$. The plot assumes $N = 10$ equally spaced coupling points, with equal coupling strengths at all points, and an anharmonicity $\omega_{2,1} - \omega_{1,0} = -0.1 \times 2\pi v/(x_2 - x_1)$. The inset shows the level structure with the relaxation rates and a drive of strength Ω_d on the $|0\rangle \leftrightarrow |2\rangle$ transition. Figure adapted from Kockum et al. (2014) with permission

2.1.3 Comparison with an Atom in Front of a Mirror

It is possible to engineer frequency-dependent relaxation rates and Lamb shifts also for small atoms. This can be achieved by placing a small atom in front of a mirror instead of in an open waveguide, a setup which has been considered in several theoretical (Meschede et al. 1990; Dorner and Zoller 2002; Beige et al. 2002; Dong et al. 2009; Koshino and Nakamura 2012; Wang et al. 2012; Tufarelli et al. 2013; Fang and Baranger 2015; Shi et al. 2015; Pichler and Zoller 2016) and experimental works (Eschner et al. 2001; Wilson et al. 2003; Dubin et al. 2007; Hoi et al. 2015; Wen et al. 2018, 2019). Here, the atomic relaxation can be enhanced or suppressed by interference with the mirror image of the atom. This setup is equivalent to a giant atom with two coupling points in a unidirectional waveguide.

However, this is the limit with a small atom in front of a mirror. In such a setup, it is not possible to increase the number of coupling points, or to have different coupling strengths at different coupling points, which means that the frequency dependence cannot be designed like for a giant atom. Furthermore, since propagation is unidirectional, it is not possible to have more advanced scattering, possible with a giant atom, where both reflection and transmission are influenced by interference between coupling points.

2.1.4   Coupling a Giant Atom to a Cavity

By introducing reflective boundary conditions at both ends of the waveguide in Fig. 1, a multimode cavity will be formed. The coupling of a giant atom to such a cavity has yet to be explored as thoroughly as the open-waveguide case. We can see that similar interference effects as in the open waveguide will come into play. It will thus, for example, be possible to arrange the coupling points such that the giant atom couples strongly to some modes of the cavity and is decoupled from other modes. This can to some extent already be achieved with a small atom, whose single coupling point can be at a node for some modes and at an antinode for others. However, we note that a recent theory proposal (Ciani and DiVincenzo 2017) uses a superconducting qubit with tunable coupling connected at multiple points to two resonators to cancel certain unwanted interaction terms while keeping desired interaction terms; it is shown that this would not have been possible with a small atom.

2.2   One Giant Atom with Time Delay

Consider a giant atom with two coupling points spaced such that it takes a time τ for light (or sound) to travel between them. In the previous section, it was assumed that τ was small compared to the relaxation time $1/\Gamma$. When this no longer is the case, the giant atom enters the non-Markovian regime, where the time evolution of the system can depend on what the system state was at an earlier time. In a giant atom, this non-Markovianity can manifest itself in revivals of the atomic population if energy is sent out from the atom at one coupling point and later is reabsorbed at another coupling point.

Four theoretical studies (Guo et al. 2017; Ask et al. 2019a; Guo et al. 2019, 2020) have explored this regime (the latter three considering more than two coupling points). In Ask et al. (2019a), it was shown that $\Gamma\tau = 1$ constitutes a sharp border for when time-delay effects become visible. When the system transitions from $\Gamma\tau < 1$ to $\Gamma\tau > 1$, the response of the giant atom to a weak coherent probe goes from showing one resonance to showing two. This is similar to the appearance of a vacuum Rabi splitting when an atom becomes strongly coupled to a cavity (the mathematical condition for the appearance of the splitting is actually exactly the same as for an atom in a multimode cavity Ask et al. 2019a; Krimer et al. 2014). In the case of the giant atom, the multiple coupling points act as a cavity when the coupling becomes strong enough or the travelling time becomes long enough.

In Guo et al. (2017), the cases $\tau > \Gamma$ and $\tau \gg \Gamma$ were studied in more detail. As τ increases, an initially excited giant atom exhibits more and more revivals of its population. In the limit of large τ, it turns out that the total energy stored in the giant atom and between its coupling points no longer decays exponentially with time t, as for a small atom, but instead decays polynomially ($\propto 1/\sqrt{t}$). Furthermore, the timescale for this decay is no longer set by the decay rate Γ, but by the travel time τ. These predictions for a giant atom with time delay were recently confirmed in an experiment (Andersson et al. 2019) (see Sect. 3.1 for more on the experimental platform used).

In Guo et al. (2019), it was shown that extending the setup from Guo et al. (2017) to more three or more coupling points enables qualitatively different phenomena: oscillating bound states. These oscillating bound states do not decay into the waveguide, but the energy oscillates persistently between the atom and the waveguide modes in-between the outermost coupling points of the atom. This result appears connected to that of Ask et al. (2019a) discussed above, and similar results have been obtained in Guo et al. (2020).

There are similarities between a giant atom with time delay and the previously studied (Dorner and Zoller 2002; Tufarelli et al. 2013; Pichler and Zoller 2016) setup with a small atom placed far from a mirror. However, in the giant-atom case scattering processes will involve both reflection and transmission, and the second-order correlation functions for these signals, calculated in Guo et al. (2017), exhibit oscillations between bunching and anti-bunching on a timescale set by τ.

2.3 Multiple Giant Atoms

When multiple small atoms are coupled to a waveguide, they can be spaced wave-length distances apart, which leads to interference effects influencing the collective behaviour of the atoms (Gu et al. 2017; Roy et al. 2017; Lehmberg 1970b, a; Lalumière et al. 2013; Zheng and Baranger 2013). Well-known examples include super- and sub-radiance (Dicke 1954; Lalumière et al. 2013), i.e., increased and decreased emission rates due to collective decay, and an effective coupling (sometimes called collective Lamb shift) between pairs of atoms, mediated by virtual photons in the transmission line (Friedberg et al. 1973; Scully and Svidzinsky 2010; Wen et al. 2019). Given this, one might wonder whether there is something left to set multiple giant atoms apart from multiple small atoms. After all, it was mainly the interference effects that separated a single giant atom from a single small atom.

In Kockum et al. (2018), the properties of multiple giant atoms were studied thoroughly and compared to those of multiple small atoms. The simplest cases considered are pictured in Fig. 4. For each of these setups, a master equation of the same form can be derived, assuming again that the travel time between coupling points is negligible:

$$
\dot{\rho} = -i \left[\omega_a' \frac{\sigma_z^a}{2} + \omega_b' \frac{\sigma_z^b}{2} + g \left(\sigma_-^a \sigma_+^b + \sigma_+^a \sigma_-^b \right), \rho \right]
$$
$$
+ \Gamma_a \mathcal{D} \left[\sigma_-^a \right] \rho + \Gamma_b \mathcal{D} \left[\sigma_-^b \right] \rho + \Gamma_{\text{coll}} \left[\left(\sigma_-^a \rho \sigma_+^b - \frac{1}{2} \left\{ \sigma_+^a \sigma_-^b, \rho \right\} \right) + \text{H.c.} \right],
$$
$$
\tag{11}
$$

where ω_j' is the transition frequency of atom j (we label the left atom a and the right atom b) including Lamb shifts, g is the strength of the exchange interaction mediated by the waveguide between the atoms, Γ_j is the individual relaxation rate of atom j, Γ_{coll} is the collective relaxation rate, and H.c. denotes Hermitian conjugate.

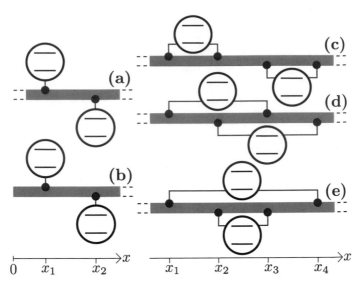

Fig. 4 Setups for two small and two giant atoms. **a** Two small atoms in an open waveguide. **b** Two small atoms in a waveguide terminated by a mirror on the left. **c** Two "separate" giant atoms, where the rightmost coupling point of the left atom is left of the leftmost coupling point of the right atom. **d** Two "braided" giant atoms, where each atom has a coupling point that lies in between the two coupling points of the other atom. **e** Two "nested" giant atoms, where the coupling points of one atom all lie in-between the coupling points of the other atom. Figure adapted from Kockum et al. (2018) with permission

Assuming that the atoms couple to the waveguide with equal strength at each coupling point, and that the distances between neigbouring coupling points are equal, corresponding to a phase shift φ, the coefficients g, Γ_j, and Γ_{coll} in Eq. (11) have simple expressions as functions of φ (Kockum et al. 2018). These functions are plotted in Fig. 5 for all the setups in Fig. 4. Looking at the individual relaxation rates (dashed curves), we see that they are always non-zero for small atoms in an open waveguide, but for setups with giant atoms there are points where $\Gamma_j = 0$, as we know from the discussion of single giant atoms in Sect. 2.1. Furthermore, at the points where $\Gamma_j = 0$, the collective relaxation rate Γ_{coll} also goes to zero. It is thus clear that setups with multiple giant atoms can be completely protected from relaxation into the waveguide.

The most remarkable feature in Fig. 5 is found when looking at the behaviour of the exchange interaction g at the points where the relaxation rates are zero. One might think that since interference effects at these points prevent the atoms from relaxing into the waveguide, it should not be possible for the waveguide to mediate interaction between the atoms. However, it turns out that g can be *non-zero* here for one of the three giant-atom setups: the braided giant atoms. This effect has recently been confirmed in experiment (Kannan et al. 2020) (see Sect. 3.2 for more on the experimental platform used).

One way to understand this protected interaction is to note that $\Gamma_j = 0$ when the phase between the coupling points of atom j is an odd integer multiple of π. The

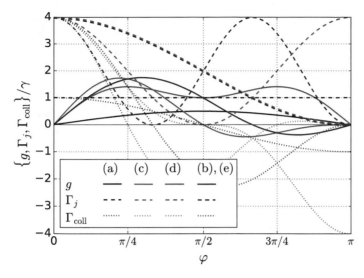

Fig. 5 Exchange interaction g (solid curves), individual relaxation rates Γ_j (dashed curves), and collective relaxation rates Γ_{coll} (dotted curves) as a function of φ for the setups in Fig. 4. The colours of the curves denote the ordering of coupling points: ab [small atoms, Fig. 4a, black], $aabb$ [separate giant atoms, Fig. 4c, blue], $abab$ [braided giant atoms, Fig. 4d, green], and $abba$ [nested giant atoms, Fig. 4e, red]. The last case is qualitatively equivalent to small atoms in front of a mirror [Fig. 4b]. For this case, there are two dashed curves (red), one for Γ_a and one for Γ_b. Figure adapted from Kockum et al. (2018) with permission

collective relaxation is due to interference between emission from coupling points of different atoms, but the sum total of these contributions is zero if the emissions from the two coupling points of one of the atoms interfere destructively. The exchange interaction arises due to emission from coupling points of one atom being absorbed at coupling points of the other atom. If the giant atoms are in the separate or nested configurations, the emissions from the two coupling points of atom b cancel if $\Gamma_b = 0$, but in the case of *braided* giant atoms, the two inner coupling points are placed in-between the coupling points of the other atom, so there is no condition forcing the contributions from the two coupling points of the other atom to interfere destructively.

We note that the protected interaction with braided giant atoms is reminiscent of the interaction between two small atoms in a waveguide with a bandgap (Kurizki 1990; Lambropoulos et al. 2000; Sundaresan et al. 2019). In that case, a bound state of photons forms around each atom that has a frequency in the bandgap, where propagation in the waveguide is impossible. The extension of these bound states decays exponentially with distance, but if two bound states overlap, the atoms can interact without decaying into the waveguide.

It is shown in Kockum et al. (2018) that the above conclusions about relations between relaxation rates and exchange interactions in giant atoms remain true even for the most general setups, with an arbitrary number of giant atoms, each having an arbitrary number of coupling points at arbitrary coordinates and with different coupling strength at each coupling point. This opens up interesting possibilities for constructing larger setups with protected exchange interaction between many giant atoms (Kockum et al. 2018).

It is also interesting that the case of two small atoms in front of a mirror, equivalent to nested giant atoms (red curves in Fig. 5), allows interaction even if one (but not both) of the atoms is prevented from relaxing into the waveguide. This has recently been confirmed in an experiment (Wen et al. 2019) with superconducting qubits in a transmission-line waveguide, and expanded upon in a connected theoretical study (Lin et al. 2019).

Finally, we note that a recent theoretical study (Karg et al. 2019) extended the treatment from giant atoms to arbitrary quantum systems, e.g., harmonic oscillators, interacting with a waveguide at multiple points. The study took into account losses in the waveguide and also considered the impact of time delays, and showed how these factors can affect the protected interaction that is possible with a nested setup.

3 Experiments with Giant Atoms

Waveguide QED can be implemented in several experimental systems (Gu et al. 2017; Roy et al. 2017), e.g., with quantum dots coupled to photonic crystal waveguides (Arcari et al. 2014), with quantum emitters coupled to plasmons in nanowires (Akimov et al. 2007; Huck and Andersen 2016), and with natural atoms coupled to optical fibres (Bajcsy et al. 2009), but the most versatile platform at the moment appears to be superconducting qubits coupled to transmission lines (Gu et al. 2017; Astafiev et al. 2010a, b; Hoi et al. 2011, 2012; van Loo et al. 2013; Hoi et al. 2013, 2015; Liu and Houck 2017; Forn-Díaz et al. 2017; Wen et al. 2018; Mirhosseini et al. 2018, 2019; Sundaresan et al. 2019; Wen et al. 2019). There are thus many systems where giant atoms could be implemented. So far, as reviewed in this section, experiments have been conducted exclusively with superconducting qubits, coupled to either surface acoustic waves (SAWs, Sect. 3.1) or transmission lines (Sect. 3.2). A theoretical proposal exists for an implementation with cold atoms in optical lattices (Sect. 3.3), and we expect that experiments will eventually be performed using more platforms.

3.1 Superconducting Qubits and Surface Acoustic Waves

Superconducting qubits (You and Nori 2011; Xiang et al. 2013; Gu et al. 2017; Kockum and Nori 2019) are electrical circuits with capacitances, inductances, and Josephson junctions (which function as non-linear inductances) that can emulate properties of natural atoms, e.g., energy-level structures and coupling to an electromagnetic field. These circuits usually have resonance frequencies ω on the order of

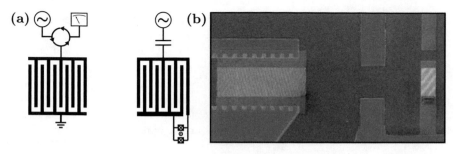

Fig. 6 Experimental implementation of a giant atom with a superconducting qubit coupled to SAWs. **a** Sketch of the experimental setup. The IDT on the left is used both to send out SAWs to the right towards the qubit and to convert reflected SAW signals from the qubit into a voltage signal that can be read out. The qubit on the right has its capacitance formed like an IDT to interact with the SAWs. The two islands of the capacitance are also connected through two Josephson junctions (boxes with crosses), which function as a non-linear inductance, making the qubit essentially an anharmonic LC oscillator. The qubit can also be driven electrically through a gate on the top. **b** False-colour image of the experimental sample. The blue parts are the IDT to the right and the qubit to the left. The yellow parts are ground planes and the electrodes connecting to the IDT. The aspect ratio of the IDT, with fingers being much longer than they are wide, collimates the SAW beam such that it travels straight towards the qubit (and also in the opposite direction). Figure from Aref et al. (2016) with permission

GHz and are cooled to low temperatures $T \ll \hbar\omega/k_B$ to prevent the thermal fluctuations interfering with quantum properties.

In 2014, an experiment (Gustafsson et al. 2014) managed to couple a superconducting qubit of the transmon type (Koch et al. 2007) to SAWs, which are vibrations that propagate on the surface of a substrate (Datta 1986; Morgan 2007). The experimental setup is shown in Fig. 6. The substrate on which the SAWs propagate is piezoelectric, which means that the vibrations acquire an electromagnetic component. Vibrations can be induced by applying an oscillating voltage across two electrodes, in the form of an interdigitated transducer (IDT), placed on the surface. If the spacing between fingers in the electrode matches the wavelength of SAWs at the frequency of the applied signal, the induced SAWs add up coherently. Conversely, propagating SAWs that arrive at the transducer induce charge on the fingers such that the vibrations are converted into a voltage signal. The crucial invention in Gustafsson et al. (2014) was to let the capacitance in the transmon qubit double as an IDT to mediate a direct coupling between qubit and SAWs. Because of the slow propagation speed of SAWs, $v \approx 3000\,\text{m/s}$, the IDT finger spacing was on the order of $d \approx 1\,\mu\text{m}$ to match the resonance frequency around $\omega \approx 5\,\text{GHz}$. As can be seen in the figure, many fingers were used in the qubit IDT, which corresponded to tens of wavelengths, making this a truly giant atom.

This first experiment with a giant atom could only probe the atom around a single frequency, since the IDT used to convert signals had a narrow bandwidth. The frequency-dependence of the qubit coupling (see Sect. 2.1.2) could therefore not be tested. However, the experimental platform with SAWs and qubits, called circuit quantum acoustodynamics (QAD) (Gustafsson et al. 2014; Aref et al. 2016; Manenti et al. 2017), has been adopted in several research groups. In their experiments (Manenti et al. 2017; Noguchi et al. 2017; Moores et al. 2018; Satzinger et al. 2018; Bolgar et al. 2018; Sletten et al. 2019; Bienfait et al. 2019), the qubit is coupled to a resonator for the SAW modes. Since the resonator is long, it has a narrow free spectral range, and the frequency-dependent coupling of the qubit is evident from how it couples with different strength to different modes. This selective coupling to modes has been used in a clever way to read out the number of phonons in a mode via the qubit (Sletten et al. 2019).

A particular advantage of the SAWs is that their slow propagation speed makes it possible to engineer a giant atom with a very long distance between coupling points. In the experiment of Andersson et al. (2019), distances exceeding 400 wavelengths were realized, corresponding to $\Gamma\tau \approx 14$, i.e., well in the non-Markovian regime discussed in Sect. 2.2.

Another recent experiment (Andersson et al. 2020) with a superconducting transmon qubit and SAWs used the possibility to engineer the relaxation rates of the first two transitions of the transmon (see Sec. 2.1.2) to enable EIT. This appears to be the first time that EIT of a propagating mechanical mode has been demonstrated.

3.2 Superconducting Qubits and Microwave Transmission Lines

Superconducting qubits are usually coupled to microwave transmission lines, or LC resonators, instead of SAWs. Also the setup with a transmission line can be used to implement giant atoms, as proposed in Kockum et al. (2014). One simply couples the transmission line to the qubit at one point, meanders the transmission line back and forth on the chip until a wavelength distance has been reached, and then connects the transmission line to the qubit once more. Due to size limitations, this approach will not allow for distances between coupling points on the order of hundreds of wavelengths or more, as is possible with SAWs. However, with the transmission line it is possible to engineer the coupling at each point and the distance between coupling points with great precision, which can be crucial for demonstrating the interference effects that lie at the heart of giant atoms.

Two recent experiments have followed this approach to implement one (Vadiraj et al. 2020) and two (Kannan et al. 2020) giant atoms. In the experiment with one giant atom, the frequency-dependent coupling shown in Fig. 2 was measured and the ability to manipulate the relaxation rates in a multilevel atom as in Fig. 3 was shown. In the experiment with two giant atoms, the decoherence-free interaction discussed in Sect. 2.3 was demonstrated.

This opens up interesting possibilities for preparing entangled many-body states in waveguide QED with many atoms, which otherwise is difficult due to the dissipation into the waveguide which always is present for small atoms (Kannan et al. 2020).

3.3 Cold Atoms in Optical Lattices

All experiments with giant atoms so far have taken place in 1D geometries at microwave frequencies and used superconducting qubits. A recent theory proposal (González-Tudela et al. 2019) shows how giant atoms instead could be implemented in higher dimensions on another platform for quantum-optics simulation: cold atoms in optical lattices. Here, one would use atoms with two internal states, each of which couples to a different optical lattice, realized by counter-propagating lasers. In one state, the atom mimics a photon moving in a lattice; in the other state, the atom mimics an atom trapped in a specific site. By rapidly modulating the relative positions of the two lattices, it is possible to engineer an effective interaction where the atomic state couples to the photonic state at multiple points (González-Tudela et al. 2019). It may be possible to achieve a similar effect with superconducting qubits coupled to several sites in a 2D lattice of superconducting resonators. While such lattices have been analysed and realized previously (Koch et al. 2010; Houck et al. 2012; Underwood et al. 2016), to the best of our knowledge it has not been suggested previously to couple one qubit to several lattice sites in such a setup.

The proposed setup with cold atoms displays rich physics with the giant atoms coupled to 2D photonic environments that have a band structure. It is possible to construct interference such that a single giant atom relaxes by only emitting its energy in certain directions. It is also possible to decouple giant atoms completely from the environment, but still have them interact by exchange interactions, like in Sect. 2.3. While this interference was possible with just two coupling points per atom in 1D, the 2D case requires at least four coupling points.

4 Conclusion and Outlook

Giant atoms are emerging as a new, interesting field of quantum optics. Following the first experimental realization and theoretical study in 2014, the field has grown quickly in the past 5 years. Theoretical investigations have been extended from one to multiple giant atoms, from 1D to higher-dimensional environments coupling to the atoms, and from the Markovian to the non-Markovian regime, where time delays between coupling points matter. These investigations have revealed remarkable properties of giant atoms, including frequency-dependent couplings and decoherence-free interactions, which are hard or impossible to realize with small atoms.

In parallel, the experimental platform for giant atoms, with SAWs coupled to superconducting qubits, has been further developed. There are now also experiments with superconducting qubits coupled to microwave transmission lines, and an experimental platform with cold atoms in optical lattices has been proposed. The experiments have confirmed many of the theoretical predictions, and also contributed with new ideas for applications of giant atoms.

Looking towards the future, we can formulate a long research agenda for giant atoms. At the heart of this agenda is the fact that giant atoms mainly differ from small atoms by the interference effects introduced by the multiple coupling points, which already has been shown to lead to new effects. It therefore seems prudent to revisit many well-known quantum-optics phenomena to see if giant atoms can enhance them or enable new physics. Below, we give a list of such projects:

- **Superradiance:** For multiple small atoms coupled to light, it is well known that quantum interference effects can give rise to enhanced light emission, superradiance, where N atoms emit light at an increased rate, proportional to N^2 (Dicke 1954; Shammah et al. 2018). The reverse process, "superabsorption", is also possible (Higgins et al. 2014; Yang et al. 2019), and may be of importance in photosynthesis and future solar cells. It is thus highly relevant to see if giant atoms can enhance superradiance and superabsorption.
- **Ultrastrong coupling:** When the strength of the coupling between light and matter starts to approach the bare resonance frequencies in the system, it is called ultrastrong (Kockum et al. 2019; Forn-Díaz et al. 2019). In this regime, the rotating-wave approximation breaks down and the number of excitations in the system is no longer conserved in the absence of drives. For a giant atom ultrastrongly coupled to an open waveguide, it would be interesting to map out the ground state of the system, since results for a small atom indicates that it should contain virtual photons clustered around each connection point (Sanchez-Burillo et al. 2014). However, ultrastrong coupling with giant atoms comes with several theoretical challenges, which make analytical results hard to achieve. For example, a giant atom with ultrastrong coupling will inevitably be in a regime where the travel time between coupling points is non-negligible (Ask et al. 2019a).
- **Generating non-classical light:** It has recently been shown that coherently driving a small atom in front of a mirror can lead to the generation of non-classical states of light with a negative Wigner function (Quijandría et al. 2018). Could a giant atom do the same?
- **Matryoshka atoms:** The topology in Fig. 4e, nested atoms, is reminiscent of a Russian matryoshka doll. Although it does not enable decoherence-free interaction like braided atoms do, it seems to have other interesting properties. If the distance between the coupling points of the outer atom is large, the outer atom could effectively act as a cavity (Ask et al. 2019b), similar to what two small atoms placed far away on either side of a central atom can do (Guimond et al. 2016). Also, preliminary results indicate that two nested giant atoms can emulate electromagnetically induced transparency in a Λ system without any external drive (Ask et al. 2019c). With many nested giant atoms, the situation is similar to having many atoms in front of a mirror. Thus, for certain inter-coupling-point distances, these giant atoms should be able to combine into fewer effective larger atoms, as can happen in the mirror case (Lin et al. 2019).
- **Chiral quantum optics:** In some waveguide-QED setups with small atoms, it is possible to realize chiral couplings, i.e., that the atoms only couple to one propagation direction in the waveguide (Lodahl et al. 2017). Although it is not

yet clear if this can be implemented in experiments with giant atoms, it seems interesting to study chiral quantum optics with giant atoms theoretically. A related question is whether interference between light propagating in a waveguide, and light taking the "shortcut" between two coupling points through a giant atom, can be used to realize an effective chiral coupling.

This was recently answered affirmatively for a setup with two atoms that are both directly coupled to each other and each coupled at its own single point to a waveguide ($\sim \lambda/4$ apart) (Guimond et al 2020).

Acknowledgements AFK acknowledges support from the Swedish Research Council (grant number 2019-03696), and from the Knut and Alice Wallenberg Foundation through the Wallenberg Centre for Quantum Technology (WACQT).

References

A.V. Akimov, A. Mukherjee, C.L. Yu, D.E. Chang, A.S. Zibrov, P.R. Hemmer, H. Park, M.D. Lukin, Generation of single optical plasmons in metallic nanowires coupled to quantum dots. Nature **450**, 402 (2007). http://www.nature.com/doifinder/10.1038/nature06230

G. Andersson, M. K. Ekström, P. Delsing, Electromagnetically induced transparency in a propagating mechanical mode. Physical Review Letters **124**, 240402 (2020)

G. Andersson, B. Suri, L. Guo, T. Aref, P. Delsing, Nonexponential decay of a giant artificial atom. Nat. Phys. **15**, 1123 (2019) http://arxiv.org/abs/1812.01302

M. Arcari, I. Söllner, A. Javadi, S. Lindskov Hansen, S. Mahmoodian, J. Liu, H. Thyrrestrup, E.H. Lee, J.D. Song, S. Stobbe, P. Lodahl, Near-unity coupling efficiency of a quantum emitter to a photonic crystal waveguide. Phys. Rev. Lett. **113**, 093603 (2014). http://link.aps.org/doi/10.1103/PhysRevLett.113.093603

T. Aref, P. Delsing, M.K. Ekström, A.F. Kockum, M.V. Gustafsson, G. Johansson, P.J. Leek, E. Magnusson, R. Manenti, Quantum acoustics with surface acoustic waves, in *Superconducting Devices in Quantum Optics*, eds. by R.H. Hadfield, G. Johansson (Springer, Berlin, 2016). http://dx.doi.org/10.1007/978-3-319-24091-6_9

A. Ask, et al., In preparation (2019b)

A. Ask, et al., In preparation (2019c)

A. Ask, M. Ekström, P. Delsing, G. Johansson, Cavity-free vacuum-Rabi splitting in circuit quantum acoustodynamics. Phys. Rev. A **99**, 013840 (2019a). http://dx.doi.org/10.1103/PhysRevA.99.013840

O.V. Astafiev, A.A. Abdumalikov, A.M. Zagoskin, Y.A. Pashkin, Y. Nakamura, J.S. Tsai, Ultimate on-chip quantum amplifier. Phys. Rev. Lett. **104**, 183603 (2010b). http://link.aps.org/doi/10.1103/PhysRevLett.104.183603

O. Astafiev, A.M. Zagoskin, A.A. Abdumalikov, Y.A. Pashkin, T. Yamamoto, K. Inomata, Y. Nakamura, J.S. Tsai, Resonance fluorescence of a single artificial atom. Science **327**, 840 (2010a). http://www.sciencemag.org/cgi/doi/10.1126/science.1181918

M. Bajcsy, S. Hofferberth, V. Balic, T. Peyronel, M. Hafezi, A.S. Zibrov, V. Vuletic, M.D. Lukin, Efficient all-optical switching using slow light within a hollow fiber. Phys. Rev. Lett. **102**, 203902 (2009). http://link.aps.org/doi/10.1103/PhysRevLett.102.203902

A. Beige, J. Pachos, H. Walther, Spontaneous emission of an atom in front of a mirror. Phys. Rev. A **66**, 063,801 (2002). http://link.aps.org/doi/10.1103/PhysRevA.66.063801

H.A. Bethe, The electromagnetic shift of energy levels. Phys. Rev. **72**, 339 (1947). http://link.aps.org/doi/10.1103/PhysRev.72.339

A. Bienfait, K.J. Satzinger, Y.P. Zhong, H.S. Chang, M.H. Chou, C.R. Conner, É. Dumur, J. Grebel, G.A. Peairs, R.G. Povey, A.N. Cleland, Phonon-mediated quantum state transfer and remote qubit entanglement. Science **364**, 368 (2019). http://www.sciencemag.org/lookup/doi/10.1126/science.aaw8415

A.N. Bolgar, J.I. Zotova, D.D. Kirichenko, I.S. Besedin, A.V. Semenov, R.S. Shaikhaidarov, O.V. Astafiev, Quantum regime of a two-dimensional phonon cavity. Phys. Rev. Lett. **120**, 223603 (2018). https://link.aps.org/doi/10.1103/PhysRevLett.120.223603

H.J. Carmichael, *Statistical Methods in Quantum Optics 1* (Springer, Berlin, 1999). http://link.springer.com/10.1007/978-3-662-03875-8

A. Ciani, D.P. DiVincenzo, Three-qubit direct dispersive parity measurement with tunable coupling qubits. Phys. Rev. B **96**, 214511 (2017). http://dx.doi.org/10.1103/PhysRevB.96.214511

C. Cohen-Tannoudji, J. Dupont-Roc, G. Grynberg, *Atom-Photon Interactions* (Wiley, Hoboken, 1998)

S. Datta, *Surface Acoustic Wave Devices* (Prentice Hall, Upper Saddle River, 1986)

R.H. Dicke, Coherence in spontaneous radiation processes. Phys. Rev. **93**, 99 (1954). http://link.aps.org/doi/10.1103/PhysRev.93.99

H. Dong, Z.R. Gong, H. Ian, L. Zhou, C.P. Sun, Intrinsic cavity QED and emergent quasinormal modes for a single photon. Phys. Rev. A **79**, 063847 (2009). http://link.aps.org/doi/10.1103/PhysRevA.79.063847

U. Dorner, P. Zoller, Laser-driven atoms in half-cavities. Phys. Rev. A **66**, 023816 (2002). http://link.aps.org/doi/10.1103/PhysRevA.66.023816

F. Dubin, D. Rotter, M. Mukherjee, C. Russo, J. Eschner, R. Blatt, Photon correlation versus interference of single-atom fluorescence in a half-cavity. Phys. Rev. Lett. **98**, 183003 (2007). http://link.aps.org/doi/10.1103/PhysRevLett.98.183003

J. Eschner, C. Raab, F. Schmidt-Kaler, R. Blatt, Light interference from single atoms and their mirror images. Nature **413**, 495 (2001). http://www.nature.com/doifinder/10.1038/35097017

Y.L.L. Fang, H.U. Baranger, Waveguide QED: Power spectra and correlations of two photons scattered off multiple distant qubits and a mirror. Phys. Rev. A **91**, 053845 (2015). http://link.aps.org/doi/10.1103/PhysRevA.91.053845

P. Forn-Díaz, J.J. García-Ripoll, B. Peropadre, J.L. Orgiazzi, M.A. Yurtalan, R. Belyansky, C.M. Wilson, A. Lupascu, Ultrastrong coupling of a single artificial atom to an electromagnetic continuum in the nonperturbative regime. Nat. Phys. **13**, 39 (2017). http://www.nature.com/doifinder/10.1038/nphys3905

P. Forn-Díaz, L. Lamata, E. Rico, J. Kono, E. Solano, Ultrastrong coupling regimes of light-matter interaction. Rev. Mod. Phys. **91**, 025005 (2019). http://dx.doi.org/10.1103/RevModPhys.91.025005

R. Friedberg, S.R. Hartmann, J.T. Manassah, Frequency shifts in emission and absorption by resonant systems of two-level atoms. Phys. Rep. **7**, 101 (1973). http://linkinghub.elsevier.com/retrieve/pii/037015737390001X

C.W. Gardiner, P. Zoller, *Quantum Noise*, 3rd edn. (Springer, Berlin, 2004)

A. González-Tudela, C. Sánchez Muñoz, J.I. Cirac, Engineering and harnessing giant atoms in high-dimensional baths: a proposal for implementation with cold atoms. Phys. Rev. Lett. **122**, 203603 (2019). https://link.aps.org/doi/10.1103/PhysRevLett.122.203603

X. Gu, A.F. Kockum, A. Miranowicz, Y.X. Liu, F. Nori, Microwave photonics with superconducting quantum circuits. Phys. Rep. **718–719**, 1–102 (2017). https://doi.org/10.1016/j.physrep.2017.10.002

P.O. Guimond, A. Roulet, H.N. Le, Scarani, V, Rabi oscillation in a quantum cavity: Markovian and non-Markovian dynamics. Phys. Rev. A **93**, 023808 (2016). http://dx.doi.org/10.1103/PhysRevA.93.023808

P. O. Guimond, B. Vermersch, M. L. Juan, A. Sharafiev, G. Kirchmair, P. Zoller, A unidirectional on-chip photonic interface for superconducting circuits. npj Quantum Inf. **6**, 32 (2020)

L. Guo, A.L. Grimsmo, A.F. Kockum, M. Pletyukhov, G. Johansson, Giant acoustic atom: a single quantum system with a deterministic time delay. Phys. Rev. A **95**, 053821 (2017). http://link.aps. org/doi/10.1103/PhysRevA.95.053821

L. Guo, A. F. Kockum, F. Marquardt, G. Johansson, Oscillating bound states for a giant atom. Phys. Rev. Res. arXiv:1911.13028 (2019)

S. Guo, Y. Wang, T. Purdy, J. Taylor, Beyond Spontaneous Emission: Giant Atom Bounded in Continuum. Phys. Rev. A **102**, 033706 (2020). https://doi.org/10.1103/PhysRevA.102.033706

M.V. Gustafsson, T. Aref, A.F. Kockum, M.K. Ekström, G. Johansson, P. Delsing, Propagating phonons coupled to an artificial atom. Science **346**, 207 (2014). http://www.sciencemag.org/cgi/ doi/10.1126/science.1257219

R. Hanson, L.P. Kouwenhoven, J.R. Petta, S. Tarucha, L.M.K. Vandersypen, Spins in few-electron quantum dots. Rev. Mod. Phys. **79**, 1217 (2007). http://link.aps.org/doi/10.1103/RevModPhys. 79.1217

S. Haroche, Nobel Lecture: Controlling photons in a box and exploring the quantum to classical boundary. Rev. Mod. Phys. **85**, 1083 (2013). https://doi.org/10.1103/RevModPhys.85.1083

K.D.B. Higgins, S.C. Benjamin, T.M. Stace, G.J. Milburn, B.W. Lovett, E.M. Gauger, Superabsorption of light via quantum engineering. Nat. Commun. **5**, 4705 (2014). http://www.nature.com/ doifinder/10.1038/ncomms5705

I.C. Hoi, A.F. Kockum, T. Palomaki, T.M. Stace, B. Fan, L. Tornberg, S.R. Sathyamoorthy, G. Johansson, P. Delsing, C.M. Wilson, Giant cross-Kerr effect for propagating microwaves induced by an artificial atom. Phys. Rev. Lett. **111**, 053601 (2013). http://link.aps.org/doi/10. 1103/PhysRevLett.111.053601

I.C. Hoi, A.F. Kockum, L. Tornberg, A. Pourkabirian, G. Johansson, P. Delsing, C.M. Wilson, Probing the quantum vacuum with an artificial atom in front of a mirror. Nat. Phys. **11**, 1045 (2015). http://www.nature.com/doifinder/10.1038/nphys3484

I.C. Hoi, T. Palomaki, J. Lindkvist, G. Johansson, P. Delsing, C.M. Wilson, Generation of nonclassical microwave states using an artificial atom in 1D open space. Phys. Rev. Lett. **108**, 263601 (2012). http://link.aps.org/doi/10.1103/PhysRevLett.108.263601

I.C. Hoi, C.M. Wilson, G. Johansson, T. Palomaki, B. Peropadre, P. Delsing, Demonstration of a single-photon router in the microwave regime. Phys. Rev. Lett. **107**, 073601 (2011). http://link. aps.org/doi/10.1103/PhysRevLett.107.073601

A.A. Houck, H.E. Türeci, J. Koch, On-chip quantum simulation with superconducting circuits. Nat. Phys. **8**, 292 (2012). http://www.nature.com/doifinder/10.1038/nphys2251

A. Huck, U.L. Andersen, Coupling single emitters to quantum plasmonic circuits. Nanophotonics **5**, 1 (2016). http://www.degruyter.com/view/j/nanoph.2016.5.issue-3/nanoph-2015-0153/nanoph-2015-0153.xml

B. Kannan, M. Ruckriegel, D. Campbell, A. F. Kockum, J. Braumüller, D. Kim, M. Kjaergaard, P. Krantz, A. Melville, B. M. Niedzielski, A. Vepsäläinen, R. Winik, J. Yoder, F. Nori, T. P. Orlando, S. Gustavsson, W. D. Oliver, Waveguide quantum electrodynamics with giant superconducting artificial atoms. Nat. **583**, 775 (2020). http://www.nature.com/articles/s41586-020-2529-9

T.M. Karg, B. Gouraud, P. Treutlein, K. Hammerer, Remote Hamiltonian interactions mediated by light. Phys. Rev. A **99**, 063829 (2019). https://link.aps.org/doi/10.1103/PhysRevA.99.063829

J. Koch, A.A. Houck, K. Le Hur, S.M. Girvin, Time-reversal-symmetry breaking in circuit-QED-based photon lattices. Phys. Rev. A **82**, 043811 (2010). http://link.aps.org/doi/10.1103/ PhysRevA.82.043811

J. Koch, T.M. Yu, J. Gambetta, A.A. Houck, D.I. Schuster, J. Majer, A. Blais, M.H. Devoret, S.M. Girvin, R.J. Schoelkopf, Charge-insensitive qubit design derived from the Cooper pair box. Phys. Rev. A **76**, 042319 (2007). http://link.aps.org/doi/10.1103/PhysRevA.76.042319

A.F. Kockum, P. Delsing, G. Johansson, Designing frequency-dependent relaxation rates and Lamb shifts for a giant artificial atom. Phys. Rev. A **90**, 013837 (2014). http://link.aps.org/doi/10.1103/ PhysRevA.90.013837

A.F. Kockum, G. Johansson, F. Nori, Decoherence-Free Interaction between Giant Atoms in Waveguide Quantum Electrodynamics. Phys. Rev. Lett. **120**, 140404 (2018). https://link.aps.org/doi/10.1103/PhysRevLett.120.140404

A.F. Kockum, A. Miranowicz, S. De Liberato, S. Savasta, F. Nori, Ultrastrong coupling between light and matter. Nat. Rev. Phys. **1**, 19 (2019). http://dx.doi.org/10.1038/s42254-018-0006-2

A.F. Kockum, F. Nori, Quantum bits with Josephson junctions, *Fundamentals and Frontiers of the Josephson Effect*, ed. by F. Tafuri (Springer, Berlin, 2019). https://doi.org/10.1007/978-3-030-20726-7

K. Koshino, Y. Nakamura, Control of the radiative level shift and linewidth of a superconducting artificial atom through a variable boundary condition. New J. Phys. **14**, 043005 (2012). http://stacks.iop.org/1367-2630/14/i=4/a=043005?key=crossref.8e8ee9c5cb468c374b4c2a56c13ab881

D.O. Krimer, M. Liertzer, S. Rotter, H.E. Türeci, Route from spontaneous decay to complex multimode dynamics in cavity QED. Phys. Rev. A **89**, 033820 (2014). https://link.aps.org/doi/10.1103/PhysRevA.89.033820

G. Kurizki, Two-atom resonant radiative coupling in photonic band structures. Phys. Rev. A **42**, 2915 (1990). https://link.aps.org/doi/10.1103/PhysRevA.42.2915

K. Lalumière, B.C. Sanders, A.F. van Loo, A. Fedorov, A. Wallraff, A. Blais, Input-output theory for waveguide QED with an ensemble of inhomogeneous atoms. Phys. Rev. A **88**, 043806 (2013). http://link.aps.org/doi/10.1103/PhysRevA.88.043806

W.E. Lamb, R.C. Retherford, Fine structure of the hydrogen atom by a microwave method. Phys. Rev. **72**, 241 (1947). http://link.aps.org/doi/10.1103/PhysRev.72.241

P. Lambropoulos, G.M. Nikolopoulos, T.R. Nielsen, S. Bay, Fundamental quantum optics in structured reservoirs. Rep. Prog. Phys. **63**, 455 (2000). http://stacks.iop.org/0034-4885/63/i=4/a=201?key=crossref.4e89934b6d1170192620f7f10999fb37

R.H. Lehmberg, Radiation from an N-atom system. II. Spontaneous emission from a pair of atoms. Phys. Rev. A **2**, 889 (1970b). http://link.aps.org/doi/10.1103/PhysRevA.2.889

R.H. Lehmberg, Radiation from an N-atom system. I. General formalism. Phys. Rev. A **2**, 883 (1970a). http://link.aps.org/doi/10.1103/PhysRevA.2.883

D. Leibfried, R. Blatt, C. Monroe, D. Wineland, Quantum dynamics of single trapped ions. Rev. Mod. Phys. **75**, 281 (2003). http://link.aps.org/doi/10.1103/RevModPhys.75.281

K.T. Lin, T. Hsu, C.Y. Lee, I.C. Hoi, G.D. Lin, Scalable collective Lamb shift of a 1D superconducting qubit array in front of a mirror. Sci. Rep. **9**, 19175 (2019). https://doi.org/10.1038/s41598-019-55545-5

G. Lindblad, On the generators of quantum dynamical semigroups. Commun. Math. Phys. **48**, 119 (1976). https://doi.org/10.1007/BF01608499

Y. Liu, A.A. Houck, Quantum electrodynamics near a photonic bandgap. Nat. Phys. **13**, 48 (2017). http://www.nature.com/doifinder/10.1038/nphys3834

P. Lodahl, S. Mahmoodian, S. Stobbe, A. Rauschenbeutel, P. Schneeweiss, J. Volz, H. Pichler, P. Zoller, Chiral quantum optics. Nature **541**, 473 (2017). http://www.nature.com/doifinder/10.1038/nature21037

R. Manenti, A.F. Kockum, A. Patterson, T. Behrle, J. Rahamim, G. Tancredi, F. Nori, P.J. Leek, Circuit quantum acoustodynamics with surface acoustic waves. Nat. Commun. **8**, 975 (2017). http://www.nature.com/articles/s41467-017-01063-9

D. Meschede, W. Jhe, E.A. Hinds, Radiative properties of atoms near a conducting plane: An old problem in a new light. Phys. Rev. A **41**, 1587 (1990). http://link.aps.org/doi/10.1103/PhysRevA.41.1587

M. Mirhosseini, E. Kim, V.S. Ferreira, M. Kalaee, A. Sipahigil, A.J. Keller, O. Painter, Superconducting metamaterials for waveguide quantum electrodynamics. Nat. Commun. **9**, 3706 (2018). http://dx.doi.org/10.1038/s41467-018-06142-z

M. Mirhosseini, E. Kim, X. Zhang, A. Sipahigil, P.B. Dieterle, A.J. Keller, A. Asenjo-Garcia, D.E. Chang, O. Painter, Cavity quantum electrodynamics with atom-like mirrors. Nature **569**, 692 (2019). http://www.nature.com/articles/s41586-019-1196-1

B.A. Moores, L.R. Sletten, J.J. Viennot, K.W. Lehnert, Cavity quantum acoustic device in the multimode strong coupling regime. Phys. Rev. Lett. **120**, 227701 (2018). https://link.aps.org/doi/10.1103/PhysRevLett.120.227701

D. Morgan, *Surface Acoustic Wave Filters*, 2nd edn. (Academic Press, Cambridge, 2007)

A. Noguchi, R. Yamazaki, Y. Tabuchi, Y. Nakamura, Qubit-assisted transduction for a detection of surface acoustic waves near the quantum limit. Phys. Rev. Lett. **119**, 180505 (2017). https://link.aps.org/doi/10.1103/PhysRevLett.119.180505

H. Pichler, P. Zoller, Photonic circuits with time delays and quantum feedback. Phys. Rev. Lett. **116**, 093601 (2016). http://link.aps.org/doi/10.1103/PhysRevLett.116.093601

F. Quijandría, I. Strandberg, G. Johansson, Steady-state generation of wigner-negative states in one-dimensional resonance fluorescence. Phys. Rev. Lett. **121**, 263603 (2018). https://link.aps.org/doi/10.1103/PhysRevLett.121.263603

D. Roy, C.M. Wilson, O. Firstenberg, Colloquium: strongly interacting photons in one-dimensional continuum. Rev. Mod. Phys. **89**, 021001 (2017). http://link.aps.org/doi/10.1103/RevModPhys.89.021001

E. Sanchez-Burillo, D. Zueco, J.J. Garcia-Ripoll, L. Martin-Moreno, Scattering in the ultrastrong regime: nonlinear optics with one photon. Phys. Rev. Lett. **113**, 263604 (2014). http://link.aps.org/doi/10.1103/PhysRevLett.113.263604

K.J. Satzinger, Y.P. Zhong, H.S. Chang, G.A. Peairs, A. Bienfait, M.H. Chou, A.Y. Cleland, C.R. Conner, É. Dumur, J. Grebel, I. Gutierrez, B.H. November, R.G. Povey, S.J. Whiteley, D.D. Awschalom, D.I. Schuster, A.N. Cleland, Quantum control of surface acoustic-wave phonons. Nature **563**, 661 (2018). http://www.nature.com/articles/s41586-018-0719-5

M.O. Scully, A.A. Svidzinsky, The Lamb shift - yesterday, today, and tomorrow. Science **328**, 1239 (2010). http://www.sciencemag.org/cgi/doi/10.1126/science.1190737

N. Shammah, S. Ahmed, N. Lambert, S. De Liberato, F. Nori, Open quantum systems with local and collective incoherent processes: efficient numerical simulations using permutational invariance. Phys. Rev. A **98**, 063815 (2018). http://arxiv.org/abs/1805.05129https://link.aps.org/doi/10.1103/PhysRevA.98.063815

T. Shi, D.E. Chang, J.I. Cirac, Multiphoton-scattering theory and generalized master equations. Phys. Rev. A **92**, 053834 (2015). http://link.aps.org/doi/10.1103/PhysRevA.92.053834

L.R. Sletten, B.A. Moores, J.J. Viennot, K.W. Lehnert, Resolving phonon Fock states in a multimode cavity with a double-slit qubit. Phys. Rev. X **9**, 021056 (2019). http://dx.doi.org/10.1103/PhysRevX.9.021056

N.M. Sundaresan, R. Lundgren, G. Zhu, A.V. Gorshkov, A.A. Houck, Interacting qubit-photon bound states with superconducting circuits. Phys. Rev. X **9**, 011021 (2019). http://dx.doi.org/10.1103/PhysRevX.9.011021

T. Tufarelli, F. Ciccarello, M.S. Kim, Dynamics of spontaneous emission in a single-end photonic waveguide. Phys. Rev. A **87**, 013820 (2013). http://link.aps.org/doi/10.1103/PhysRevA.87.013820

D.L. Underwood, W.E. Shanks, A.C.Y. Li, L. Ateshian, J. Koch, A.A. Houck, Imaging photon lattice states by scanning defect microscopy. Phys. Rev. X **6**, 021044 (2016). https://link.aps.org/doi/10.1103/PhysRevX.6.021044

A. M. Vadiraj, A. Ask, T. G. McConkey, I. Nsanzineza, C. W. Sandbo Chang, A. F. Kockum, C. M. Wilson, Engineering the level structure of a giant artificial atom in waveguide quantum electrodynamics. arXiv:2003.14167 (2020)

A.F. van Loo, A. Fedorov, K. Lalumiere, B.C. Sanders, A. Blais, A. Wallraff, Photon-mediated interactions between distant artificial atoms. Science **342**, 1494 (2013). http://www.sciencemag.org/cgi/doi/10.1126/science.1244324

D.F. Walls, G.J. Milburn, *Quantum Optics*, 2nd edn. (Springer, Berlin, 2008). http://link.springer.com/10.1007/978-3-540-28574-8

Y. Wang, J. Minár, G. Hétet, V. Scarani, Quantum memory with a single two-level atom in a half cavity. Phys. Rev. A **85**, 013823 (2012). http://link.aps.org/doi/10.1103/PhysRevA.85.013823

P.Y. Wen, A.F. Kockum, H. Ian, J.C. Chen, F. Nori, I.C. Hoi, Reflective amplification without population inversion from a strongly driven superconducting qubit. Phys. Rev. Lett. **120**, 063603 (2018). https://link.aps.org/doi/10.1103/PhysRevLett.120.063603

P.Y. Wen, K.T. Lin, A.F. Kockum, B. Suri, H. Ian, J.C. Chen, S.Y. Mao, C.C. Chiu, P. Delsing, F. Nori, G.D. Lin, I.C. Hoi, Large collective Lamb shift of two distant superconducting artificial atoms. Phys. Rev. Lett. **123**, 233602 (2019). https://doi.org/10.1103/PhysRevLett.123.233602

M.A. Wilson, P. Bushev, J. Eschner, F. Schmidt-Kaler, C. Becher, R. Blatt, U. Dorner, Vacuum-field level shifts in a single trapped ion mediated by a single distant mirror. Phys. Rev. Lett. **91**, 213602 (2003). http://link.aps.org/doi/10.1103/PhysRevLett.91.213602

Z.L. Xiang, S. Ashhab, J.Q. You, F. Nori, Hybrid quantum circuits: superconducting circuits interacting with other quantum systems. Rev. Mod. Phys. **85**, 623 (2013). http://link.aps.org/doi/10.1103/RevModPhys.85.623

D. Yang, S.H. Oh, J. Han, G. Son, J. Kim, J. Kim, K. An, Observation of superabsorption by correlated atoms (2019). http://arxiv.org/abs/1906.06477

J.Q. You, F. Nori, Atomic physics and quantum optics using superconducting circuits. Nature **474**, 589 (2011). http://dx.doi.org/10.1038/nature10122

H. Zheng, H.U. Baranger, Persistent quantum beats and long-distance entanglement from waveguide-mediated interactions. Phys. Rev. Lett. **110**, 113601 (2013). http://link.aps.org/doi/10.1103/PhysRevLett.110.113601

Topics in Mathematics

Extended Divisibility Relations for Constraint Polynomials of the Asymmetric Quantum Rabi Model

Cid Reyes-Bustos

Abstract The quantum Rabi model (QRM) is widely regarded as one of the fundamental models of quantum optics. One of its generalizations is the asymmetric quantum Rabi model (AQRM), obtained by introducing a symmetry-breaking term depending on a parameter $\varepsilon \in \mathbb{R}$ to the Hamiltonian of the QRM. The AQRM was shown to possess degeneracies in the spectrum for values $\epsilon \in 1/2\mathbb{Z}$ via the study of the divisibility of the so-called constraint polynomials. In this article, we aim to provide further insight into the structure of Juddian solutions of the AQRM by extending the divisibility properties and the relations between the constraint polynomials with the solution of the AQRM in the Bargmann space. In particular we discuss a conjecture proposed by Masato Wakayama.

Keywords Quantum Rabi models · Degenerate eigenvalues · Constraint polynomials · Juddian solutions

1 Introduction

The *quantum Rabi model* (QRM) is one of the basic models in quantum optics, describing the interaction between a two-level atom and a light field. Its Hamiltonian H_{Rabi} is given by

$$H_{\mathrm{Rabi}} = \omega a^\dagger a + g(a + a^\dagger)\sigma_x + \Delta\sigma_z,$$

where $a^\dagger$ and a are the creation and annihilation operators of the quantum harmonic oscillator, σ_x, σ_z are the Pauli matrices

C. Reyes-Bustos (✉)
Department of Mathematical and Computing Science, School of Computing, Tokyo Institute of Technology, 2-12-1 Ookayama, Meguro-ku, Tokyo 152-8550, Japan
e-mail: reyes@c.titech.ac.jp

© The Author(s) 2021　　149
T. Takagi et al. (eds.), *International Symposium on Mathematics,*
Quantum Theory, and Cryptography, Mathematics for Industry 33,
https://doi.org/10.1007/978-981-15-5191-8_13

$$\sigma_x = \begin{bmatrix} 0 & 1 \\ 1 & 0 \end{bmatrix}, \qquad \sigma_z = \begin{bmatrix} 1 & 0 \\ 0 & -1 \end{bmatrix},$$

$\omega > 0$ is the classical frequency of light field (modeled by a quantum harmonic oscillator), $2\Delta > 0$ is the energy difference of the two-level system and $g > 0$ is the interaction strength between the two systems. In our discussion we have set $\hbar = 1$ with no loss of generality. The QRM has a $\mathbb{Z}/2\mathbb{Z}$-symmetry that allows a decomposition $H_{\text{Rabi}} = H_{+\Delta} \oplus H_{-\Delta}$ for Hamiltonians $H_{\pm\Delta}$ acting on appropriate subspaces of the Hilbert space in which H_{Rabi} acts. Degeneracies are then found to naturally appear between one eigenvalue of $H_{+\Delta}$ and one eigenvalue of $H_{-\Delta}$. The parameters (g, Δ, ω) of the QRM are classified into *parameter regimes* according to the static and dynamic properties of the resulting energy levels and their solutions (see Xie et al. 2017 for discussion on parameter regimes).

Recent developments in experimental physics (Maissen et al. 2014, Yoshi-hara et al. 2017) have managed to realize parameter regimes (including the non-perturbative ultrastrong coupling and the deep strong coupling regimes) where approximated models, such as the Jaynes–Cummings model, can no longer describe the physical properties of the QRM. These developments, along with the prospect of applications to areas such as quantum information technologies (see Haroche and Raimond 2008; Yoshihara et al. 2017) have made the study of the properties of the QRM and its spectrum an important topic in physics. At the same time, there has been interest in the research of the mathematical aspects of the QRM and its gen-eralizations (see, for example, Reyes-Bustos and Wakayama 2017; Sugiyama 2018; Wakayama 2017).

The *asymmetric quantum Rabi model* (AQRM) is one of these generalizations. The Hamiltonian of the AQRM is obtained by introducing a nontrivial interaction term that breaks the $\mathbb{Z}/2\mathbb{Z}$-symmetry in the Hamiltonian of the QRM. Concretely, its Hamiltonian is given by

$$H_{\text{Rabi}}^{\varepsilon} = \omega a^{\dagger} a + \Delta \sigma_z + g \sigma_x (a^{\dagger} + a) + \varepsilon \sigma_x,$$

with $\varepsilon \in \mathbb{R}$. In general, this model loses the $\mathbb{Z}/2\mathbb{Z}$-symmetry of the QRM making the presence of degeneracies a nontrivial question and, in particular, there appears to be no way to define invariant subspaces (called parity subspaces in the case of the QRM) whose solutions constitute degeneracies (or crossings).

However, and contrary to this intuition, degenerate states were discovered in numerical experiments for the case $\varepsilon = \frac{1}{2}$ by Li and Batchelor in (2015). Later, Masato Wakayama in (2017) proved the existence in general for the case $\varepsilon = \frac{1}{2}$ and conjectured the existence of degenerate states for the general half-integer ε case in terms of divisibility of constraint polynomials. The conjecture was recently proved affirmatively for the general case by Kazufumi Kimoto, Masato Wakayama and the author in (2017). The presence of degenerate solutions for half-integer parameter hints at the possibility of a hidden symmetry in the AQRM, as it has been discussed in Semple and Kollar (2017), Wakayama (2017).

In order to describe how the degeneracies in the spectrum of the AQRM appear, we introduce the constraint polynomials.

Definition 1 Let $N \in \mathbb{Z}_{\geq 0}$. The polynomials $P_k^{(N,\varepsilon)}(x, y)$ of degree $k \in \mathbb{Z}_{\geq 0}$ are defined recursively by

$$P_0^{(N,\varepsilon)}(x, y) = 1, \qquad P_1^{(N,\varepsilon)}(x, y) = x + y - 1 - 2\varepsilon,$$
$$P_k^{(N,\varepsilon)}(x, y) = (kx + y - k(k + 2\varepsilon))P_{k-1}^{(N,\varepsilon)}(x, y) - k(k-1)(N - k + 1)x P_{k-2}^{(N,\varepsilon)}(x, y).$$

The polynomial $P_N^{(N,\varepsilon)}(x, y)$ is called *constraint polynomial* and its defining property is that if the parameters $g, \Delta > 0$ satisfy the *constraint equation*

$$P_N^{(N,\varepsilon)}((2g)^2, \Delta^2) = 0,$$

then $\lambda = N + \varepsilon - g^2$ is an eigenvalue of $H_{\text{Rabi}}^{\varepsilon}$. Any eigenvalue of the AQRM arising from the zeros of the constraint polynomials in this way is called *Juddian eigenvalue*.

The original conjecture proposed in Wakayama (2017) is summarized in the following theorem.

Theorem 2 (Kimoto et al. 2017) *For $N, \ell \in \mathbb{Z}_{\geq 0}$, we have*

$$P_{N+\ell}^{(N+\ell,-\frac{\ell}{2})}(x, y) = A_N^{(\ell)}(x, y) P_N^{(N,\frac{\ell}{2})}(x, y), \tag{1}$$

for a polynomial $A_N^{(\ell)}(x, y) \in \mathbb{Z}[x, y]$. In addition, for $\ell, N \in \mathbb{Z}_{\geq 0}$ the polynomial $A_N^{(\ell)}(x, y)$ has no zeros for $x, y > 0$.

In other words, since the constraint polynomials at both sides of (1) correspond to the same eigenvalue, we see that any Juddian eigenvalue of the AQRM is degenerate when the parameter ε is half-integer. The proof of Theorem 2 is done by studying certain determinant expressions satisfied by the constraint polynomials.

In the same paper Wakayama (2017) (see also Reyes-Bustos and Wakayama 2017), a second conjecture was presented. This time the polynomials involved are not the constraint polynomials, but the intermediate polynomials $P_k^{(N,\varepsilon)}(x, y)$. Since these polynomials are also related to solutions of the eigenvalue problem of the QRM, the study of this conjecture may provide some new insight into the relation between solutions of the QRM.

Conjecture 3 (Wakayama 2017) *Let $N, \ell, k \in \mathbb{Z}_{\geq 0}$. There are polynomials $A_k^{(N,\ell)}(x, y)$ and $B_k^{(N,\ell)}(x, y)$ in $\mathbb{Z}[x, y]$ such that*

$$P_{k+\ell}^{(N+\ell,-\frac{\ell}{2})}(x, y) = A_k^{(N,\ell)}(x, y) P_k^{(N,\frac{\ell}{2})}(x, y) + B_k^{(N,\ell)}(x, y)$$

with $B_N^{(N,\ell)}(x, y) = B_0^{(N,\ell)} = 0$. Furthermore, we have $A_k^{(N,\ell)}(x, y) > 0$ for $x, y > 0$.

It is important to notice that the way it was described in Wakayama (2017), the conjecture has not a unique solution. We discuss the issue in Sect. 3 and by extending the divisibility properties of the constraint polynomials, we give a candidate solution to the conjecture above. In addition, we describe the relation of the constraint polynomials with the coefficient solutions of the eigenvalue problem of AQRM in the Bargmann space picture.

Finally, we remark that there have been recent efforts to define regime parameters of the QRM using information from the energy levels of the solutions and not just the dynamic properties (see Rossatto et al. 2017). This approach is based on knowledge on the parameters for which exceptional solutions appear (for instance, the zeros of constraint polynomials). We expect that the results given here for constraint polynomials may provide some further insight for the studies in this direction.

2 The Confluent Picture of the Asymmetric Quantum Rabi Model

In this section we introduce the asymmetric quantum Rabi model (AQRM) and the realization of its eigenvalue problem in the Bargmann space, equivalent to a system of linear confluent Heun differential equations. After that we see that the coefficients of the solutions of the AQRM are expressed in terms of the constraint polynomials and other related polynomials. A good reference for Bargmann space methods is Schweber (1967).

The Bargmann space $\mathcal{H}_{\mathcal{B}}$ is the space of complex functions $f : \mathbb{C} \to \mathbb{C}$ holomorphic everywhere in the complex plane satisfying

$$\|f\|_{\mathcal{B}} = \left(\frac{1}{\pi} \int_{\mathbb{C}} |f(z)|^2 e^{-|z|^2} dx dy \right)^{1/2} < \infty$$

for $z = x + iy$ and where $dx dy$ is the Lebesgue measure in $\mathbb{C} \simeq \mathbb{R}^2$.

An important property of the Bargmann space is that it contains entire functions f having asymptotic expansion of the form

$$f(z) = e^{\alpha_1 z} z^{-\alpha_0} (c_0 + c_1 z^{-1} + c_2 z^{-2} + \cdots), \tag{2}$$

as $z \to \infty$ (see Braak 2011b). In particular, normal solutions of differential equations having an unramified singular point of rank 2 at infinity are included.

The Bargmann space $\mathcal{H}_{\mathcal{B}}$ is seen to be a Hilbert space unitarily equivalent to $L^2(\mathbb{R})$ and the realization of the creation and annihilation operators is given by

$$a \to \partial_z, \qquad a^{\dagger} \to z,$$

where we use ∂_z to denote $\frac{\partial}{\partial z}$.

Recall that the Hamiltonian $H_{\text{Rabi}}^{\varepsilon}$ of the AQRM is given by

$$H_{\text{Rabi}}^{\varepsilon} = \omega a^{\dagger} a + \Delta \sigma_z + g \sigma_x (a^{\dagger} + a) + \varepsilon \sigma_x. \tag{3}$$

Without loss of generality, we set $\omega = 1$ for the remainder of the paper. Thus, when $H_{\text{Rabi}}^{\varepsilon}$ is realized as an operator acting on $\mathcal{H}_{\mathcal{B}} \otimes \mathbb{C}^2$, the Hamiltonian $H_{\text{Rabi}}^{\varepsilon}$ is given by

$$\tilde{H}_{\text{Rabi}}^{\varepsilon} := \begin{bmatrix} z \partial_z + \Delta & g(z + \partial_z) + \varepsilon \\ g(z + \partial_z) + \varepsilon & z \partial_z - \Delta \end{bmatrix}.$$

Then, the time-independent Schrödinger equation $H_{\text{Rabi}}^{\varepsilon} \varphi = \lambda \varphi$ ($\lambda \in \mathbb{R}$) is equivalent to the system of first-order differential equations

$$\tilde{H}_{\text{Rabi}}^{\varepsilon} \psi = \lambda \psi, \quad \psi = \begin{bmatrix} \psi_1(z) \\ \psi_2(z) \end{bmatrix},$$

where eigenfunctions of $H_{\text{Rabi}}^{\varepsilon}$ associated with a given eigenvalue $\lambda \in \mathbb{R}$ correspond to solutions $\psi_i \in \mathcal{H}_{\mathcal{B}}$, $i = 1, 2$.

The eigenvalue problem of the AQRM is then reduced to finding entire functions $\psi_1, \psi_2 \in \mathcal{H}_{\mathcal{B}}$, and real number λ satisfying

$$\begin{cases} (z \partial_z + \Delta) \psi_1 + (g(z + \partial_z) + \varepsilon) \psi_2 = \lambda \psi_1, \\ (g(z + \partial_z) + \varepsilon) \psi_1 + (z \partial_z - \Delta) \psi_2 = \lambda \psi_2. \end{cases}$$

Now, by setting $\phi_{\pm} = \psi_1 \pm \psi_2$, we get

$$\begin{cases} (z + g) \dfrac{d}{dz} \phi_+ + (gz + \varepsilon - \lambda) \phi_+ + \Delta \phi_- = 0, \\ (z - g) \dfrac{d}{dz} \phi_- - (gz + \varepsilon + \lambda) \phi_- + \Delta \phi_+ = 0. \end{cases} \tag{4}$$

We note that the system (4) is equivalent to a second-order confluent Heun differential equation with an (unramified) irregular singular point at $z = \infty$ in addition to regular singular points at $z = \pm g$ (c.f. Braak 2016). Therefore, by the discussion above and (2), any entire solution ψ of (4) is actually $\psi \in \mathcal{H}_{\mathcal{B}} \otimes \mathbb{C}^2$. This is a key property used to prove the integrability in Braak (2011a).

Notice also that by applying the substitution $z \to -z$, we obtain the alternative system

$$\begin{cases} (z + g) \dfrac{d}{dz} \bar{\phi}_- + (gz + \varepsilon - \lambda) \bar{\phi}_- + \Delta \bar{\phi}_+ = 0, \\ (z - g) \dfrac{d}{dz} \bar{\phi}_+ - (gz + \varepsilon + \lambda) \bar{\phi}_+ + \Delta \bar{\phi}_- = 0 \end{cases} \tag{5}$$

where $\bar{\phi}_{\pm}(z) = \phi_{\pm}(-z)$. Furthermore, the two systems are equivalent under the transformation $\varepsilon \to -\varepsilon$.

Setting $x = \lambda + g^2$, the solutions around the singularity $z = g$ (for $x \pm \varepsilon \notin \mathbb{Z}$) are given by

$$\phi_+(z) = e^{-gz} \sum_{n=0}^{\infty} \frac{\Delta K_n^-}{x - \varepsilon - n}(z + g)^n, \qquad \phi_-(z) = e^{-gz} \sum_{n=0}^{\infty} K_n^-(z + g)^n, \qquad (6)$$

and by the symmetry mentioned above, the other set of solutions is given by

$$\bar{\phi}_-(z) = e^{gz} \sum_{n=0}^{\infty} \frac{\Delta K_n^+}{x - \varepsilon - n}(z + g)^n, \qquad \bar{\phi}_+(z) = e^{gz} \sum_{n=0}^{\infty} K_n^+(z + g)^n, \qquad (7)$$

related by $\phi_+(z) = \bar{\phi}_+(-z)$ and $\phi_-(z) = \bar{\phi}_-(-z)$. For $n \in \mathbb{Z}_{\geq 0}$, define the functions $f_n^{\pm} = f_n^{\pm}(x, g, \Delta, \varepsilon)$ by

$$f_n^{\pm}(x, g, \Delta, \varepsilon) = 2g + \frac{1}{2g}\left(n - x \pm \varepsilon + \frac{\Delta^2}{x - n \pm \varepsilon}\right). \qquad (8)$$

The coefficients $K_n^{\pm}(x) = K_n^{\pm}(x, g, \Delta, \varepsilon)$ are then given by the recurrence relation

$$nK_n^{\pm}(x) = f_{n-1}^{\pm}(x, g, \Delta, \varepsilon)K_{n-1}^{\pm}(x) - K_{n-2}^{\pm}(x) \quad (n \geq 1) \qquad (9)$$

with initial condition $K_{-1}^{\pm} = 0$ and $K_0^{\pm} = 1$.

The solutions (6) (resp. (7)) in general do not represent entire solutions. The condition for the solutions to be entire is given by the G-function. Next, we recall the definition of the G-function and refer the reader to Braak (2011a, 2011b) for the full details.

Definition 4 The G-function for the Hamiltonian $H_{\text{Rabi}}^{\varepsilon}$ is defined as

$$G_{\varepsilon}(x; g, \Delta) := \Delta^2 \bar{R}^+(x; g, \Delta, \varepsilon)\bar{R}^-(x; g, \Delta, \varepsilon) - R^+(x; g, \Delta, \varepsilon)R^-(x; g, \Delta, \varepsilon)$$

where

$$R^{\pm}(x; g, \Delta, \varepsilon) = \sum_{n=0}^{\infty} K_n^{\pm}(x)g^n \quad \text{and} \quad \bar{R}^{\pm}(x; g, \Delta, \varepsilon) = \sum_{n=0}^{\infty} \frac{K_n^{\pm}(x)}{x - n \pm \varepsilon}g^n,$$

$$(10)$$

whenever $x \mp \varepsilon \notin \mathbb{Z}_{\geq 0}$, respectively.

The main property of the G-function (see, for example, Braak 2011a) is that for a fixed tuple of parameters (g, Δ, ε), the zeros x_n of $G_{\varepsilon}(x; g, \Delta)$ correspond to eigenvalues $\lambda_n = x_n - g^2$ of $H_{\text{Rabi}}^{\varepsilon}$ with $x_n \neq N \pm \varepsilon$ for any integer $N \in \mathbb{Z}$. Any such eigenvalue is called a *regular eigenvalue* of the QRM. More precisely, if x

is a zero of the G-function, the solutions (6) can be analytically continued to the whole plane, and thus constitute solutions of the eigenvalue problem for the given eigenvalue $\lambda = x - g^2$.

In general, not every eigenvalue of the AQRM is regular. An eigenvalue that is not regular is called *exceptional eigenvalue*. Equivalently, exceptional eigenvalues are those of the form $\lambda = N \pm \varepsilon - g^2$. If the power series in the solution for an exceptional eigenvalue is terminating (i.e., is a polynomial), it is called *Juddian*, otherwise it is called *non-Juddian exceptional* eigenvalue. We recall from the introduction that Juddian eigenvalues are those that arise from zeros of the constraint polynomials. We also remark that the exceptional eigenvalues are closely related to the poles of the G-function, and refer the reader to Kimoto et al. (2017), Li and Batchelor (2015) for more information on exceptional eigenvalues.

After the preparations, we relate the coefficients of the solutions (resp. the G-function), with constraint polynomials. For brevity, we set $c_k^{(\varepsilon)} = k(k + 2\varepsilon)$ and $\lambda_k = k(k - 1)(N - k + 1)$. Then the polynomial $P_k^{(N,\varepsilon)}(x, y)$ is the determinant of a $k \times k$ tridiagonal matrix

$$P_k^{(N,\varepsilon)}(x, y) = \det(\mathbf{I}_k y + \mathbf{A}_k^{(N)} x + \mathbf{U}_k^{(\varepsilon)}) \tag{11}$$

where $\mathbf{I}_k$ is the identity matrix of size k and

$$\mathbf{A}_k^{(N)} = \text{tridiag} \begin{bmatrix} i & 0 \\ \lambda_{i+1} & \end{bmatrix}_{1 \le i \le k}, \quad \mathbf{U}_k^{(\varepsilon)} = \text{tridiag} \begin{bmatrix} -c_i^{(\varepsilon)} & 1 \\ 0 & \end{bmatrix}_{1 \le i \le k},$$

where we use the notation

$$\text{tridiag} \begin{bmatrix} a_i & b_i \\ c_i & \end{bmatrix}_{1 \le i \le n} := \begin{bmatrix} a_1 & b_1 & 0 & 0 & \cdots & 0 \\ c_1 & a_2 & b_2 & 0 & \cdots & 0 \\ 0 & c_2 & a_3 & b_3 & \cdots & 0 \\ \vdots & \ddots & \ddots & \ddots & \ddots & \vdots \\ 0 & \cdots & 0 & c_{n-2} & a_{n-1} & b_{n-1} \\ 0 & \cdots & 0 & 0 & c_{n-1} & a_n \end{bmatrix}.$$

The relation between the Nth coefficient of the G-function and the constraint polynomials is seen in the next lemma.

Lemma 5 (Kimoto et al. 2017) *Let $N \in \mathbb{Z}_{\ge 0}$. For $g > 0$, the relation*

$$(N!)^2 (2g)^N K_N^-(N + \varepsilon; g, \Delta, \varepsilon) = P_N^{(N,\varepsilon)}((2g)^2, \Delta^2) \tag{12}$$

holds. In addition, if $\varepsilon = \ell/2$ ($\ell \in \mathbb{Z}$), it also holds that

$$((N + \ell)!)^2 (2g)^{N+\ell} K_{N+\ell}^+(N + \ell/2; g, \Delta, \ell/2) = P_{N+\ell}^{(N+\ell, -\ell/2)}((2g)^2, \Delta^2).$$

From this point of view, the constraint polynomials are multiples of the coefficients of the solutions of the associated equation system of differential equations for $x = N + \varepsilon$. This fact is important since it allows us to relate the residues at the poles of the G-function with the presence or absence of exceptional solutions (see Kimoto et al. 2017, Propositions 5.3, 5.5 and 5.6).

We proceed to generalize the result above to all the coefficients of the G-function. First, we note a simple but important relation between the coefficients $K_n^-(N + \varepsilon; g, \Delta, \varepsilon)$ and $K_n^-(n + \varepsilon; g, \Delta, \varepsilon)$ of the G-functions and the corresponding relation between constraint polynomials.

Lemma 6 *For $N, n \in \mathbb{Z}_{\geq 0}$ with $n \leq N$,*

$$K_n^-(N + \varepsilon; g, \Delta, \varepsilon) = K_n^-(n + \varepsilon; g, \Delta, \varepsilon) + q_0(g, \Delta, \varepsilon, n, N),$$

where $(2g)^n q_0(g, \Delta, \varepsilon, n, N) \in \mathbb{Z}[g, \Delta, \varepsilon, n, N]$ and

$$q_0(g, \Delta, \varepsilon, N, N) = q_0(g, \Delta, \varepsilon, n, n) = 0.$$

Moreover,
$$P_k^{(N,\varepsilon)}(x, y) = P_k^{(k,\varepsilon)}(x, y) + \bar{q}_0(g, \Delta, \varepsilon, n, N),$$

where $\bar{q}_0(g, \Delta, \varepsilon, n, N) \in \mathbb{Z}[g, \Delta, \varepsilon, n, N]$ and $\bar{q}_0(g, \Delta, \varepsilon, N, N) = \bar{q}_0$ $(g, \Delta, \varepsilon, n, n) = 0$.

Proof We give the proof for the polynomials $P_k^{(N,\varepsilon)}(x, y)$ as the proof for the coefficients $K_n^-(N + \varepsilon; g, \Delta, \varepsilon)$ is done in a completely analogous way. In the determinant expression (11) for $P_k^{(N,\varepsilon)}(x, y)$, in each term $\lambda_i = i(i - 1)(N - i + 1)$, we write $N = k + (N - k)$ and then factor out the terms including $N - k$ by the multilinearity of the determinant. This gives the result. □

Next, we relate the coefficients of the solutions at $\bar{x} = N + \varepsilon$ with the constraint polynomials $P_n^{(n,\varepsilon)}(x, y)$. In the lemma below, for $a \in \mathbb{C}$ and $n \in \mathbb{Z}_{\geq 0}$, $(a)_n = a(a + 1) \cdots (a + n - 1)$ is the Pochhammer symbol.

Lemma 7 *For $N, n \in \mathbb{Z}_{\geq 0}$ with $n \leq N$, we have*

$$n!(N - n + 1)_n(2g)^n K_n^-(N + \varepsilon; g, \Delta, \varepsilon) = P_n^{(n,\varepsilon)}((2g)^2, \Delta^2) + q_1(x, y; N, n, \varepsilon),$$

with $q_1(x, y; N, n, \varepsilon) \in \mathbb{Z}[x, y, N, n, \varepsilon]$ such that $q_1(x, y; N, N, \varepsilon) = q_1(x, y; n, n, \varepsilon) = 0$.

Proof For $n \leq N$, define the auxiliary polynomials $P_k^{(N,n,\varepsilon)}(x, y)$ by the three-term recurrence relation

$$P_k^{(N,n,\varepsilon)}(x, y) = ((N - n + k)x + y - (N - n + k)^2 - 2(N - n + k)\varepsilon) P_{k-1}^{(N,n,\varepsilon)}(x, y)$$
$$- (N - n + k)(N - n + k - 1)(n - k + 1)x P_{k-2}^{(N,n,\varepsilon)}, \tag{13}$$

with initial conditions $P_0^{(N,n,\varepsilon)}(x, y) = 1$ and

$$P_1^{(N,n,\varepsilon)}(x, y) = (N - n + 1)x + y - (N - n + 1)^2 - 2(N - n + 1)\varepsilon.$$

Note that setting $n = N$ gives $P_k^{(N,N,\varepsilon)}(x, y) = P_k^{(N,\varepsilon)}(x, y)$.

Next, the determinant form (or continuant) of the three-term recurrence relation for the coefficients $K_n^-(x; g, \Delta, \varepsilon)$ is given by

$$K_n^-(x; g, \Delta, \varepsilon) = \frac{1}{n!} \det \begin{pmatrix} f_{n-1}^-(x) & 1 & 0 & \cdots & 0 & 0 \\ n-1 & f_{n-2}^-(x) & 1 & \cdots & 0 & 0 \\ \vdots & \vdots & & \ddots & \cdots & \vdots & \vdots \\ 0 & 0 & 0 & \cdots & 1 & f_0^-(x) \end{pmatrix},$$

where we factored $\frac{1}{k}$ from each of the rows. Next, we see that

$$\begin{aligned} f_k^-(N + \varepsilon) &= 2g + \frac{1}{2g}\left(k - N - 2\varepsilon + \frac{\Delta^2}{N - k}\right) \\ &= \frac{1}{(2g)(N - k)}\left((2g)^2(N - k) - (N - k)^2 - 2\varepsilon(N - k) + \Delta^2\right) \\ &= \frac{1}{(2g)(N - k)}h(k, g, \Delta), \end{aligned}$$

with $h(k, g, \Delta)$ defined implicitly. Thus, we obtain the expression

$$\begin{aligned} K_n^-(N + \varepsilon; g, \Delta, \varepsilon) &= \frac{1}{n!(2g)^2(N - n + 1)_n} \\ &\times \text{tridiag}\begin{bmatrix} h(n - i, g, \Delta) & (2g)^2(N - n + i)(N - n + i + 1)(n - i) \\ 1 & \end{bmatrix}_{1 \le i \le n}, \end{aligned}$$

and we verify that the three-term recurrence relation corresponding to this determinant is exactly the one defining the polynomials $P_k^{(N,n,\varepsilon)}(x, y)$ above, with $x = (2g)^2$ and $y = \Delta^2$. Thus, we have proved that

$$n!(N - n + 1)_n(2g)^n K_n^-(N + \varepsilon; g, \Delta, \varepsilon) = P_n^{(N,n,\varepsilon)}((2g)^2, \Delta^2).$$

The result then follows by factoring out the elements containing $N - n$ from the determinant associated with the three-term recurrence relation (13). $\qquad \square$

From Lemmas 6 and 7, we immediately have the following Corollary giving several expressions for the coefficients in terms of the polynomials $P_n^{(N,\varepsilon)}(x, y)$.

Corollary 8 *For $N, n \in \mathbb{Z}_{\geq 0}$ with $n \leq N$, we have*

$$P_n^{(N,\varepsilon)}((2g)^2, \Delta^2) = (n!)^2(2g)^n K_n^-(N + \varepsilon; g, \Delta, \varepsilon) + q_2(g^2, \Delta^2, n, N),$$

where $q_2(g^2, \Delta^2, n, N) \in \mathbb{Z}[g^2, \Delta^2, N, n, \varepsilon]$ such that

$$q_2(g^2, \Delta^2, n, n) = q_2(g^2, \Delta^2, N, N) = 0.$$

Furthermore, we have

$$P_n^{(N,\varepsilon)}((2g)^2, \Delta^2) = (n!)^2 (2g)^n K_n^-(n + \varepsilon; g, \Delta, \varepsilon) + \bar{q}_2(g^2, \Delta^2, n, N),$$

with $\bar{q}_2(g^2, \Delta^2, n, N)$ satisfying the same properties as $q_2(g^2, \Delta^2, n, N)$

Using the results above, we can give an expression of the solutions of the confluent picture of the AQRM in terms of constraint polynomials. To see this, we notice that for $n \in \mathbb{Z}_{\geq 0}$, the following identity holds

$$P_n^{(x,\varepsilon)}((2g)^2, \Delta^2) = (n!)(x - n + 1)_n (2g)^n K_n^-(x + \varepsilon; g, \Delta, \varepsilon) + (x - n)q_n(g^2, \Delta^2, x), \quad (14)$$

where $x \notin \mathbb{Z}_{\geq 0}$ and $q_n(g^2, \Delta^2, x)$ is a polynomial with integer coefficients.

Next, we see that the solutions (6), (7) or the functions $R^\pm$, $\bar{R}^\pm$ appearing in the definition of the G-function can be expressed in terms of constraint polynomials. For instance, we have

$$R^-(x + \varepsilon; g, \Delta, \varepsilon) = \sum_{n=0}^{\infty} \frac{P_n^{(x,\varepsilon)}((2g)^2, \Delta^2)}{(n!)(x - n + 1)_n (2g)^n} + \sum_{n=0}^{\infty} \frac{(x - n)q_n(g^2, \Delta^2, x)}{(n!)(x - n + 1)_n (2g)^n}.$$

From this expression (and the corresponding ones for R^+, $\bar{R}^\pm$) it is possible to give an alternate method for computing the residues at the poles of the G-function to the one in Kimoto et al. (2017).

3   Extended Divisibility Properties for Constraint and Related Polynomials

In this section we return to Conjecture 3, originally presented in Wakayama (2017) (see also Reyes-Bustos and Wakayama 2017). As mentioned in the introduction, in its current form, the conjecture may not have a unique solution. Indeed, let $A_k^{(N,\varepsilon)}(x, y)$, $B_k^{(N,\varepsilon)}(x, y)$ and $\bar{A}_k^{(N,\varepsilon)}(x, y)$, $\bar{B}_k^{(N,\varepsilon)}(x, y)$ be two pairs of polynomials satisfying the conditions of the conjecture. Moreover, if the coefficients of $\frac{1}{2}\left(A_k^{(N,\varepsilon)}(x, y) + \bar{A}_k^{(N,\varepsilon)}(x, y)\right)$ and $\frac{1}{2}\left(B_k^{(N,\varepsilon)}(x, y) + \bar{B}_k^{(N,\varepsilon)}(x, y)\right)$ are integers, then these polynomials also satisfy the conditions of the conjecture as long as the polynomial $\frac{1}{2}\left(A_k^{(N,\varepsilon)}(x, y) + \bar{A}_k^{(N,\varepsilon)}(x, y)\right)$ has the positivity condition.

To get a better understanding of the divisibility structure, we extend some of the results given in Kimoto et al. (2017) and give a proposal for a solution of the conjecture that is compatible with the case of the constraint polynomials. In particular, we show how to obtain a family of solutions to the conjecture by using a method related to the one discussed above.

First, we recall a simple lemma on diagonalization that we use in the proofs below.

Lemma 9 (Kimoto et al. 2017) *For $1 \leq k \leq N$, the eigenvalues of $\mathbf{A}_k^{(N)}$ are $\{1, 2, \ldots, k\}$ and the eigenvectors are given by the columns of the lower triangular matrix $\mathbf{E}_k^{(N)}$ given by*

$$(\mathbf{E}_k^{(N)})_{i,j} = (-1)^{i-j} \binom{i}{j} \frac{(i-1)!(N-j)!}{(j-1)!(N-i)!},$$

for $1 \leq i, j \leq k$.

Proof We have to check that $(\mathbf{A}_k^{(N)} \mathbf{E}_k^{(N)})_{i,j} = j(\mathbf{E}_k^{(N)})_{i,j}$ for every i, j. By definition, we see that

$$(\mathbf{A}_k^{(N)} \mathbf{E}_k^{(N)})_{i,j} = j(\mathbf{E}_k^{(N)})_{i,j} \iff (j-i)(\mathbf{E}_k^{(N)})_{i,j} = \lambda_i (\mathbf{E}_k^{(N)})_{i-1,j}$$
$$\iff (j-i)\binom{i}{j} = -i\binom{i-1}{j},$$

and the last equality is easily verified. $\qquad\square$

Next, we see that in general the polynomials $P_k^{(N,\varepsilon)}(x, y)$ are expressed as the determinant of a tridiagonal matrix plus a rank-one matrix.

Proposition 10 *Let $k \in \mathbb{Z}_{\geq 0}$, then*

$$P_k^{(N,\varepsilon)}(x, y) = \det\left(\mathbf{I}_k y + \mathbf{D}_k x + \mathbf{C}_k^{(N,\varepsilon)} + \mathbf{e}_k{}^T \mathbf{u}_k^{(N)}\right),$$

where $\mathbf{I}_k$ is the identity matrix, $\mathbf{D}_k = \mathrm{diag}(1, 2, \ldots, k)$, and $\mathbf{C}_k^{(N,\varepsilon)}$ is the tridiagonal matrix given by

$$\mathbf{C}_k^{(N,\varepsilon)} = \mathrm{tridiag}\begin{bmatrix} -i(2(N-i)+1+2\varepsilon) & 1 \\ i(i+1)c_{N-i}^{(\varepsilon)} & \end{bmatrix}_{1 \leq i \leq k},$$

$\mathbf{e}_k \in \mathbb{R}^k$ is the kth standard basis vector, and $\mathbf{u}_k^{(N)} \in \mathbb{R}^k$ is given entrywise by

$$\left(\mathbf{u}_k^{(N)}\right)_j = (-1)^{k-j+2} \binom{k+1}{j} \frac{k!(N-j)!}{(j-1)!(N-k-1)!}.$$

Proof By Lemma 9, the eigenvalues of $\mathbf{A}_k^{(N)}$ are $\{1, 2, \ldots, k\}$ and the eigenvectors are given by the columns of the lower triangular matrix $\mathbf{E}_k^{(N)}$ given by

$$(\mathbf{E}_k^{(N)})_{i,j} = (-1)^{i-j} \binom{i}{j} \frac{(i-1)!(N-j)!}{(j-1)!(N-i)!}.$$

Then, it suffices to verify that

$$\mathbf{U}_k^{(\varepsilon)}\mathbf{E}_k^{(N)} = \mathbf{E}_k^{(N)}\mathbf{C}_k^{(N,\varepsilon)} + \mathbf{E}_k^{(N)}\mathbf{e}_k{}^T\mathbf{u}_k^{(N)}. \tag{15}$$

Note that the kth column of $\mathbf{E}_k^{(N)}$ is $\mathbf{e}_k$, therefore the last summand reduces to $\mathbf{e}_k{}^T\mathbf{u}_k^{(N)}$.
    For $i, j \le k$, set

$$d_{ij} = (-1)^{i-j}\binom{i}{j}\frac{(i-1)!(N-j)!}{(j-1)!(N-i)!},$$

then, by using the elementary identities

$$j(j+1)c_{N-j}^{(\varepsilon)}d_{i,j+1} = -(i-j)(N-j+2\varepsilon)d_{ij},$$

$$d_{i+1,j} - d_{i,j-1} = (i^2 + j^2 + ij - j - iN - jN)d_{ij},$$

we see that

$$-c_i^{(\varepsilon)}d_{ij} + d_{i+1,j} + j(2(N-j)+1+2\varepsilon)d_{ij} - d_{i,j-1} - j(j+1)c_{N-j}^{(\varepsilon)}d_{i,j+1} = 0. \tag{16}$$

For $i, j \le k$, we have $d_{ij} = (\mathbf{E}_\mathbf{k}^{(\mathbf{N},\varepsilon)})_{i,j}$ and (16) directly gives (15) for $1 \le j \le k$ and
$1 \le i \le k-1$. For $i = k$, equation (16) reads

$$(\mathbf{U}_k^{(\varepsilon)}\mathbf{E}_k^{(N)} - \mathbf{E}_k^{(N)}\mathbf{C}_k^{(N,\varepsilon)})_{k,j} = -d_{k+1,j},$$

and the right-hand side is equal to the ith entry of $\mathbf{u}_k^{(N)}$, as desired. $\square$

Note that when $k = N$, by the definition of the entries, the vector $\mathbf{u}_k^{(N)}$ is equal
to the zero vector, and the proposition above reduces to Proposition 4.2 of Kimoto
et al. (2017).

Corollary 11 *Let $k \in \mathbb{Z}_{\ge 0}$, then*

$$P_k^{(N,\varepsilon)}(x, y) = \det\left(\mathbf{I}_k y + \mathbf{D}_k x + \mathbf{C}_k^{(N,\varepsilon)}\right) + R_k^{(N,\varepsilon)}(x, y),$$

for a polynomial $R_k^{(N,\varepsilon)} \in \mathbb{R}[x, y]$ with $R_N^{(N,\varepsilon)}(x, y) = 0$.

Note that the polynomial $R_k^{(N,\varepsilon)}$ satisfies the condition expected to be satisfied by
the polynomial $B_k^{(N,\ell)}(x, y)$ of the conjecture. Moreover, the polynomials described
by the determinant expression of a tridiagonal matrix

$$\det\left(\mathbf{I}_k y + \mathbf{D}_k x + \mathbf{C}_k^{(N,\varepsilon)}\right)$$

are exactly the polynomials $Q_k^{(N,\varepsilon)}(x, y)$ of Remark 3.6 of Kimoto et al. (2017).

Proof It is well-known that if $\mathbf{A}$ is a square matrix, then

$$\det(\mathbf{A} + \mathbf{v}^T \mathbf{u}_k^{(N)}) = \det(\mathbf{A}) + {}^T\mathbf{v}\,\mathrm{adj}(\mathbf{A})\mathbf{u}_k^{(N)},$$

where $\mathrm{adj}(A)$ is the adjugate matrix, the transpose of the matrix of cofactors of A. Applying this result along with Proposition 10, we get the determinant expression. Furthermore, we see that

$$R_k^{(N,\varepsilon)}(x, y) = {}^T\mathbf{e}_k \,\mathrm{adj}\left(\mathbf{I}_k y + \mathbf{D}_k x + \mathbf{C}_k^{(N,\varepsilon)}\right)\mathbf{u}_k^{(N)}$$

is a polynomial, since $\det\left(\mathbf{I}_k y + \mathbf{D}_k x + \mathbf{C}_k^{(N,\varepsilon)}\right)$ is clearly a polynomial. As mentioned above, $\mathbf{u}_k^{(N)} = 0$ when $N = k$, and thus the second claim follows. $\square$

Remark 12 The polynomial $R_k^{(N,\varepsilon)}(x, y)$ is given explicitly by

$$R_k^{(N,\varepsilon)}(x, y) = -\sum_{j=0}^{k-1}(-1)^{k-j}\binom{k+1}{j+1}\frac{k!(N-(j+1))!}{j!(N-(k+1))!}P_j^{(N,\varepsilon)}(x, y).$$

In particular, this expression can be interpreted as the Fourier expansion of the polynomial $R_k^{(N,\varepsilon)}(x, y)$ with respect to the family of generalized orthogonal polynomials $\left\{P_k^{(N,\varepsilon)}(x, y)\right\}_{k\geq 0}$ (compare with Remark 7.2 in Kimoto et al. 2017). Here, generalized orthogonal polynomials (with respect to the variable y) are used in the sense of Brezinski (1980).

It also follows that

$$Q_k^{(N,\varepsilon)}(x, y) = \sum_{j=0}^{k}(-1)^{k-j}\binom{k+1}{j+1}\frac{k!(N-(j+1))!}{j!(N-(k+1))!}P_j^{(N,\varepsilon)}(x, y), \qquad (17)$$

and since $Q_k^{(N,\varepsilon)}(x, y)$ are polynomials given by the determinant of a tridiagonal matrix, we immediately see that the right-hand side of (17) satisfy the three-term recurrence relation

$$\begin{aligned}Q_k^{(N,\varepsilon)}(x, y) =&(kx + y - k(2(N + 1 - k) - 1 + 2\varepsilon))Q_{k-1}^{(N,\varepsilon)}(x, y)\\ &- k(k-1)(N+1-k)(N+1-k+2\varepsilon)Q_{k-2}^{(N,\varepsilon)}(x, y),\end{aligned}$$

which should be contrasted with Definition 1.

We note one more interesting consequence of equation (17). Setting vectors

$$\begin{aligned}{}^T\boldsymbol{P}_k^{(N,\varepsilon)}(x, y) &= (P_0^{(N,\varepsilon)}(x, y), P_1^{(N,\varepsilon)}(x, y), \ldots, P_{k-1}^{(N,\varepsilon)}(x, y))\\ {}^T\boldsymbol{Q}_k^{(N,\varepsilon)}(x, y) &= (Q_0^{(N,\varepsilon)}(x, y), Q_1^{(N,\varepsilon)}(x, y), \ldots, Q_{k-1}^{(N,\varepsilon)}(x, y)),\end{aligned}$$

we verify that

$$Q_k^{(N,\varepsilon)}(x, y) = \mathbf{E}_k^{(N)} P_k^{(N,\varepsilon)}(x, y),$$

where $\mathbf{E}_k^{(N)}$ is the matrix of Lemma 9. These identities and the relation with orthogonal polynomials are part of a forthcoming paper by the author Reyes-Bustos (2019).

For completeness, we note the case $k = N$ of the corollary above, which reduces to the result given in Kimoto et al. (2017), is used to show, among other things, that for a fixed $x \in \mathbb{R}$ (resp. $y \in \mathbb{R}$) all the roots with respect to y (resp. x) of the constraint polynomial $P_N^{(N,\varepsilon)}(x, y)$ are real when $\varepsilon > -1/2$ (see Theorem 3.6 of Kimoto et al. 2017).

Corollary 13 *Let* $N \in \mathbb{Z}_{\geq 0}$. *We have*

$$P_N^{(N,\varepsilon)}(x, y) = \det\left(\mathbf{I}_N y + \mathbf{D}_N x + \mathbf{S}_N^{(N,\varepsilon)}\right),$$

where $\mathbf{D}_N$ *is the diagonal matrix of Proposition 10 and* $\mathbf{S}_N^{(N,\varepsilon)}$ *is the symmetric matrix given by*

$$\mathbf{S}_N^{(N,\varepsilon)} = \text{tridiag}\left[\begin{matrix} -i(2(N - i) + 1 + 2\varepsilon) & \sqrt{i(i + 1)c_{N-i}^{(\varepsilon)}} \\ \sqrt{i(i + 1)c_{N-i}^{(\varepsilon)}} & \end{matrix}\right]_{1 \leq i \leq N}.$$

Proof Consider the case $k = N$ in Proposition 10. Notice that the matrices $\mathbf{I}_N y + \mathbf{D}_N x + \mathbf{C}_N^{(N,\varepsilon)}$ and $\mathbf{I}_N y + \mathbf{D}_N x + \mathbf{S}_N^{(N,\varepsilon)}$ are tridiagonal. By comparing the off diagonal elements, we see that the two determinants are equal. □

Similar to the case $N = k$, when the parameter ε is half-integer, we have special divisibility properties for the polynomials $P_k^{(N,\varepsilon)}(x, y)$ obtained by factoring the determinant expression.

Proposition 14 *Let* $\ell, k \in \mathbb{Z}_{\geq 0}$, *then*

$$P_{k+\ell}^{(N+\ell,-\frac{\ell+N-k}{2})}(x, y) = \bar{A}_k^{(N,\ell)}(x, y) P_k^{(N,\frac{\ell+N-k}{2})}(x, y) + \bar{B}_k^{(N,\ell)}(x, y)$$

with $\bar{B}_N^{(N,\ell)}(x, y) = 0$. *Moreover, the polynomial* $\bar{A}_k^{(N,\ell)}(x, y)$ *is given by*

$$\bar{A}_k^{(N,\ell)}(x, y) = \frac{(k + \ell)!}{k!} \det \text{tridiag}\left[\begin{matrix} x + \frac{y}{k+i} + 2i - 1 + k - N - \ell & 1 \\ c_{-i}^{(\frac{N+\ell-k}{2})} & \end{matrix}\right]_{1 \leq i \leq \ell}.$$

As can be easily seen from the definition, and as we have already considered above in (14), the variable N in the constraint polynomial can be taken to assume real values, in other words, we can assume that it is a free variable. In this way, this result, along with Theorem 16 below, can be interpreted as divisibility modulo $N - k$, that is,

$$P_{k+\ell}^{(N+\ell,-\frac{\ell+N-k}{2})}(x, y) \equiv \bar{A}_k^{(N,\ell)}(x, y) P_k^{(N,\frac{\ell+N-k}{2})}(x, y) \quad (\mathrm{mod}\ N - k).$$

We make this assumption in the remainder of this section to simplify the proofs.

Proof We begin with the determinant expression of Corollary 11 for the polynomial $P_{k+\ell}^{(N+\ell,-\frac{\ell+N-k}{2})}(x, y)$, that is

$$P_{k+\ell}^{(N+\ell,-\frac{\ell+N-k}{2})}(x, y) = \det\left(\mathbf{I}_{k+\ell}y + \mathbf{D}_{k+\ell}x + \mathbf{C}_{k+\ell}^{(N+\ell,-\frac{\ell+N-k}{2})}\right) + q_{k+\ell}(x, y),$$

where $q_{k+\ell}(x, y)$ is a polynomial divisible by $N - k$. The tridiagonal matrix $\mathbf{C}_{k+\ell}^{(N+\ell,-\frac{\ell+N-k}{2})}$ is given by

$$\mathbf{C}_{k+\ell}^{(N+\ell,-\frac{\ell+N-k}{2})} = \mathrm{tridiag}\begin{bmatrix} -i(-2i+1+\ell+N+k) & 1 \\ i(i+1)(N+\ell-i)(k-i) & \end{bmatrix}_{1 \le i \le k+\ell}.$$

Note that when $i = k$, the off-diagonal element $i(i+1)(N+\ell-i)(k-i)$ vanishes and $\det\left(\mathbf{I}_{k+\ell}y + \mathbf{D}_{k+\ell}x + \mathbf{C}_{k+\ell}^{(N+\ell,-\frac{\ell+N-k}{2})}\right)$ can be computed as the product of the determinant of a $k \times k$ matrix and the determinant of an $\ell \times \ell$ matrix.

Let us first consider the determinant of the $\ell \times \ell$-matrix factor. It is given by

$$\det \mathrm{tridiag}\begin{bmatrix} y + (k+i)x - (k+i)(-2(k+i)+1+\ell+N+k) & 1 \\ (k+i)(k+i+1)(N+\ell-k-i)(-i) & \end{bmatrix}_{1 \le i \le \ell}$$

which is easily seen to be equal to

$$\bar{A}_k^{(N,\ell)}(x, y) = \frac{(k+\ell)!}{k!} \det \mathrm{tridiag}\begin{bmatrix} x + \frac{y}{k+i} + 2i - 1 + k - N - \ell & 1 \\ c_{-i}^{(\frac{N+\ell-k}{2})} & \end{bmatrix}_{1 \le i \le \ell}.$$

Let us denote by $q(x, y; N, \ell, k)$ the remaining factor, that is,

$$q(x, y; N, \ell, k) = \det \mathrm{tridiag}\begin{bmatrix} ix + y - i(-2i+1+\ell+N+k) & 1 \\ i(i+1)(N+\ell-i)(k-i) & \end{bmatrix}_{1 \le i \le k}.$$

By Corollary 11, we have

$$P_k^{(N,\frac{\ell+N-k}{2})}(x, y) - R_k^{(N,\frac{\ell+N-k}{2})}$$

$$= \det \mathrm{tridiag}\begin{bmatrix} ix + y - i(3N - 2i + 1 + \ell - k) & 1 \\ i(i+1)(N-i)(2N-i+\ell-k) & \end{bmatrix}_{1 \le i \le k},$$

the right-hand side can be written as

$$\text{det tridiag} \begin{bmatrix} ix + y - i(-2i + 1 + \ell + N + k + 2(N - k)) & 1 \\ i(i + 1)(k - i + (N - k))(N + \ell - i + (N - k)) & \end{bmatrix}_{1 \le i \le k},$$

and noticing that entrywise, the entries of the matrix of the determinant differ to those in the determinant expression of $q(x, y; N, \ell, k)$ only by factors of $N - k$, we obtain

$$q(x, y; N, \ell, k) = P_k^{(N, \frac{\ell+N-k}{2})}(x, y) + q'(x, y; N, \ell, k)$$

for a polynomial $q'(x, y; N, \ell, k)$ satisfying $q'(x, y; N, \ell, N) = 0$. This completes the proof. $\qquad\qquad\square$

In order to consider the result for the desired parameter $\varepsilon = \ell/2$, we need the following lemma.

Lemma 15 *Let $k \in \mathbb{Z}_{\geq 0}$ and $\delta \in \mathbb{R}$. Then, we have*

$$P_k^{(N, \varepsilon + \delta)}(x, y) = P_k^{(N, \varepsilon)}(x, y) + 2\delta q_k^{(N, \varepsilon)}(x, y)$$

for some polynomial $q^{(N, \varepsilon)}(x, y) \in \mathbb{R}[x, y]$.

Proof It is clear that $q_0^{(N, \varepsilon)}(x, y) = 0$ and $q_1^{(N, \varepsilon)}(x, y) = 1$. Then, assume that it holds for all $i \leq k$ for some $k \in \mathbb{Z}_{\geq 0}$. We have,

$$
\begin{aligned}
P_k^{(N, \varepsilon + a)}(x, y) &= (kx + y - c_k^{(\varepsilon + a)}) P_{k-1}^{(N, \varepsilon + a)}(x, y) - \lambda_k x P_{k-2}^{((N, \varepsilon + a))}(x, y) \\
&= P_k^{(N, \varepsilon)}(x, y) - 2ka P_{k-1}^{(N, \varepsilon)}(x, y) + 2a(kx + y - c_k^{(\varepsilon + a)}) q_{k-1}^{(N, \varepsilon)} \\
&\quad - 2a\lambda_k x q_{k-2}^{(N, \varepsilon)}(x, y) \\
&= P_k^{(N, \varepsilon)}(x, y) + 2a q_k^{(N, \varepsilon)}(x, y)
\end{aligned}
$$

and the result follows by induction. $\qquad\qquad\square$

Finally, we give a particular solution to Conjecture 3.

Theorem 16 *Let $\ell, k \in \mathbb{Z}_{\geq 0}$, then*

$$P_{k+\ell}^{(N+\ell, -\frac{\ell}{2})}(x, y) = A_k^{(N, \ell)}(x, y) P_k^{(N, \frac{\ell}{2})}(x, y) + B_k^{(N, \ell)}(x, y)$$

with $B_N^{(N, \ell)}(x, y) = 0$. Moreover, the polynomial $A_k^{(N, \ell)}(x, y)$ is given by

$$A_k^{(N, \ell)}(x, y) = \frac{(k + \ell)!}{k!} \text{det tridiag} \begin{bmatrix} x + \frac{y}{k+i} + 2i - 1 - \ell & 1 \\ c_{-i}^{(\frac{\ell}{2})} & \end{bmatrix}_{1 \le i \le \ell}.$$

Note that the polynomial $A_k^{(N, \ell)}(x, y)$ does not depend on the parameter N. Because of this, positivity follows trivially from the result for the polynomials $A_k^{(\ell)}(x, y)$ given in Kimoto et al. (2017). That is, we have $A_k^{(N, \ell)}(x, y) > 0$ for $x, y > 0$.

Proof First, by using Lemma 15 above on the polynomials at both sides of Proposition 14, it is easy to see that

$$P_{k+\ell}^{(N+\ell,-\frac{\ell}{2})}(x, y) = \bar{A}_k^{(N,\ell)}(x, y) P_k^{(N,\frac{\ell}{2})}(x, y) + \bar{C}_k^{(N,\ell)}(x, y)$$

for some polynomial $\bar{C}_k^{(N,\ell)}(x, y)$ satisfying $\bar{C}_N^{(N,\ell)}(x, y) = 0$. Note that the matrices in the determinant expressions of $\bar{A}_k^{(N,\ell)}(x, y)$ and $A_k^{(N,\ell)}(x, y)$ differ entrywise at most by factor of $N - k$, therefore

$$A_k^{(N,\ell)}(x, y) = \bar{A}_k^{(N,\ell)}(x, y) + (N - k)q^{(N,\ell)}(x, y)$$

for some polynomial $q^{(N,\ell)}(x, y) \in \mathbb{Z}[x, y]$ completing the proof. $\qquad \square$

It is important to mention that Theorem 16 may be proved by defining directly

$$B_k^{(N,\ell)}(x, y) = P_{k+\ell}^{(N+\ell,-\frac{\ell}{2})}(x, y) - A_k^{(\ell)}(x, y) P_k^{(N,\frac{\ell}{2})}(x, y),$$

and appealing to the results of Kimoto et al. (2017). However, in the proof above we wanted to emphasize how the polynomial $A_k^{(\ell)}(x, y)$ appears naturally by extending the main results of Kimoto et al. (2017).

Let us now return to the discussion on Conjecture 3 started at the beginning of the section. For an arbitrary (nonzero) polynomial $p(x, y)$, by setting

$$\hat{A}_k^{(\ell)}(x, y) = A_k^{(\ell)}(x, y) + k(N - k)p(x, y)$$

we verify the relation

$$P_{k+\ell}^{(N+\ell,-\frac{\ell}{2})}(x, y) = \hat{A}_k^{(N,\ell)}(x, y) P_k^{(N,\frac{\ell}{2})}(x, y) + \hat{B}_k^{(N,\ell)}(x, y),$$

with

$$\hat{B}_k^{(N,\ell)}(x, y) = B_k^{(N,\ell)}(x, y) - k(N - k)p(x, y) P_k^{(N,\frac{\ell}{2})}(x, y),$$

giving another solution to the conjecture as long as

$$\hat{A}_k^{(\ell)}(x, y) > 0$$

for $x, y > 0$ and $0 \le k \le N$. Therefore, this method gives a family of solutions of the conjecture related to the particular solution $A_k^{(\ell)}(x, y)$. It would be desirable to consider the problem of characterizing all the solutions to the problem posed in Conjecture 3 or in other words to consider the problem of finding the solutions with minimal degree for $\hat{B}_k^{(N,\ell)}(x, y)$ (or $\hat{A}_k^{(\ell)}(x, y)$) while retaining the condition of positivity of $\hat{A}_k^{(\ell)}(x, y)$. We note that the method for showing the positivity of the polynomial $A_k^{(\ell)}(x, y)$ in Kimoto et al. (2017) cannot be extended in general to the polynomial $\hat{A}_k^{(\ell)}(x, y)$ described here.

As a conclusion, we leave the question of Conjecture 3 open, but change the problem from one of existence to one of characterization of solutions according to the discussion above.

Problem 17 Characterize all pairs of solutions $A_k^{(N,\ell)}(x, y)$ and $B_k^{(N,\ell)}(x, y)$ of Conjecture 3. Alternatively, describe the "minimal" solutions according to certain criteria (e.g., degree).

4 Open Problems

To complement Problem 17, in this section we describe some open problems related with constraint polynomials and Juddian solutions of the AQRM and the QRM.

4.1 Number of Exceptional Solutions of the AQRM

For fixed $\Delta > 0$ and $N \in \mathbb{Z}_{\geq 0}$, the number of values of $g > 0$ such that $\lambda = N \pm \varepsilon - g^2$ is a Juddian solution is, by the results in Li and Batchelor (2015) (see also Kimoto et al. 2017), exactly $N - k$, where k is the integer satisfying

$$k(k + 2\varepsilon) \leq \Delta^2 < (k + 1)(k + 1 + 2\varepsilon).$$

This gives a complete answer to the problem of counting the number of Juddian solutions for fixed Δ when g is allowed to vary. From the G-functions for non-Juddian exceptional eigenvalues (called T-function in Kimoto et al. 2017), it is not difficult to obtain a condition on Δ for the existence solution for non-Juddian exceptional solutions for the case of the QRM, but such an estimate provides no information on the exact number of non-Juddian exceptional solutions and no further results in this direction are known.

A different problem in the same line is to determine, for a fixed $g, \Delta > 0$, the number of exceptional solutions present in the spectrum of $H_{\text{Rabi}}^{\varepsilon}$. For the case of Juddian eigenvalues, it corresponds to finding all the $N \in \mathbb{Z}_{\geq 0}$ such that

$$P_N^{(N,\varepsilon)}((2g)^2, \Delta^2) = 0,$$

for a given $g, \Delta > 0$. We recall here that since the polynomials $P_N^{(N,\varepsilon)}((2g)^2, \Delta^2)$ do not constitute a family of orthogonal polynomials in the usual sense (i.e., with respect to the variables $x = (2g)^2$ or $y = \Delta^2$) with the exception of the case $\Delta = 0$, there is almost no information known about the relation between their zeros. The same problem can be posed for non-Juddian exceptional eigenvalues but as in the Juddian case, there are no results in this direction.

4.2 Classification of Parameter Regimes

The parameter regimes for the QRM are defined according to different observed properties of the QRM, specially its dynamic properties, and whether the model can be approximated by simpler models (like the Jaynes–Cummings models). However, as remarked in Rossatto et al. (2017), the characterization of the coupling regimes is not universally agreed and there is a need for a more specific criterion.

In the same paper, the authors give a new proposal for characterization on the coupling regimes of the QRM that depends not only on the parameters of the system but also on the energy levels of the system. This new classification is based on the study of approximate exceptional solutions of eigenvalue problem of the QRM. The new classification has the advantage of giving precise differentiation between the coupling regimes based on observations made by the authors on the statical and dynamical properties of the QRM in these regimes.

For instance, in this proposal the *perturbative ultrastrong coupling regime* (pUSC) roughly corresponds to combinations of parameters g, ω, Δ and eigenvalues λ lying to the left of the first Juddian solution in the spectral curve graph. The *perturbative deep strong coupling regime* (pDSC) is similarly defined by the combination of parameters g, ω, Δ and eigenvalues λ lying past a boundary curve (in the (λ, g)-plane) after the last Juddian solution (or the first non-Juddian solution). The *non-perturbative ultrastrong-deep strong coupling regime* (npUSC-DSC) would then correspond to the remaining region in the (λ, g)-plane.

Thus, it is important to estimate the parameters corresponding to the first and last Juddian solution for each level N, and also the first non-Juddian exceptional solution for the level N, in order to describe the boundaries between the parameter regimes in an effective way. In a more general sense, it would be interesting to have an estimate for the distribution of the zeros of constraint polynomials and constraint functions for non-Juddian exceptional eigenvalues.

Acknowledgements This work was supported by JST CREST Grant Number JPMJCR14D6, Japan. The author would like to thank the anonymous referee for some crucial comments related to the proposed solution of Conjecture 3.

References

D. Braak, Integrability of the Rabi Model. Phys. Rev. Lett. **107**, 100401 (2011a)
D. Braak, *Online Supplement of "Integrability of the Rabi Model"* (2011b)
D. Braak, Analytical solutions of basic models in quantum optics, in *Applications + Practical Conceptualization + Mathematics = fruitful Innovation, Proceedings of the Forum of Mathematics*

for Industry 2014, vol. 11, Mathematics for Industry, ed. by R. Anderssen, et al. (Springer, Berlin, 2016), pp. 75–92

C. Brezinski, *Padé-Type Approximation and General Orthogonal Polynomials* (Springer Basel AG, Basel, 1980)

S. Haroche, J.M. Raimond, *Exploring the Quantum - Atoms, Cavities and Photons* (Oxford University Press, Oxford, 2008)

K. Kimoto, C. Reyes-Bustos, M. Wakayama, Determinant expressions of constraint polynomials and degeneracies of the asymmetric quantum Rabi model. *International Mathematics Research Notices,* Published Online: 20 April 2020. (2020)

Z.-M. Li, M.T. Batchelor, Algebraic equations for the exceptional eigen spectrum of the generalized Rabi model. J. Phys. A: Math. Theor. **48**, 454005 (13pp) (2015)

C. Maissen, G. Scalari, F. Valmorra, M. Beck, J. Faist, S. Cibella, R. Leoni, C. Reichl, C. Charpentier, W. Wegscheider, Ultrastrong coupling in the near field of complementary split-ring resonators. Phys. Rev. B **90**, 205309 (2014)

C. Reyes-Bustos, *Residual structure for the family of constraint polynomials for the quantum Rabi model* (In preparation)

C. Reyes-Bustos, M. Wakayama, Spectral degeneracies in the asymmetric quantum Rabi model, in *Mathematical Modelling for Next-Generation Cryptography, vol. 29, Mathematics for Industry*, ed. by T. Takagi, et al. (Springer, Berlin, 2017), pp. 117–137

D.Z. Rossatto, C.J. Villa-Bôas, M. Sanz, E. Solano, Spectral classification of coupling regimes in the quantum Rabi model. Phys. Rev. A **96**, 013849 (2017)

S. Schweber, On the application of Bargmann Hilbert spaces to dynamical problems. Ann. Phys. **41**, 205–229 (1967)

J. Semple, M. Kollar, Asymptotic behavior of observables in the asymmetric quantum Rabi model. J. Phys. A: Math. Theor. **51**, 044002 (2017)

S. Sugiyama, Spectral zeta functions for the quantum Rabi models. Nagoya Math. J. **229**, 52–98 (2018)

M. Wakayama, Symmetry of asymmetric quantum Rabi models. J. Phys. A: Math. Theor. **50**, 174001 (22pp) (2017)

Q.-T. Xie, H.-H. Zhong, M.T. Batchelor, C.-H. Lee, The quantum Rabi model: solutions and dynamics. J. Phys. A: Math. Theor. **50**, 113001 (2017)

F. Yoshihara, T. Fuse, S. Ashhab, K. Kakuyanagi, S. Saito, K. Semba, Supercondunting qubit oscillator circuit beyond the ultrastrong-coupling regime. Nat. Phys. **13**, 44–47 (2017)

Generalized Group–Subgroup Pair Graphs

Kazufumi Kimoto

Abstract A regular finite graph is called a Ramanujan graph if its zeta function satisfies an analog of the Riemann Hypothesis. Such a graph has a small second eigenvalue so that it is used to construct cryptographic hash functions. Typically, explicit family of Ramanujan graphs are constructed by using Cayley graphs. In the paper, we introduce a generalization of Cayley graphs called generalized group–subgroup pair graphs, which are a generalization of group–subgroup pair graphs defined by Reyes-Bustos. We study basic properties, especially spectra of them.

Keywords Cayley graphs · Spectra of graphs · Group–subgroup pair graphs · Group actions · Homogeneity · Representation theory · Characters

1 Introduction

A k-regular finite graph is called a *Ramanujan graph* if its zeta function satisfies an analog of the *Riemann hypothesis*. This condition is equivalent to say that every nontrivial (i.e. $\neq \pm k$) eigenvalue of the graph is less than or equal to $2\sqrt{k-1}$. Thus the second largest eigenvalue in absolute value of a Ramanujan graph is small, and this means that it has a large isoperimetric constant (i.e. it is an *expander graph*), so that random walks on such a graph rapidly converge to the uniform distribution as the number of walk steps tends to infinity. Consequently, as an application to cryptography, Ramanujan graphs can be used to construct cryptographic hash functions (see Charles et al. (2009), in which hash functions are constructed from LPS graphs Lubotzky et al. (1988) and Pizer graphs (1990)).

In order to construct (a family of) Ramanujan graphs, the *Cayley graphs* are an important tool; a Cayley graph is a graph whose vertex set is a finite group, and the adjacency of vertices is described in terms of the multiplication of the group. In fact,

K. Kimoto (✉)
Department of Mathematical Sciences, Faculty of Science,
University of the Ryukyus, 1 Senbaru Nishihara, cho Okinawa 903-0213, Japan
e-mail: kimoto@math.u-ryukyu.ac.jp

T. Takagi et al. (eds.), *International Symposium on Mathematics,
Quantum Theory, and Cryptography*, Mathematics for Industry 33,
https://doi.org/10.1007/978-981-15-5191-8_14

169

most of the known explicit constructions of infinite families of Ramanujan graphs are given as Cayley graphs, and the construction is based on deep results in number theory associated with the group (for instance, the construction of the LPS graphs due to Lubotzky et al. (1988) is based on the Ramanujan–Petersson conjecture on automorphic forms).

Thus it is natural to consider the generalization of Cayley graphs to enlarge the possibility to produce Ramanujan graphs and/or expander families. *Group–subgroup pair graphs* (or pair graph for short) Reyes-Bustos (2016), which are defined for a triplet (G, H, S) of a finite group G, a subgroup $H \subset G$ and a suitable subset $S \subset G$, are one of such attempts. A pair graph is regular in special cases and provides interesting examples of Ramanujan graphs. However, we can construct regular pair graphs only when $[G : H] \leq 2$. The purpose of this paper is to give a generalization of group–subgroup pair graphs, which can provide Ramanujan graph even when $[G : H] > 2$. A generalized pair graph is a graph defined for a pair (G, H) of a group and its subgroup together with a suitable family $\mathcal{S}$ of subsets in G. We study basic properties, especially spectra of them.

Here is the brief description on the organization of the paper: In Sect. 2, we recall basic conventions on graphs. In Sect. 3, we recall the definitions of Cayley graphs and group–subgroup pair graphs, and give several examples of them. In Sect. 4, we introduce the notion of homogeneity of a graph. In Sect. 5, we give a generalization of group–subgroup pair graphs. In Sect. 6, we describe the spectra of generalized group–subgroup pair graphs.

1.1 Conventions

For a matrix A, A^* is the transposed complex conjugate of A, and $\mathrm{Tr}(A)$ is the trace of A. The n by n identity matrix is denoted by I_n.

For a group G, we use the symbol e to indicate the identity element of G. We denote by χ^ρ the *character* of a given representation ρ of G: $\chi^\rho(x) = \mathrm{Tr}(\rho(x))$ for $x \in G$. The *unitary dual* of G (i.e. the set of all equivalence classes of unitary irreducible representations of G) is denoted by $\widehat{G}$. The *dual group* of G is defined to be $G^* = \mathrm{Hom}(G, \mathbb{C}^\times)$. We often identify G^* with the subset

$$\{\pi \in \widehat{G} \mid \deg \pi = 1\} \subset \widehat{G}$$

consisting of 1-dimensional representations of G via the bijection $\pi \mapsto \chi^\pi$. When G is abelian, we have $G^* = \widehat{G}$. We denote by $\mathbf{1}$ the trivial character of G (i.e. $\mathbf{1}(x) = 1$, $x \in G$).

2 Preliminaries

In what follows, a graph is always assumed to be *finite*, *undirected* and *simple* otherwise stated.

Let $X = (V, E)$ be a graph. The number of vertices $|V|$ and edges $|E|$ are called the *order* and *size* of the graph, respectively. We often write $x \sim y$ to indicate that two vertices x and y are adjacent, i.e. $xy \in E$. We denote by $\mathcal{N}(x)$ the *neighborhood* of x: $\mathcal{N}(x) = \{y \in V \mid x \sim y\}$. The *degree* $\deg(x)$ of a vertex x is the number of edges incident to x. If X is simple, then $\deg(x)$ is equal to $|\mathcal{N}(x)|$.

We call X a *k-regular* graph if $\deg(x) = k$ for every $x \in V$. We introduce two generalizations of this notion for later use. Suppose that V has a partition $V = V_1 \sqcup \cdots \sqcup V_m$.

(1) If the degree is constant on each subset V_i, say d_i, then we call X a $(d_1, \ldots, d_m)$-*regular* graph.

(2) If

$$d_{ij} := \left| \{ y \in V_j \mid x \sim y \} \right| \quad (x \in V_i)$$

depends only on i and j, then we say X is a *D-regular* graph, where $D = (d_{ij})_{1 \le i, j \le m}$. Notice that if

$$\sum_{i=1}^{m} d_{ir} = \sum_{j=1}^{m} d_{rj} =: d_r \quad (r = 1, \ldots, m),$$

then X is $(d_1, \ldots, d_m)$-regular ($\deg(x) = d_i$ for any $x \in V_i$).

Numbering the vertices, say $V = \{v_1, \ldots, v_N\}$ ($N = |V|$), we define the *adjacency matrix* $A = A_X$ of X by

$$A = (a_{ij})_{1 \le i, j \le N}, \quad a_{ij} = \begin{cases} 1 & v_i \sim v_j, \\ 0 & \text{otherwise.} \end{cases}$$

A depends on the choice of numbering of V, however, it is uniquely determined up to conjugation by permutation matrices. An eigenvalue of A is called an eigenvalue of the graph X. We denote by $\mathrm{Spec}(X)$ the multiset consisting of eigenvalues of X. If X is k-regular, then k is the largest eigenvalue of X, and every eigenvalue of X lies in the interval $[-k, k]$. We put

$$\lambda(X) := \max \left\{ |\lambda| \mid \lambda \in \mathrm{Spec}(X),\ \lambda \ne \pm k \right\}.$$

X is called a *Ramanujan graph* if

$$\lambda(X) \le 2\sqrt{k - 1}.$$

Remark 1 This condition $\lambda(X) \leq 2\sqrt{k-1}$ is equivalent to the analog of the Riemann hypothesis

$$\zeta_X(q^{-s})^{-1} = 0 \quad (q = k - 1) \implies \mathrm{Re}(s) = \frac{1}{2}$$

for the Ihara zeta function

$$\zeta_X(u) = \prod_{[P]} (1 - u^{\nu(P)})^{-1}$$

of X, where $[P]$ runs over all the "primes" in X, and $\nu(P)$ is the "length" of P. See, for example, Terras (2011) for detail.

Remark 2 It is known that the second largest eigenvalue λ_1 of X satisfies

$$\lambda_1 > 2\sqrt{k-1} - \frac{2\sqrt{k-1} - 1}{m}$$

when $\mathrm{diam}(X) \geq 2m + 2 \geq 4$, where $\mathrm{diam}(X)$ denotes the diameter of X Nilli (1991).

Remark 3 The notion of Ramanujan graphs is extended to non-regular graphs in several cases. For instance, a (p, q)-regular bipartite graph X is called *Ramanujan bigraph* if

$$\left| \sqrt{p-1} - \sqrt{q-1} \right| \leq \lambda(X) \leq \sqrt{p-1} + \sqrt{q-1}.$$

See, for example, Feng and Li (1996), Hashimoto (1989).

Example 1 The cycle graph C_n of order n is a 2-regular graph, and its eigenvalues are given by $2\cos\frac{2j\pi}{n}$ $(j = 0, 1, \ldots, n-1)$, which are all less than or equal to $2 = 2\sqrt{2-1}$. Hence C_n is Ramanujan for any $n \geq 3$.

3 Cayley Graphs and Group–Subgroup Pair Graphs

We briefly recall the basics of the Cayley graphs and group–subgroup pair graphs. We refer to Fulton and Harris (1991) for basic facts on representation theory.

3.1 Cayley Graphs

Definition 1 Let G be a group and $S \subset G$ be a symmetric generating set, that is, $S^{-1} = S$ and $\langle S \rangle = G$. The *Cayley graph* $\mathrm{Cay}(G, S)$ is a graph whose vertex set is G and two vertices $x, y \in G$ are adjacent if and only if $y = xs$ for some $s \in S$.

Let $\mathcal{R}$ be the *left regular representation* of G, which is the permutation representation induced from the left translation. Explicitly, if we index the elements in G as $G = \{g_1, \ldots, g_N\}$ ($N = |G|$), then $\mathcal{R}(g)$ ($g \in G$) can be realized as a matrix whose (i, j)-entry is $\delta(g_i^{-1} g g_j)$, where $\delta(x)$ is 1 if $x = e$ and 0 otherwise. Then the adjacency matrix $\mathcal{A}$ of $\mathrm{Cay}(G, S)$ is given by

$$\mathcal{A} = \sum_{s \in S} \mathcal{R}(s).$$

Since the irreducible decomposition of $\mathcal{R}$ is given by

$$\mathcal{R} \sim \bigoplus_{\pi \in \hat{G}} \pi^{\oplus \deg \pi},$$

there exists a certain unitary matrix U such that

$$U^* \mathcal{R}(g) U = \bigoplus_{\pi \in \hat{G}} \pi(g)^{\oplus \deg \pi}.$$

It follows that

$$U^* \mathcal{A} U = \bigoplus_{\pi \in \hat{G}} \left(\sum_{s \in S} \pi(s) \right)^{\oplus \deg \pi},$$

and hence the characteristic polynomial of the adjacency matrix $\mathcal{A}$ is written as

$$\det(x\, I_N - \mathcal{A}) = \prod_{\pi \in \hat{G}} \det\left(x\, I_{\deg \pi} - \sum_{s \in S} \pi(s) \right)^{\deg \pi}.$$

When G is abelian, every irreducible representation of G is 1-dimensional and we have

$$\mathrm{Spec}(\mathrm{Cay}(G, S)) = \left\{ \sum_{s \in S} \varphi(s) \,\middle|\, \varphi \in G^* \right\}.$$

Example 2 Let $G = D_n = \langle s, t \rangle$ be the dihedral group of degree $2n$ ($s^n = t^2 = e$, $tst = s^{-1}$). Take a symmetric generating subset $S = \{s, s^{-1}, t\}$. Then the Cayley graph $\mathrm{Cay}(G, S)$ is a 3-regular graph which is isomorphic to the Cartesian product of the path graph P_1 of length 1 and the cycle graph C_n of length n (Fig. 1). The following are the pictures of $\mathrm{Cay}(G, S)$ for $n = 5, 6, 7, 8$:
    The eigenvalues of $\mathrm{Cay}(G, S)$ are given by

$$2 \cos \frac{2j\pi}{n} \pm 1 \quad (j = 0, 1, \ldots, n - 1).$$

We see that $\mathrm{Cay}(G, S)$ is no longer Ramanujan if $2 \cos \frac{2\pi}{n} + 1 > 2\sqrt{2}$ or $n \geq 16$.

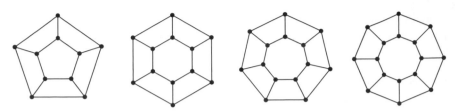

Fig. 1 $\mathrm{Cay}(D_n, S)$ for $n = 5, 6, 7, 8$

3.2 Group–Subgroup Pair Graphs

Definition 2 (Reyes-Bustos (2016)) Let G be a group, H a subgroup of G and $S \subset G$ such that $S_0 = S \cap H$ is symmetric (i.e. $S_0^{-1} = S_0$). The *group–subgroup pair graph* (or *pair graph* for short) $\mathcal{G}(G, H, S)$ is a graph whose vertex set is G and two vertices $x, y \in G$ are adjacent if and only if there exist $h \in H$ and $s \in S$ such that $\{x, y\} = \{h, hs\}$.

Remark 4 If $G = H = \langle S \rangle$, then $\mathcal{G}(G, G, S) = \mathrm{Cay}(G, S)$. If $[G : H] = 2$ and $S_0 = \varnothing$, then $\mathcal{G}(G, H, S)$ is bipartite.

Example 3 If $H = \{e\}$ and $S = G \setminus \{e\}$, then $\mathcal{G}(G, H, S)$ is the *star graph* $K_{1,k}$ (with $|G| = k + 1$). For instance, the pair graph for $G = \mathbb{Z}_8 = \mathbb{Z}/8\mathbb{Z}$, $H = \{0\}$ and $S = \mathbb{Z}_8 \setminus \{0\}$ is

$$\mathcal{G}(\mathbb{Z}_8, \{0\}, \mathbb{Z}_8 \setminus \{0\}) = K_{1,7} =$$

Here we summarize several elementary facts on pair graphs (see Reyes-Bustos (2016) for the proof). Assume that H is a subgroup of G with index $k + 1$ and order n. Put $N = |G| = (k + 1)n$ for short. Fix a set $\{x_0 = e, x_1, x_2, \ldots, x_k\}$ of representatives of the right cosets in G modulo H:

$$G = \bigsqcup_{i=0}^{k} V_i, \qquad V_i := H x_i,$$

and put $S_i = H x_i \cap S$. We also put $d_i = |S_i|$ and $d = |S|$. We denote by $\mathcal{A}$ the adjacency matrix for $\mathcal{G}(G, H, S)$, and by λ_i $(i = 0, 1, \ldots, N - 1)$ the eigenvalues of $\mathcal{G}(G, H, S)$ which are ordered in decreasing order: $\lambda_0 \geq \lambda_1 \geq \cdots \geq \lambda_{N-1}$.

- We have

$$\deg(v) = \begin{cases} d & v \in V_0 = H, \\ d_i & v \in V_i \ (i = 1, \ldots, k). \end{cases}$$

In particular, $\mathcal{G}(G, H, S)$ is regular if and only if $k = 0$ or $k = 1$ and $S_0 = \emptyset$.

- $\mathcal{G}(G, H, S)$ is a D-regular graph for

$$
D = \begin{pmatrix} d_0 \ d_1 \ \dots \ d_k \\ d_1 \\ \vdots & & O \\ d_k \end{pmatrix}.
$$

- $\mathcal{G}(G, H, S)$ is bipartite if and only if $S_0 = \emptyset$. The bipartition of G is then given by V_0 and $\bigcup_{i=1}^{k} V_i$.
- $\mathcal{G}(G, H, S)$ is connected if and only if $S_i \neq \emptyset$ for all $i \geq 1$ and $S_0 \cup \bigcup_{i=1}^{k} S_i S_i^{-1}$ generates H (Theorem 3.3 in Reyes-Bustos (2016)).
- $\mathcal{G}(G, H, S)$ has eigenvalues (called *trivial eigenvalues*; see Theorem 5.1 in Reyes-Bustos (2016))

$$
\mu_\pm = \frac{1}{2} \left(d_0 \pm \left(d_0^2 + 4 \sum_{i=1}^{k} d_i^2 \right)^{1/2} \right).
$$

μ_+ is the largest eigenvalue, and it is simple if $\mathcal{G}(G, H, S)$ is connected. For any eigenvalue λ of $\mathcal{G}(G, H, S)$ other than $\pm\lambda_0$, we have $|\lambda| < \lambda_0$.

- When $[G : H] = 2$, $\mathcal{G}(G, H, S)$ is Ramanujan if $|S| \geq n + 2 - 2\sqrt{n}$.

When the subgroup H is *abelian*, the eigenvalues of $\mathcal{G}(G, H, S)$ can be expressed in terms of *group characters* of H as follows.

Theorem 1 (Kimoto, 2018, Theorem 3) *If H is abelian, then the eigenvalues of $\mathcal{G}(G, H, S)$ are given by*

$$
\lambda_{\varphi,\pm} = \frac{1}{2} \left(\sum_{h \in H_0} \varphi(h) \pm \left(\left(\sum_{h \in H_0} \varphi(h) \right)^2 + 4 \sum_{j=1}^{k} \left| \sum_{h \in H_i} \varphi(h) \right|^2 \right)^{1/2} \right) \quad (\varphi \subset H^*)
$$

and zeros whose multiplicity is at least $(k - 1)n$. Here $H_i := S_i x_i^{-1} \subset H$.

4 Homogeneity

We introduce a simple notion concerning the symmetry of a graph. Let $X = (V, E)$ be a graph. Assume that a group G acts on V. We say that X is G-*homogeneous* if $x \sim y$ implies $gx \sim gy$ for any $g \in G$. This is equivalent to say that G is embedded in the *graph automorphism group* $\mathrm{Aut}(X)$ of the graph X. We see that $\mathcal{N}(gx) = g\mathcal{N}(x)$ and hence $\deg(x) = \deg(gx)$ for any $x \in V$ and $g \in G$. In particular, if $G \curvearrowright V$ is transitive (i.e. for any $x, y \in V$, we can find $g \in G$ such that $y = gx$), then X is regular.

Remark 5 X is Aut(X)-homogeneous.

Remark 6 A G-homogeneous graph X is vertex-transitive (i.e. for any $x, y \in V$, there exists a graph isomorphism f such that $y = f(x)$) if $G \curvearrowright V$ is transitive.

Example 4 A Cayley graph $X = \text{Cay}(G, S)$ is G-homogeneous by the natural left translation $(g, x) \mapsto gx$. X is $G \times G$-homogeneous via $((g_1, g_2), x) \mapsto g_1 x g_2^{-1}$ if and only if S is *normal* or G-conjugate invariant (i.e. $gSg^{-1} = S$ for all $g \in G$) or S is a union of several conjugacy classes of G. In such a case, we have

$$\det(x\, I_N - \mathcal{A}) = \prod_{\pi \in \widehat{G}} \left(x - \frac{1}{\deg \pi} \sum_{s \in S} \chi^\pi(s) \right)^{(\deg \pi)^2}$$

by Schur's lemma since $\sum_{s \in S} \pi(s)$ commutes with every $\pi(g)$ $(g \in G)$ for each $\pi \in \widehat{G}$. Here χ^π is the character of π.

Example 5 A pair graph $X = \mathcal{G}(G, H, S)$ is H-homogeneous.

Proposition 1 *Let $X = (V, E)$ be a graph with a group action $G \curvearrowright V$ which is free (i.e. stabilizer of any $v \in V$ is trivial) and transitive. Then $X \cong \text{Cay}(G, S)$ for a certain $S \subset G$.*

Proof We have $\mathcal{N}(gv) = g\mathcal{N}(v)$ for each $g \in G$ and $s \in S$. There exists $S \subset G$ such that $\mathcal{N}(v) = \{sv \mid s \in S\}$. It is straightforward to check that $X \cong \text{Cay}(G, S)$. $\qquad\square$

We roughly observe that the spectra $\text{Spec}(X)$ of a graph X tends to be simple if X is equipped with a large symmetry. Pair graphs can be regarded as a class of graphs which have weakened but nontrivial symmetry (or homogeneity) compared to Cayley graphs.

In the following section, we introduce a generalization of pair graphs, which are free but non-transitive H-homogeneous graphs.

5 Generalized Group–Subgroup Pair Graph

5.1 Definition

Let G be a finite group and H its subgroup of index $k + 1$. For later use, we put $N = |G|, n = |H|$ (hence we have $N = (k + 1)n$). Fix a collection of representatives $\{x_0 = e, x_1, \ldots, x_k\}$ of $H \backslash G$ and put $V_i = Hx_i$ $(i = 0, 1, \ldots, k)$. Let $\mathcal{S} = \{S_{ij}\}_{i,j=0}^k$ be a family of subsets in G such that

(1) $S_{ij} \subset V_i^{-1} V_j = x_i^{-1} H x_j$,
(2) $e \notin S_{ij}$,
(3) $S_{ij}^{-1} = S_{ji}$.

For two vertices $x, y \in G$, we connect these two by an edge if and only if $y = xs$ for some $s \in S_{ij}$ when $x \in V_i$ and $y \in V_j$ ($i, j = 0, 1, \ldots, k$). We denote this graph by $\mathcal{G}(G, H, \mathcal{S})$, and call such a graph a *generalized group–subgroup pair graph*, or simply *generalized pair graph*. Put

$$
D = \begin{pmatrix}
d_{00} & d_{01} & \cdots & d_{0k} \\
d_{10} & d_{11} & \cdots & d_{1k} \\
\vdots & \vdots & \ddots & \vdots \\
d_{k0} & d_{k1} & \cdots & d_{kk}
\end{pmatrix}
$$

with $d_{ij} = |S_{ij}|$. Notice that D is symmetric. We also put

$$
d_s = \sum_{j=0}^{k} d_{sj} = \sum_{i=0}^{k} d_{is} \quad (s = 0, 1, \ldots, k).
$$

Then $\mathcal{G}(G, H, \mathcal{S})$ is a D-regular and $(d_0, d_1, \ldots, d_k)$-regular graph. Thus, if every row sum and column sum of D is equal to d, then $\mathcal{G}(G, H, \mathcal{S})$ is d-regular. By the definition, we readily see that the following lemma holds.

Lemma 1 $\mathcal{G}(G, H, \mathcal{S})$ *is H-homogeneous, that is, $x \sim y$ implies $hx \sim hy$ for any $x, y \in G$ and $h \in H$.*

When $k = 1$ or $[G : H] = 2$, H is normal and $G/H \cong \mathbb{Z}/2\mathbb{Z}$, and hence it follows that

$$
S_{00}, S_{11} \subset V_0, \quad S_{01}, S_{10} \subset V_1.
$$

In this case, $\mathcal{G}(G, H, \mathcal{S})$ is (d_0, d_1)-biregular, and it is regular if $|S_{00}| = |S_{11}|$.

Remark 7 When $S_{ii} = \varnothing$ ($i = 0, 1, \ldots, k$), then $\mathcal{G}(G, H, \mathcal{S})$ is a *multi-partite graph*.

5.2 Examples

Example 6 Let $X = (V, E)$ be a graph of order $k + 1$ with $V = \{0, 1, \ldots, k\}$, and $\mathcal{A} = (a_{ij})_{0 \leq i, j \leq k}$ be its adjacency matrix. Take a group $G = \{x_0, x_1, \ldots, x_k\}$ of order $k + 1$, and put $H = \{e\}$ and

$$
S_{ij} = \begin{cases}
\varnothing & a_{ij} = 0, \\
\{x_i^{-1} x_j\} & a_{ij} = 1.
\end{cases}
$$

Then $\mathcal{G}(G, H, \mathcal{S}) \cong X$. Thus any finite graph is captured in the framework of generalized pair graphs (with trivial symmetry).

Example 7 Let G be a finite group, H its subgroup of index $k+1$ and $S \subset G$ a subset such that $S \cap H$ is symmetric. Fix a collection of representatives $\{x_0 = e, x_1, \ldots, x_k\}$ of $H \backslash G$ and put $V_i = H x_i$ $(i = 0, 1, \ldots, k)$. Define

$$S_{0i} = S \cap V_i, \quad S_{i0} = S_{0i}^{-1} \quad (i = 0, 1, \ldots, k),$$
$$S_{ij} = \varnothing \quad (i \neq 0, j \neq 0).$$

Then $\mathcal{G}(G, H, \mathcal{S})$ is reduced to the original group–subgroup pair graph $\mathcal{G}(G, H, S)$.

Example 8 Let $G = D_n = \langle s, t \rangle$ be the dihedral group of degree $2n$. We take $H = \langle s \rangle$ and $x_0 = e$, $x_1 = t$. Put

$$S_{00} = \{s, s^{-1}\}, \quad S_{01} = S_{10} = \{t\}, \quad S_{11} = \{s^2, s^{-2}\}.$$

Then $\mathcal{G}(G, H, \mathcal{S})$ is a $\begin{pmatrix} 2 & 1 \\ 1 & 2 \end{pmatrix}$-regular graph (and hence it is 3-regular). The following are the pictures of $\mathcal{G}(G, H, \mathcal{S})$ for $n = 5, 6, 7, 8$ (Fig. 2): when $n = 5$, $\mathcal{G}(G, H, \mathcal{S})$ is isomorphic to the Petersen graph (the leftmost one in the picture above). These four examples are Ramanujan graphs:

$$\det(x I - A) = \begin{cases} (x-3)(x-1)^5(x+2)^4 & n = 5, \\ (x-3)(x-1)x^2(x+2)^2(x^2-5)(x^2-2)^2 & n = 6, \\ (x-3)(x-1)(x^6+2x^5-6x^4-10x^3+10x^2+11x-1)^2 & n = 7, \\ (x-3)(x-1)(x^2-5)(x^2+2x-1)^2(x^4-4x^2+1)^2 & n = 8 \end{cases}$$

and

$$\lambda(X) \approx \begin{cases} 2 & n = 5, \\ 2.2361 & n = 6, \\ 2.3319 & n = 7, \\ 2.4142 & n = 8, \end{cases}$$

which are less than $2\sqrt{2} \approx 2.8284$. In general, the eigenvalues of $\mathcal{G}(G, H, \mathcal{S})$ are given by

$$\cos \frac{2\pi j}{n} + \cos \frac{4\pi j}{n} \pm \sqrt{\left(\cos \frac{2\pi j}{n} - \cos \frac{4\pi j}{n}\right)^2 + 1} \quad (j = 0, 1, \ldots, n-1).$$

$\mathcal{G}(G, H, \mathcal{S})$ is Ramanujan whenever $n \leq 23$, and is not Ramanujan when $n \geq 24$.

Example 9 Let $G = D_n$ be the dihedral group of degree $2n$, and we take $H = \langle s \rangle$ and $x_0 = e$, $x_1 = t$. Put

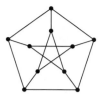
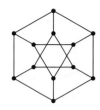
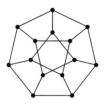

Fig. 2 $\mathcal{G}(G, H, \mathcal{S})$ for $n = 5, 6, 7, 8$

$$S_{00} = \{s, s^{-1}\}, \quad S_{01} = S_{10} = \{st, s^{-1}t\}, \quad S_{11} = \{s^2, s^{-2}\}.$$

Then $\mathcal{G}(G, H, \mathcal{S})$ is a $\begin{pmatrix} 2 & 2 \\ 2 & 2 \end{pmatrix}$-regular graph (and hence it is 4-regular). The following are the pictures of $\mathcal{G}(G, H, \mathcal{S})$ for $n = 5, 6, 7, 8$ (Fig. 3): these four examples are Ramanujan graphs:

$$\det(x\,I - \mathcal{A}) = \begin{cases} x(x-4)(x^4 + 2x^3 - 4x^2 - 5x + 5)^2 & n = 5, \\ x^3(x-4)(x+2)^2(x^2-8)(x^2-2)^2 & n = 6, \\ x(x-4)(x^6 + 2x^5 - 8x^4 - 15x^3 + 14x^2 + 28x + 7)^2 & n = 7, \\ x^3(x-4)(x+2)^2(x^2-8)(x^4 - 6x^2 + 4)^2 & n = 8 \end{cases}$$

and

$$\lambda(X) \approx \begin{cases} 2.4667 & n = 5, \\ 2.8284 & n = 6, \\ 2.6377 & n = 7, \\ 2.8284 & n = 8, \end{cases}$$

which are less than $2\sqrt{3} \approx 3.4641$. In general, the eigenvalues of $\mathcal{G}(G, H, \mathcal{S})$ are given by

$$\cos\frac{2\pi j}{n} + \cos\frac{4\pi j}{n} \pm \sqrt{\left(\cos\frac{2\pi j}{n} - \cos\frac{4\pi j}{n}\right)^2 + 4\cos^2\frac{2\pi j}{n}} \quad (j = 0, 1, \ldots, n-1).$$

$\mathcal{G}(G, H, \mathcal{S})$ is Ramanujan whenever $n \leq 15$, and is not Ramanujan when $n \geq 16$.

In general, when $[G : H] = 2$, take $S_{00} \subset H = Hx_0$ such that $S_{00}^{-1} = S_{00}$ and $S_{01} \subset Hx_1$. We also take a nontrivial group automorphism f of H. Put $S_{11} = f(S_{00})$ and $S_{10} = S_{01}^{-1}$. Then we get a regular graph $\mathcal{G}(G, H, \mathcal{S})$.

Fig. 3 $\mathcal{G}(G, H, \mathcal{S})$ for $n = 5, 6, 7, 8$

6 Spectra of $\mathcal{G}(G, H, \mathcal{S})$

6.1 Adjacency Matrix of $\mathcal{G}(G, H, \mathcal{S})$

Let $\mathcal{A}$ be the adjacency matrix of $\mathcal{G}(G, H, \mathcal{S})$. For a concrete description of $\mathcal{A}$, we write $H = \{h_0, \ldots, h_{n-1}\}$ with $h_0 = e$, and put $g_{ni+j} = h_j x_i$ for $i = 0, \ldots, k$ and $j = 0, \ldots, n - 1$. Thus we have $G = \{g_0, g_1, \ldots, g_{N-1}\}$. Then $\mathcal{A}$ is of the form

$$\mathcal{A} = \begin{pmatrix} \mathcal{A}_{00} & \mathcal{A}_{01} & \ldots & \mathcal{A}_{0k} \\ \mathcal{A}_{10} & \mathcal{A}_{11} & \ldots & \mathcal{A}_{1k} \\ \vdots & \vdots & \ddots & \vdots \\ \mathcal{A}_{k0} & \mathcal{A}_{k1} & \ldots & \mathcal{A}_{kk} \end{pmatrix},$$

where each block $\mathcal{A}_{pq}$ ($0 \leq p, q \leq k$) is given by

$$(\mathcal{A}_{pq})_{ij} = \begin{cases} 1 & h_i^{-1} h_j \in H_{pq} := x_p S_{pq} x_q^{-1}, \\ 0 & \text{otherwise.} \end{cases}$$

We notice that we can express each $\mathcal{A}_{pq}$ as

$$\mathcal{A}_{pq} = \sum_{s \in H_{pq}} \mathcal{R}_H(s),$$

where $\mathcal{R}_H$ is the left regular representation of H.

6.2 When H is abelian

If H is *abelian*, then $\mathcal{R}_H$ is a direct sum of all inequivalent 1-dimensional (irreducible) representations of H, that is, there exists a certain unitary matrix U such that

$$U^* \mathcal{R}_H(h) U \sim \bigoplus_{\varphi \in H^*} \varphi(h).$$

Hence

$$U^* \mathcal{A}_{pq} U = \sum_{s \in H_{pq}} \bigoplus_{\varphi \in H^*} \varphi(s).$$

Since $\{U^* \mathcal{A}_{pq} U\}_{p,q}$ commutes with each other, we have the following theorem.

Theorem 2 *Assume that H is an abelian subgroup of G. The adjacency matrix $\mathcal{A}$ of the generalized pair graph $\mathcal{G}(G, H, \mathcal{S})$ is given by*

$$\det(x\, I_N - \mathcal{A}) = \prod_{\varphi \in H^*} \det(x\, I_{k+1} - \mathcal{A}_\varphi),$$

where $\mathcal{A}_\varphi$ with $\varphi \in H^*$ is given by

$$\mathcal{A}_\varphi = \left(\sum_{s \in H_{ij}} \varphi(s) \right)_{0 \le i, j \le k}.$$

Remark 8 When $H = \{e\}$, we see that $H^* = \{\mathbf{1}\}$ and $\mathcal{A}_\mathbf{1} = \mathcal{A}$. Thus the theorem above is trivial.

Remark 9 Notice that $\mathcal{A}_\mathbf{1} = D$. It follows that the eigenvalues of D are also eigenvalues of $\mathcal{G}(G, H, \mathcal{S})$ if H is abelian. It is natural to ask the relation between $\mathrm{Spec}(\mathcal{A})$ and $\mathrm{Spec}(D)$ when H is non-abelian. We leave this as a future problem.

Remark 10 When $\mathcal{G}(G, H, \mathcal{S})$ is a pair graph, that is, $\mathcal{A}_{ij} = O$ if $i \ne 0$ and $j \ne 0$, we have

$$\det(x\, I_N - \mathcal{A}) = x^{(k-1)n} \det\left(x^2 I_n - x\, \mathcal{A}_{00} - \sum_{j=1}^{k} \mathcal{A}_{0j} \mathcal{A}_{j0} \right)$$

without any assumption on H. If H is abelian, then Theorem 1 follows immediately from the equation above.

6.3 Petersen Extension

Let G be a group, H be a subgroup of G with index 2 and $X := \mathrm{Cay}(H, S)$ be a k-regular Cayley graph. Assume that $G = H \cup Hw$ with $w \in G$. Take a group endomorphism $\sigma \in \mathrm{End}(H)$. Notice that $X' := \mathrm{Cay}(H, \sigma(S)) \cong X$ if σ is an *automorphism*. Put

$$S_{00} = S, \quad S_{11} = \sigma(S), \quad S_{01} = \{w\}, \quad S_{10} = \{w^{-1}\}.$$

Fig. 4 Cay(H, S) and its
Petersen extension
$\mathcal{G}(G, H, S)$

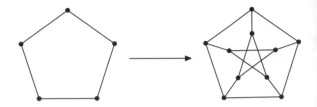

Then $\widetilde{X} = \mathcal{G}(G, H, S)$ is a $(k + 1)$-regular H-homogeneous graph. We call this the *Petersen extension* of Cay(H, S). The adjacency matrix $\widetilde{A}$ of $\widetilde{X}$ is given by

$$\widetilde{A} = \begin{pmatrix} A & I_n \\ I_n & A' \end{pmatrix},$$

where A and A' are the adjacency matrices of X and X', and it follows that

$$\det(x\, I_{2n} - \widetilde{A}) = \det(x^2 I_n - x(A + A') + AA' - I_n).$$

Example 10 When $G = D_5 = \langle s, t \rangle$, $H = \langle s \rangle$, $S = \{s, s^{-1}\}$, $w = t$ and $\sigma \in \mathrm{Aut}(H)$ is given by $\sigma(h) = h^2$ ($h \in H$), the Petersen extension $\mathcal{G}(G, H, S)$ of Cay(H, S) is the *Petersen graph* (Fig. 4).

Remark 11 If σ is the identity map of H (i.e. $X' = X$), then the Petersen extension $\mathcal{G}(G, H, S)$ of Cay(H, S) is just a Cartesian product of Cay(H, S) and the path graph $P_1 = \bullet\!\!-\!\!\bullet$.

In general, it is not true that the Petersen extension $\widetilde{X}$ of $X = \mathrm{Cay}(H, S)$ is Ramanujan when X is Ramanujan. Thus we propose the following problem.

Problem 1 Characterize the quintuple (G, H, S, w, σ) such that both Cay(H, S) and its Petersen extension with w and σ are Ramanujan.

6.3.1   Examples: Dihedral case

We look at the case where $G = D_n = \langle s, t \rangle$, $H = \langle s \rangle$ and $w = t$, for instance. In this case, an endomorphism σ of H is given by $\sigma(h) = h^l$ for certain $l \in \mathbb{Z}$, and $\sigma \in \mathrm{Aut}(H)$ if and only if $\gcd(n, l) = 1$. We also notice that $wSw^{-1} = tSt = S$ for any symmetric generating subset S of H.

Let $X_{n,l} := \mathcal{G}(G, H, S)$ be the Petersen extension of Cay(H, S) defined by w and $\sigma : H \ni h \mapsto h^l \in H$. Then, the family S is given by

$$S_{00} = S, \quad S_{01} = S_{10} = \{t\}, \quad S_{11} = \{s^l \mid s \in S\}.$$

For each character $\varphi \in H^*$, define

$$\alpha_\varphi := \sum_{s \in S} \varphi(s), \quad \beta_\varphi := \sum_{s \in S} \varphi(s^l).$$

By Theorem 2, we see that

$$\det(x\, I_{2n} - A) = \prod_{\varphi \in H^*} \det(x\, I_2 - A_\varphi), \quad A_\varphi = \begin{pmatrix} \alpha_\varphi & 1 \\ 1 & \beta_\varphi \end{pmatrix},$$

where A is the adjacency matrix of $X_{n,l}$. Hence the eigenvalues of $X_{n,l}$ are given by

$$\frac{\alpha_\varphi + \beta_\varphi \pm \sqrt{(\alpha_\varphi - \beta_\varphi)^2 + 4}}{2} \quad (\varphi \in H^*).$$

Example 11 If $n \geq 3$ and $S = \{s, s^{-1}\}$, then

$$\alpha_\varphi = e^{\frac{2\pi i j}{n}} + e^{-\frac{2\pi i j}{n}} = 2\cos\frac{2\pi j}{n},$$

$$\beta_\varphi = e^{\frac{2l\pi i j}{n}} + e^{-\frac{2l\pi i j}{n}} = 2\cos\frac{2l\pi j}{n}$$

for $\varphi \in H^*$ given by $\varphi(s) = e^{\frac{2\pi i j}{n}}$. Thus the eigenvalues of $X_{n,l}$ are calculated as

$$\cos\frac{2\pi j}{n} + \cos\frac{2l\pi j}{n} \pm \sqrt{\left(\cos\frac{2\pi j}{n} - \cos\frac{2l\pi j}{n}\right)^2 + 1} \quad (j = 0, 1, \ldots, n-1).$$

We can numerically check that

(1) if $n \leq 53$ and $n \neq 48$, then there exists l such that $X_{n,l}$ is Ramanujan,
(2) if $n \geq 54$ or $n = 48$, then $X_{n,l}$ is not Ramanujan for any choice of l.

When n is *odd* and $\gcd(n, l) = 1$ (i.e. $\sigma \in \mathrm{Aut}(H)$), then we see that

(1) if $n \leq 53$ and $n \neq 45$, then there exists l such that $X_{n,l}$ is Ramanujan,
(2) if $n \geq 55$ or $n = 45$, then $X_{n,l}$ is not Ramanujan for any choice of l.

Example 12 If $n = 2m \geq 4$ is *even* and $S = \{s, s^m, s^{-1}\}$, then

$$\alpha_\varphi = e^{\frac{2\pi i j}{n}} + e^{\frac{2m\pi i j}{n}} + e^{-\frac{2\pi i j}{n}} = (-1)^j + 2\cos\frac{2\pi j}{n},$$

$$\beta_\varphi = e^{\frac{2l\pi i j}{n}} + e^{\frac{2lm\pi i j}{n}} + e^{-\frac{2l\pi i j}{n}} = (-1)^{lj} + 2\cos\frac{2l\pi j}{n}$$

for $\varphi \in H^*$ given by $\varphi(s) = e^{\frac{2\pi i j}{n}}$. We can numerically check that

(1) if $m \leq 29$ ($n \leq 58$), then there exists l such that $X_{n,l}$ is Ramanujan,

(2) if $m \geq 30$ ($n \geq 60$), then $X_{n,l}$ is not Ramanujan for any choice of l.

Example 13 If $n \geq 5$ and $S = \{s, s^2, s^{-1}, s^{-2}\}$, then

$$\alpha_\varphi = e^{\frac{2\pi i j}{n}} + e^{\frac{4\pi i j}{n}} + e^{-\frac{2\pi i j}{n}} + e^{-\frac{4\pi i j}{n}} = 2\cos\frac{2\pi j}{n} + 2\cos\frac{4\pi j}{n},$$

$$\beta_\varphi = e^{\frac{2l\pi i j}{n}} + e^{\frac{4l\pi i j}{n}} + e^{-\frac{2l\pi i j}{n}} + e^{-\frac{4l\pi i j}{n}} = 2\cos\frac{2l\pi j}{n} + 2\cos\frac{4l\pi j}{n}$$

for $\varphi \in H^*$ given by $\varphi(s) = e^{\frac{2\pi i j}{n}}$. We can numerically check that

(1) if $n \leq 33$, then there exists l such that $X_{n,l}$ is Ramanujan,
(2) if $n \geq 34$, then $X_{n,l}$ is not Ramanujan for any choice of l.

Remark 12 In the paper, we discuss the construction of graphs when a finite group G and its subgroup H are given. It would be also interesting to consider the situation where finite groups G, H and an epimorphism $p\colon G \twoheadrightarrow H$ are given (i.e. H is a quotient group of G).

Acknowledgements This work was supported by JST CREST Grant Number JPMJCR14D6, Japan. The author would like to thank the anonymous reviewer for his/her comments and suggestions.

References

D. Charles, K. Lauter, E. Goren, Cryptographic hash functions from expander graphs. J. Cryptol. **22**, 93–113 (2009)

K. Feng, W.-C.W. Li, Spectra of hypergraphs and applications. J. Number Theory **60**(1), 1–22 (1996)

W. Fulton, J. Harris, Representation Theory: A First Course. Graduate Texts in Mathematics, vol. 129 (Springer, New York, 1991)

K. Hashimoto, Zeta functions of finite graphs and representations of p-adic groups. Automorphic forms and geometry of arithmetic varieties. In: Advanced Studies in Pure Mathematics, vol. 15 (Academic Press, Boston, MA, 1989), pp. 211–280

K. Kimoto, Spectra of Group-Subgroup Pair Graphs, in Mathematical Modelling for Next-Generation Cryptography, Math. Ind. eds. by T. Takagi et al., vol. 29 (Springer, Singapore, 2018)

A. Lubotzky, R. Phillips, P. Sarnak, Ramanujan graphs. Combinatorica **8**, 261–277 (1988)

A. Nilli, On the second eigenvalue of a graph. Discrete Math. **91**, 207–210 (1991)

A.K. Pizer, Ramanujan graphs and Hecke operators. Bull. Am. Math. Soc. (N.S.) **23**, 127–137 (1990)

C. Reyes-Bustos, Cayley-type graphs for group-subgroup pairs. Linear Algebra Appl. **488**, 320–349 (2016)

A. Terras, *Zeta Functions of Graphs—A stroll through the Garden* (Cambridge Univ Press, Cambridge, 2011)

Post-Quantum Cryptography

A Survey of Solving SVP Algorithms and Recent Strategies for Solving the SVP Challenge

Masaya Yasuda

Abstract Recently, lattice-based cryptography has received attention as a candidate of post-quantum cryptography (PQC). The essential security of lattice-based cryptography is based on the hardness of classical lattice problems such as the shortest vector problem (SVP) and the closest vector problem (CVP). A number of algorithms have been proposed for solving SVP exactly or approximately, and most of them are useful also for solving CVP. In this paper, we give a survey of typical algorithms for solving SVP from a mathematical point of view. We also present recent strategies for solving the Darmstadt SVP challenge in dimensions higher than 150.

Keywords Shortest vector problem (SVP) · Enumeration · Sieve · Lattice basis reduction · LLL · BKZ · Random sampling · Sub-sieving

1 Introduction

There has recently been a substantial amount of research for large-scale quantum computers. On the other hand, if such computers were built, they could break currently used public-key cryptosystems such as the RSA cryptosystem and the elliptic curve cryptography. (See Shor 1994 for Shor's quantum algorithms.) In order to prepare information security systems to be able to resist quantum computing, the US National Institute of Standards and Technology (NIST) began a process to develop new standards for PQC in 2015 and called for proposals in 2016. It has rapidly accelerated to research lattice-based cryptography as a candidate of PQC. Specifically, at the submission deadline of the end of November 2017 for the call, NIST received more than 20 proposals of lattice-based cryptosystems. Among them, more than 10 proposals were allowed for Round 2 submissions around the end of January 2019. (See the web page of NIST 2016.) The security of such proposals relies on the hard-

M. Yasuda (✉)
Institute of Mathematics for Industry, Kyushu University, 744 Motooka, Nishi-ku Fukuoka 819–0395, Japan
e-mail: yasuda@imi.kyushu-u.ac.jp

© The Author(s) 2021

T. Takagi et al. (eds.), *International Symposium on Mathematics, Quantum Theory, and Cryptography*, Mathematics for Industry 33, https://doi.org/10.1007/978-981-15-5191-8_15

ness of cryptographic lattice problems such as learning with errors (LWE) and NTRU. Such problems are reduced to approximate-SVP or approximate-CVP. (For example, see Albrecht et al. 2018 for details.) Therefore, it is becoming more important to understand classical lattice problems for evaluating the security of lattice-based PQC candidates.

For a positive integer n, a (full-rank) *lattice* L in $\mathbb{R}^n$ is the set of all integral linear combinations of linearly independent vectors $\mathbf{b}_1, \ldots, \mathbf{b}_n$ in $\mathbb{R}^n$. (The set of the $\mathbf{b}_i$'s is called a *basis* of L.) Given a basis of a lattice L, SVP asks to find the non-zero shortest vector in L. In this paper, we give a survey of typical algorithms for solving SVP from a mathematical point of view. These algorithms can be classified into two categories, depending on whether they solve SVP exactly or approximately. Exact-SVP algorithms perform an exhaustive search for an integer combination of the basis vectors $\mathbf{b}_i$'s to find the non-zero shortest lattice vector $\mathbf{v} = \sum_{i=1}^{n} v_i \mathbf{b}_i \in L$, and their cost is expensive. In contrast, approximate-SVP algorithms are much faster than exact algorithms, but they find short lattice vectors, not necessarily the shortest ones. However, exact- and approximate-SVP algorithms are complementary. For example, exact algorithms apply an approximation algorithm as a preprocessing to reduce their expensive cost, while several approximate-SVP algorithms call many times an exact algorithm in low dimension as a subroutine to find a very short lattice vector. In this paper, we also introduce recent strategies for solving the Darmstadt SVP challenge Darmstadt (2010), in which sample lattice bases are presented in order to test algorithms solving SVP. In particular, these strategies combine approximate- and exact-SVP algorithms to efficiently solve SVP in high dimensions such as $n \geq 150$.

Notation. The symbols $\mathbb{Z}$, $\mathbb{Q}$, and $\mathbb{R}$ denote the ring of integers, the field of rational numbers, and the field of real numbers, respectively. Let $\lfloor z \rceil$ denote the rounding integer of an integer z. We represent all vectors in *column format*. For $\mathbf{a} = (a_1, \ldots, a_n)^\top \in \mathbb{R}^n$, let $\|\mathbf{a}\|$ denote its Euclidean norm. For $\mathbf{a} = (a_1, \ldots, a_n)^\top$ and $\mathbf{b} = (b_1, \ldots, b_n)^\top$, let $\langle \mathbf{a}, \mathbf{b} \rangle$ denote the inner product $\sum_{i=1}^{n} a_i b_i$. Denote by $V_n(R)$ the volume of the n-dimensional ball of radius $R > 0$ centered at the origin. In particular, we let $v_n = V_n(1)$ denote the volume of the unit ball. Then $V_n(R) = v_n R^n$ and

$$v_n = \frac{\pi^{n/2}}{\Gamma(1 + n/2)} \sim \frac{1}{\sqrt{\pi n}} \left(\frac{2\pi e}{n} \right)^{n/2}$$

using Stirling's formula, where $\Gamma(s) = \int_0^\infty t^{s-1} e^{-t} dt$ denotes the Gamma function.

2 Mathematical Background

In this section, we introduce basic definitions and properties on lattices, and present famous lattice problems whose hardness ensures the essential security of lattice-based cryptography. (For example, see Galbraith 2012, Part IV or Nguyen 2009 for more details.)

2.1 Lattices and Their Bases

For a positive integer n, let $\mathbf{b}_1, \ldots, \mathbf{b}_n$ be n linearly independent (column) vectors in $\mathbb{R}^n$. The set of all integral linear combinations of the $\mathbf{b}_i$'s is a (full-rank) *lattice*

$$L = \mathcal{L}(\mathbf{b}_1, \ldots, \mathbf{b}_n) = \left\{ \sum_{i=1}^{n} v_i \mathbf{b}_i : v_i \in \mathbb{Z} \text{ for all } 1 \leq i \leq n \right\}$$

of dimension n with basis $\mathbf{B} = (\mathbf{b}_1, \ldots, \mathbf{b}_n) \in \mathbb{R}^{n \times n}$. (A basis is regarded not only as a set of vectors, but also as a matrix whose column vectors span a lattice.) Every lattice has infinitely many bases if $n \geq 2$; if two bases $\mathbf{B}_1$ and $\mathbf{B}_2$ span the same lattice, then there exists an $n \times n$ unimodular matrix $\mathbf{U} \in \mathrm{GL}_n(\mathbb{Z})$ with $\mathbf{B}_1 = \mathbf{B}_2 \mathbf{U}$. The *volume* of L is defined as $\mathrm{vol}(L) = |\det(\mathbf{B})|$, independent of the choice of bases.

The *Gram–Schmidt orthogonalization* for an (ordered) basis $\mathbf{B}$ is the orthogonal family $\mathbf{B}^* = (\mathbf{b}_1^*, \ldots, \mathbf{b}_n^*) \in \mathbb{R}^{n \times n}$, recursively defined by $\mathbf{b}_1^* = \mathbf{b}_1$ and for $2 \leq i \leq n$

$$\mathbf{b}_i^* = \mathbf{b}_i - \sum_{j=1}^{i-1} \mu_{i,j} \mathbf{b}_j^*, \text{ where } \mu_{i,j} = \frac{\langle \mathbf{b}_i, \mathbf{b}_j^* \rangle}{\|\mathbf{b}_j^*\|^2} \text{ for } 1 \leq j < i \leq n.$$

Notice that the Gram–Schmidt vectors $\mathbf{b}_i^*$'s depend on the order of basis vectors in $\mathbf{B}$. For convenience, set $\mu = (\mu_{i,j}) \in \mathbb{R}^{n \times n}$ where let $\mu_{i,j} = 0$ for all $i < j$ and $\mu_{k,k} = 1$ for all k. Then $\mathbf{B} = \mathbf{B}^* \mu$, and thus $\mathrm{vol}(L) = \prod_{i=1}^{n} \|\mathbf{b}_i^*\|$ from the orthogonality of Gram–Schmidt vectors. For $2 \leq \ell \leq n$, let π_ℓ denote the orthogonal projection over the orthogonal supplement of the $\mathbb{R}$-vector space $\langle \mathbf{b}_1, \ldots, \mathbf{b}_{\ell-1} \rangle_{\mathbb{R}}$ as

$$\pi_\ell : \mathbb{R}^n \longrightarrow \langle \mathbf{b}_1, \ldots, \mathbf{b}_{\ell-1} \rangle_{\mathbb{R}}^{\perp} = \langle \mathbf{b}_\ell^*, \ldots, \mathbf{b}_n^* \rangle_{\mathbb{R}}, \quad \pi_\ell(\mathbf{x}) = \sum_{i=\ell}^{n} \frac{\langle \mathbf{x}, \mathbf{b}_i^* \rangle}{\|\mathbf{b}_i^*\|^2} \mathbf{b}_i^*.$$

Every projection map depends on a basis. We also set $\pi_1 = \mathrm{id}$ for convenience.

2.2 Successive Minima, Hermite's Constants, and Gaussian Heuristic

For every $1 \leq i \leq n$, the ith *successive minimum* of an n-dimensional lattice L, denoted by $\lambda_i(L)$, is defined as the minimum of $\max_{1 \leq j \leq i} \|\mathbf{v}_j\|$ over all i linearly independent vectors $\mathbf{v}_1, \ldots, \mathbf{v}_i \in L$. In particular, the first minimum $\lambda_1(L)$ is the norm of the shortest non-zero vector in L. We clearly have $\lambda_1(L) \leq \lambda_2(L) \leq \cdots \leq \lambda_n(L)$ by definition. Moreover, for any basis $\mathbf{B} = (\mathbf{b}_1, \ldots, \mathbf{b}_n)$ of L, its Gram–Schmidt vectors satisfy $\lambda_i(L) \geq \min_{i \leq j \leq n} \|\mathbf{b}_j^*\|$ for every $1 \leq i \leq n$. (See Bremner 2011, Proposition 3.14 for proof.)

Hermite (1850) first proved that the quantity $\frac{\lambda_1(L)^2}{\text{vol}(L)^{2/n}}$ is upper bounded over all lattices L of dimension n. Its supremum over all lattices of dimension n is called *Hermite's constant* of dimension n, denoted by γ_n. This implies $\lambda_1(L) \leq \sqrt{\gamma_n}\text{vol}(L)^{1/n}$ for any lattice L of dimension n. As its extension, it satisfies

$$\left(\prod_{i=1}^{r} \lambda_i(L)\right)^{1/r} \leq \sqrt{\gamma_n}\text{vol}(L)^{1/n} \text{ for } 1 \leq r \leq n.$$

This is known as Minkowski's second theorem. (See Martinet 2013, Chap. 2 for proof.) It is important to know the value of γ_n in order to obtain an upper bound of $\lambda_1(L)$; Minkowski's convex body theorem implies $\gamma_n \leq 4v_n^{-2/n}$. (See Martinet 2013, Chap. 2 for proof.) This shows that

$$\lambda_1(L) \leq 2v_n^{-1/n}\text{vol}(L)^{1/n} \tag{1}$$

for any lattice L of dimension n. Moreover, it satisfies $\gamma_n \leq 1 + \frac{n}{4}$ from well-known formulas for v_n. It is very difficult to find the exact value of γ_n, and such values are known for only a few integers n. However, every γ_n is known as essentially linear in n. It also satisfies Mordell's inequality $\gamma_n \leq \gamma_k^{(n-1)/(k-1)}$ for any $n \geq k \geq 2$. (See Nguyen 2009 for more details on Hermite's constants.)

Given a lattice L of dimension n and a measurable set S in $\mathbb{R}^n$, the *Gaussian Heuristic* predicts that the number of vectors in $L \cap S$ is roughly equal to $\text{vol}(S)/\text{vol}(L)$. By applying the ball of radius $\lambda_1(L)$ centered at the origin in $\mathbb{R}^n$, it leads to the prediction of the norm of the shortest non-zero vector in L. Specifically, the expectation of $\lambda_1(L)$ according to the Gaussian Heuristic is given by

$$\text{GH}(L) = v_n^{-1/n}\text{vol}(L)^{1/n} \sim \sqrt{\frac{n}{2\pi e}}\text{vol}(L)^{1/n}.$$

This is tight compared to Eq. (1). Note that this is only a heuristic. But for "random" lattices, $\lambda_1(L)$ is asymptotically equal to $\text{GH}(L)$ with overwhelming probability Ajtai (1996).

2.3 Introduction to Lattice Problems

The most famous lattice problem is given below.

? The Shortest Vector Problem (SVP)

Given a basis $\mathbf{B} = (\mathbf{b}_1, \ldots, \mathbf{b}_n)$ of a lattice L, find the shortest non-zero vector in L, that is, a vector $\mathbf{s} \in L$ such that $\|\mathbf{s}\| = \lambda_1(L)$.

It was proven by Ajtai (1996) that SVP is NP-hard under randomized reductions. SVP can be relaxed by an approximate factor: *Given a basis of a lattice L and an approximation factor* $f \geq 1$, *find a non-zero vector* $\mathbf{v}$ *in L such that* $\|\mathbf{v}\| \leq f\lambda_1(L)$. Approximate-SVP is exactly SVP when $f = 1$. It is unlikely that one can efficiently solve approximate-SVP within quasi-polynomial factors in n, while approximate-SVP within a factor $\sqrt{n/\log(n)}$ is unlikely to be NP-hard. (See Nguyen 2009 for more details.)

Another famous lattice problem is given below.

? The Closest Vector Problem (CVP)

Given a basis $\mathbf{B} = (\mathbf{b}_1, \ldots, \mathbf{b}_n)$ of a lattice L and a target vector $\mathbf{t}$, find a vector in L closest to $\mathbf{t}$, that is, a vector $\mathbf{v} \in L$ such that the distance $\|\mathbf{t} - \mathbf{v}\|$ is minimized.

CVP is at least as hard as SVP. As in the case of SVP, we can define an approximate variant of CVP by an approximate factor. Approximate-CVP is also at least as hard as approximate-SVP with the same factor. From a practical point of view, both are considered equally hard, due to Kannan's embedding technique Kannan (1987) which can transform approximate-CVP into approximate-SVP. (See also Galbraith 2012 for the embedding.)

The security of modern lattice-based cryptosystems is based on the hardness of cryptographic lattice problems, such as the LWE and the NTRU problems. (For example, see NIST 2016 for NIST post-quantum candidates.) Such lattice problems are reduced to approximate-SVP or approximate-CVP. (For example, see Albrecht et al. 2018 for details.)

3 Solving SVP Algorithms

In this section, we present typical algorithms for solving SVP. These algorithms can be classified into two categories, depending on whether they solve SVP *exactly* or *approximately*. However, both categories are *complementary*; exact algorithms first apply an approximation algorithm as a preprocessing to reduce their cost, while blockwise algorithms (e.g., the BKZ algorithm presented below) call many times an exact algorithm in low dimension as a subroutine to find a very short lattice vector.

3.1 Exact-SVP Algorithms

Exact-SVP algorithms find the non-zero shortest lattice vector, but they are expensive. These algorithms perform an exhaustive search of all short vectors, whose number is exponential in the dimension (in the worst case). These algorithms can be split in two categories; polynomial-space algorithms and exponential-space algorithms.

3.1.1 Polynomial-Space Exact Algorithms: Enumeration

They are based on *enumeration*, which dates back to the early 1980s with work by Pohst (1981), Kannan (1983), and Fincke–Pohst (1985). Enumeration is simply an exhaustive search for an integer combination of the basis vectors such that the lattice vector is the shortest. An enumeration algorithm takes as input an enumeration radius $R > 0$ and a basis $\mathbf{B} = (\mathbf{b}_1, \ldots, \mathbf{b}_n)$ of a lattice L, and outputs all non-zero vectors $\mathbf{s}$ in L such that $\|\mathbf{s}\| \leq R$ (if exists). The radius R is taken as an upper bound of $\lambda_1(L)$, like $\sqrt{\gamma_n} \mathrm{vol}(L)^{1/n}$, to find the shortest non-zero lattice vector. It goes through the enumeration tree formed by all vectors in the projected lattices $\pi_n(L)$, $\pi_{n-1}(L), \cdots, \pi_1(L) = L$ with norm at most R. More precisely, the enumeration tree is a tree of depth n, and for each $1 \leq k \leq n+1$, the nodes at depth $n + 1 - k$ are all the vectors in the projected lattice $\pi_k(L)$ with norm at most R. In particular, the root of the tree is the zero vector because $\pi_{n+1}(L) = \{\mathbf{0}\}$. The parent of a node $\mathbf{u} \in \pi_k(L)$ at depth $n + 1 - k$ is the node $\pi_{k+1}(\mathbf{u})$ at depth $n - k$. The child nodes are arranged in order of norms.

Here we introduce the basic idea of the Schnorr–Euchner algorithm Schnorr and Euchner (1994), which is a depth first search of the enumeration tree to find all leaves in practice. (cf. Kannan's algorithm 1983 is asymptotically superior in the running time, but it is not competitive in practice due to a substantial overhead of recursive procedures. See also Micciancio and Walter 2014 for such discussion.) We represent the shortest non-zero vector as $\mathbf{s} = v_1\mathbf{b}_1 + \cdots + v_n\mathbf{b}_n \in L$ for some unknown integers v_i's. With Gram–Schmidt information of $\mathbf{B}$, it is rewritten as

$$\mathbf{s} = \sum_{i=1}^{n} v_i \left(\mathbf{b}_i^* + \sum_{j=1}^{i-1} \mu_{i,j}\mathbf{b}_j^* \right) = \sum_{j=1}^{n} \left(v_j + \sum_{i=j+1}^{n} \mu_{i,j}v_i \right) \mathbf{b}_j^*.$$

Due to the orthogonality of Gram–Schmidt vectors $\mathbf{b}_j^*$'s, the squared norms of projections of the vector $\mathbf{s}$ are given as for every $1 \leq k \leq n$

$$\|\pi_k(\mathbf{s})\|^2 = \sum_{j=k}^{n} \left(v_j + \sum_{i=j+1}^{n} \mu_{i,j}v_i \right)^2 \|\mathbf{b}_j^*\|^2.$$

If $\mathbf{s}$ is a leaf of the enumeration tree, then its projections all satisfy $\|\pi_k(\mathbf{s})\|^2 \leq R^2$ for all $1 \leq k \leq n$. These n inequalities together with above equations enable to perform an exhaustive search for the integral coordinates $v_n, v_{n-1}, \ldots, v_1$ of $\mathbf{s}$:

$$\left(v_k + \sum_{i=k+1}^{n} \mu_{i,k}v_i \right)^2 \leq \frac{R^2 - \sum_{j=k+1}^{n} \left(v_j + \sum_{i=j+1}^{n} \mu_{i,j}v_i \right)^2 \|\mathbf{b}_j^*\|^2}{\|\mathbf{b}_k^*\|^2} \tag{2}$$

for every $1 \leq k \leq n$. We start with $k = n$ in Eq. (2), that is, $0 \leq v_n \leq \frac{R}{\|\mathbf{b}_n^*\|}$, because we can restrict to "positive" nodes due to the symmetry of the enumeration tree. Choosing a candidate of v_n, we move to the next index $k = n - 1$ in Eq. (2), that is, $(v_{n-1} + \mu_{n,n-1}v_n)^2 \leq \frac{R^2 - v_n^2\|\mathbf{b}_n^*\|^2}{\|\mathbf{b}_{n-1}^*\|^2}$ to find a candidate of v_{n-1}. By repeating this procedure, assume that the integers $v_n, \ldots, v_{k+1}$ are found for some $1 < k < n$. Then Eq. (2) enables to compute an interval I_k such that $v_k \in I_k$, and thus to perform an exhaustive search for the integer v_k. A depth first search of the tree corresponds to enumerating the interval from its middle, namely, a zig-zag search like

$$v_k = \lceil c_k \rfloor, \ \lceil c_k \rfloor \pm 1, \ \lceil c_k \rfloor \pm 2, \ \cdots,$$

where $c_k = -\sum_{i=k+1}^{n} \mu_{i,k} v_i$. The basic Schnorr–Euchner enumeration algorithm Schnorr and Euchner (1994) is as below (see Gama et al. 2010, Algorithm 2 for the algorithm with some improvements).

Algorithm: The basic Schnorr–Euchner enumeration Schnorr and Euchner (1994)

Input: A basis $\mathbf{B} = (\mathbf{b}_1, \ldots, \mathbf{b}_n)$ of a lattice L and a radius R with $\lambda_1(L) \leq R$
Output: The shortest non-zero vector $\mathbf{s} = \sum_{i=1}^{n} v_i \mathbf{b}_i$ in L
1: Compute Gram–Schmidt information $\mu_{i,j}$ and $\|\mathbf{b}_i^*\|^2$ of $\mathbf{B}$
2: $(\rho_1, \ldots, \rho_{n+1}) = \mathbf{0}$, $(v_1, \ldots, v_n) = (1, 0, \ldots, 0)$, $(c_1, \ldots, c_n) = \mathbf{0}$, $(w_1, \ldots, w_n) = \mathbf{0}$
3: $k = 1$, last_nonzero $= 1$ // largest i for which $v_i \neq 0$
4: **while** true **do**
5: $\quad \rho_k \leftarrow \rho_{k+1} + (v_k - c_k)^2 \cdot \|\mathbf{b}_k^*\|^2$ // $\rho_k = \|\pi_k(\mathbf{s})\|^2$
6: $\quad$ **if** $\rho_k \leq R^2$ **then**
7: $\quad\quad$ **if** $k = 1$ **then** $R^2 \leftarrow \rho_k$, $\mathbf{s} \leftarrow \sum_{i=1}^{n} v_i \mathbf{b}_i$; // update the squared radius
8: $\quad\quad$ **else** $k \leftarrow k - 1$, $c_k \leftarrow -\sum_{i=k+1}^{n} \mu_{i,k} v_i$, $v_k \leftarrow \lceil c_k \rfloor$, $w_k \leftarrow 1$;
9: $\quad$ **else**
10: $\quad\quad$ $k \leftarrow k + 1$ // going up the tree
11: $\quad\quad$ **if** $k = n + 1$ **then return** $\mathbf{s}$;
12: $\quad\quad$ **if** $k \geq$ last_nonzero **then** last_nonzero $\leftarrow k$, $v_k \leftarrow v_k + 1$;
13: $\quad\quad$ **else**
14: $\quad\quad\quad$ **if** $v_k > c_k$ **then** $v_k \leftarrow v_k - w_k$; **else** $v_k \leftarrow v_k + w_k$; // zig-zag search
15: $\quad\quad\quad$ $w_k \leftarrow w_k + 1$
16: $\quad\quad$ **end if**
17: $\quad$ **end if**
18: **end while**

The running time of the enumeration algorithm fully depends on the total number of tree nodes N. An estimate of N can be derived from the Gaussian Heuristic. More precisely, the number of nodes at level ℓ is exactly half the number of vectors in the projected lattice $\pi_{n+1-\ell}(L)$ with norm at most R. Since $\text{vol}(\pi_{n+1-\ell}(L)) =$

$\prod_{i=n+1-\ell}^{n} \|\mathbf{b}_i^*\|$, the Gaussian Heuristic predicts the number of nodes at level ℓ scanned by the Schnorr–Euchner algorithm to be close to

$$H_\ell \approx \frac{1}{2} \cdot \frac{V_\ell(R)}{\prod_{i=n+1-\ell}^{n} \|\mathbf{b}_i^*\|}.$$

Then $N \approx \sum_{\ell=1}^{n} H_\ell$. For a "good" basis (reduced by LLL or BKZ, introduced in the next subsection), we have $\|\mathbf{b}_i^*\|/\|\mathbf{b}_{i+1}^*\| \approx q$ for some constant q. This is called the *geometric series assumption (GSA)*,[1] first introduced by Schnorr (2003). The constant q depends on the reduction algorithm. For example, we experimentally have $q \approx 1.04$ by LLL and $q \approx 1.025$ by BKZ with blocksize 20 for high-dimensional lattices (see Gama and Nguyen 2008 for details.) Now we take the enumeration radius $R = \sqrt{\gamma_n}\mathrm{vol}(L)^{1/n}$, which is optimal in the worst case. With the constant q, we estimate

$$H_\ell \approx \frac{q^{(n-\ell)(n-1)/2} V_\ell(\sqrt{\gamma_n})}{2q^{(n-\ell-1)(n-\ell)/2}} = q^{\ell(n-\ell)/2} 2^{O(n)}.$$

since we can roughly estimate $V_\ell(\sqrt{\gamma_n}) = 2^{O(n)}$ from $\sqrt{\gamma_n} = \Theta\left(\sqrt{n}\right)$ Gama et al. (2010). The right-hand term is maximized for $\ell = \frac{n}{2}$, and it is less than $q^{n^2/8}2^{O(n)}$. Thus the maximum of H_ℓ is super-exponential in n and is reached for $\ell \approx \frac{n}{2}$. (See Gama et al. 2010, Fig. 1 for the actual number of nodes, which is very close to this prediction.) Since smaller q is obtained for a more reduced basis, it shows that the more reduced the input basis is, the less are the nodes in the enumeration tree, and the cheaper enumeration cost.

It is possible to obtain substantial speedups using *pruning* techniques by Gama et al. (2010). Their idea is tempting not to enumerate all the tree nodes, by discarding certain branches. (See Aono et al. 2018 for a lower bound of the time complexity of pruned enumeration.) However, it decreases the success probability to find the shortest non-zero lattice vector $\mathbf{s}$. For instance, one might intuitively hope that $\|\pi_{n/2}(\mathbf{s})\|^2 \lesssim \|\mathbf{s}\|^2/2$, which is more restrictive than the inequality $\|\pi_{n/2}(\mathbf{s})\|^2 \le \|\mathbf{s}\|^2$. Formally, pruning replaces each of the n inequalities $\|\pi_k(\mathbf{s})\|^2 \le R^2$ by $\|\pi_k(\mathbf{s})\|^2 \le R_{n+1-k}^2$, where $R_1 \le \cdots \le R_n = R$ are n real numbers defined by a pruning strategy. A pruning parameter is set in the fplll library The FPLLL development team (2016), and a pruning function for setting R_i's is implemented in the progressive BKZ library Aono et al. (2016).

3.1.2 Exponential-Space Exact Algorithms: Sieve

These algorithms have a better asymptotic running time, but they all require exponential space $2^{\Theta(n)}$. The first algorithm of this kind is the randomized sieve algorithm proposed by Ajtai, Kumar, and Sivakumar (AKS) Ajtai et al. (2001). The AKS

---

[1]This assumption states that for a reduced basis $\mathbf{B} = (\mathbf{b}_1, \ldots, \mathbf{b}_n)$, the plots of its Gram–Schmidt log-norms $\log \|\mathbf{b}_i^*\|$ for $1 \le i \le n$ are on a straight line. (For example, see Schnorr 2003, Fig. 1.)

algorithm outputs the shortest lattice vector with overwhelming probability, and its asymptotic complexity is much better than deterministic enumeration algorithms with $2^{O(n^2)}$ time complexity. The main idea is as follows (see also Nguyen 2008, Sect. 3 or Nguyen 2009): Given a lattice L of dimension n, consider a ball S centered at the origin and of radius r with $\lambda_1(L) \leq r \leq O(\lambda_1(L))$. Then $\#(L \cap S) = 2^{O(n)}$ based on the Gaussian Heuristic. If we could perform an exhaustive search for all vectors in $L \cap S$, we could find the shortest lattice vector within $2^{O(n)}$ polynomial-time operations. Enumeration enables to perform an exhaustive search of $L \cap S$, but it requires to go through all the vectors in the union set $\widetilde{S} = \bigcup_{k=1}^{n} (\pi_k(L) \cap S)$, whose total number is much larger than $\#(L \cap S)$. In contrast, the AKS algorithm performs a randomized sampling of $L \cap S$, without going through the set $\widetilde{S}$. If it was uniformly sampled over $L \cap S$, a short lattice vector would be included in N samples with probability close to 1 for $N \gg \#(L \cap S)$. Unfortunately, it is unclear whether the uniform property is satisfied by the AKS sampling. However, it can be shown that there exists a vector $\mathbf{w} \in L \cap S$ such that $\mathbf{w}$ and $\mathbf{w} + \mathbf{s}$ can be sampled with non-zero probability for some shortest lattice vector $\mathbf{s}$. Thus the shortest lattice vector is obtained by computing the shortest difference of any pairs of the N sampled vectors in $L \cap S$.

There are several heuristic variants of the AKS algorithm with time complexity $2^{O(n)}$ and space complexity exponential in n for an n-dimensional lattice L Bai et al. (2016), Herold and Kirshanova (2017), Micciancio and Voulgaris (2010), Nguyen (2008). Given a basis of L, these algorithms build databases of lattice vectors with norms at most $R \cdot \mathrm{GH}(L)$ for a small constant $R > 0$ such as $R^2 = \frac{4}{3}$. In generic sieves, it is checked whether the sum or the difference of any pair of vectors in databases becomes shorter. The basic sieve algorithm is as below.

Algorithm: The basic sieve

Input: A basis $\mathbf{B} = (\mathbf{b}_1, \ldots, \mathbf{b}_n)$ of a lattice L and a size parameter $N = \left(\frac{4}{3}\right)^{n/2+o(n)}$
Output: A database of N short vectors in L
1: Take a set D of N random vectors in L (with norm at most $2^n \mathrm{vol}(L)^{1/n}$)
2: **while** $\exists (\mathbf{v}, \mathbf{w}) \in D^2$ such that $\|\mathbf{v} + \mathbf{w}\| < \|\mathbf{v}\|$ (resp., $\|\mathbf{v} - \mathbf{w}\| < \|\mathbf{v}\|$) **do**
3:    $\mathbf{v} \leftarrow \mathbf{v} + \mathbf{w}$ (resp., $\mathbf{v} \leftarrow \mathbf{v} - \mathbf{w}$) // update vectors in the database D
4: **end while**
5: **return** D

In Step 1 of the above algorithm, the initialization of the database D can be performed by first computing an LLL-reduced basis (see the next subsection for the LLL reduction), and taking random small integral combinations of the basis vectors. (A natural idea is to use a stronger reduction algorithm such as BKZ in order to generate shorter initial vectors.) The Nguyen–Vidick sieve (2008) finds pairs of vectors $(\mathbf{v}_1, \mathbf{v}_2)$ from D, whose sum or difference gives a shorter vector, that is, $\|\mathbf{v}_1 \pm \mathbf{v}_2\| < \max_{\mathbf{v} \in D} \|\mathbf{v}\|$. Once such a pair is found, the longest vector from the database gets replaced by $\mathbf{v}_1 \pm \mathbf{v}_2$. The database size is a priori fixed to the asymptotic heuristic minimum $2^{0.2075n+O(n)}$ in order to find enough such pairs. The running time is quadratic in the database size. The Gauss sieve (2010) is a variant of the Nguyen–Vidick sieve with substantial improvements; the main improvement is to divide the database into two parts, the so-called "list " part and the "queue" part. Both parts are separately sorted by Euclidean norm in order to make early reduction likely. In updating vectors, the queue part enables to avoid considering the same pair several times. The running time and the database size for the Gauss sieve are asymptotically the same as for the Nguyen–Vidick sieve, but its performance is better in practice. The 3-sieve Bai et al. (2016), Herold and Kirshanova (2017) searches for triples of lattice vectors whose sum gives a shorter vector. (cf. the Nguyen–Vidick and the Gauss algorithms are a kind of 2-sieve.) There are more possible triples than pairs to shorten vectors in the database, but a search for such triples is more costly. (Filtering techniques Herold and Kirshanova 2017 are required to speed up such a search.) Several tricks and techniques have been proposed to improve sieve algorithms, such as the SimHash technique Charikar (2002), Ducas (2018), Fitzpatrick et al. (2014). Several practical sieve algorithms also have been implemented in the fplll library The FPLLL development team (2016).

3.2 Approximate-SVP Algorithms

These algorithms are much faster than exact algorithms, but they output short lattice vectors, not necessarily the shortest ones.

3.2.1 LLL Reduction

The first efficient approximate-SVP algorithm is the celebrated algorithm by Lenstra, Lenstra, and Lovász (LLL) Lenstra et al. (1982). Nowadays it is known as the most famous algorithm of *lattice basis reduction*, which finds a lattice basis with short and nearly orthogonal basis vectors. Such a basis is called *reduced* or *good*. We introduce the notion of LLL reduction. Let $\mathbf{B} = (\mathbf{b}_1, \ldots, \mathbf{b}_n)$ be a basis of a lattice L, and $\mathbf{B}^* = (\mathbf{b}_1^*, \ldots, \mathbf{b}_n^*)$ its Gram–Schmidt vectors with coefficients $\mu_{i,j}$. For a parameter $\frac{1}{4} < \delta < 1$, the basis $\mathbf{B}$ is called δ-*LLL-reduced* if it satisfies two conditions: (i) (Size-reduction condition) $|\mu_{i,j}| \leq \frac{1}{2}$ for all $1 \leq j < i \leq n$. (ii) (Lovász' condition) $\delta \|\mathbf{b}_{k-1}^*\|^2 \leq \|\pi_{k-1}(\mathbf{b}_k)\|^2$ for all $2 \leq k \leq n$. This can be rewritten as $\|\mathbf{b}_k^*\|^2 \geq (\delta - \mu_{k,k-1}^2)\|\mathbf{b}_{k-1}^*\|^2$. Any δ-LLL-reduced basis satisfies the below properties (see Bremner 2011 for proof):

- $\|\mathbf{b}_1\| \leq \alpha^{(n-1)/4}\mathrm{vol}(L)^{1/n}$, where $\alpha = \frac{4}{4\delta-1} > \frac{4}{3}$.
- $\|\mathbf{b}_i\| \leq \alpha^{(n-1)/2}\lambda_i(L)$ for $1 \leq i \leq n$, and $\prod_{i=1}^{n} \|\mathbf{b}_i\| \leq \alpha^{n(n-1)/4}\mathrm{vol}(L)$.

Given any basis of L, the LLL algorithm finds a δ-LLL-reduced basis of L. As seen from the above second property, it can solve approximate-SVP with factor $\alpha^{(n-1)/2}$. The basic LLL algorithm is given below (see also Galbraith 2012, Chap. 17 or Nguyen 2009).

Algorithm: The basic LLL Lenstra et al. (1982)

Input: A basis $\mathbf{B} = (\mathbf{b}_1, \ldots, \mathbf{b}_n)$ of a lattice L, and a reduction parameter $\frac{1}{4} < \delta < 1$
Output: A δ-LLL-reduced basis $\mathbf{B}$ of L
1: Compute Gram–Schmidt information $\mu_{i,j}$ and $\|\mathbf{b}_i^*\|^2$ of the input basis $\mathbf{B}$
2: $k \leftarrow 2$
3: **while** $k \leq n$ **do**
4:   Size-reduce $\mathbf{B} = (\mathbf{b}_1, \ldots, \mathbf{b}_n)$ // At each k, we recursively change $\mathbf{b}_k \leftarrow \mathbf{b}_k - \lfloor \mu_{k,j} \rceil \mathbf{b}_j$ for $1 \leq j \leq k-1$ (e.g., see Galbraith 2012, Algorithm 24)
5:   **if** $(\mathbf{b}_{k-1}, \mathbf{b}_k)$ satisfies Lovász' condition **then**
6:     $k \leftarrow k + 1$
7:   **else**
8:     Swap $\mathbf{b}_k$ with $\mathbf{b}_{k-1}$, and update Gram–Schmidt information of $\mathbf{B}$
9:     $k \leftarrow \max(k-1, 2)$
10:   **end if**
11: **end while**

In the LLL algorithm, a pair of adjacent basis vectors $(\mathbf{b}_{k-1}, \mathbf{b}_k)$ is swapped if it does not satisfy Lovász' condition. Thus the output basis is δ-LLL-reduced if the algorithm terminates. The quantity $\text{Pot}(\mathbf{B}) = \prod_{i=1}^{n-1} \|\mathbf{b}_i^*\|^{2(n-i)}$ is called the *potential* of a basis $\mathbf{B}$. Every swap in the LLL algorithm decreases the potential of an input basis by a factor at least $\delta < 1$. (cf. the size-reduction procedure does not change the potential.) This guarantees the termination of the LLL algorithm in polynomial time in n. Furthermore, the LLL algorithm is applicable also for linearly dependent vectors to remove their linear dependency. (See Bremner 2011, Chap. 6, Cohen 2013, Sect. 2.6.4, Pohst 1987 or Sims 1994, Sect. 8.7 for details.)

3.2.2 Variants of LLL

LLL with Deep Insertions (DeepLLL)

This variant is a straightforward generalization of LLL, in which *non-adjacent* basis vectors can be changed. Specifically, a basis vector $\mathbf{b}_k$ is inserted between $\mathbf{b}_{i-1}$ and $\mathbf{b}_i$ as $\sigma_{i,k}(\mathbf{B}) = (\ldots, \mathbf{b}_{i-1}, \mathbf{b}_k, \mathbf{b}_i, \ldots, \mathbf{b}_{k-1}, \mathbf{b}_{k+1}, \ldots)$, called a *deep insertion*, if the so-called deep exchange condition $\|\pi_i(\mathbf{b}_k)\|^2 < \delta\|\mathbf{b}_i^*\|^2$ is satisfied for $\frac{1}{4} < \delta < 1$. In this case, the new GSO vector at the ith position is given by $\pi_i(\mathbf{b}_k)$, strictly shorter than the old GSO vector $\mathbf{b}_i^*$. A basis $\mathbf{B} = (\mathbf{b}_1, \ldots, \mathbf{b}_n)$ is called δ-*DeepLLL-reduced* if it satisfies two conditions: (i) it is size-reduced, (ii) $\|\pi_i(\mathbf{b}_k)\|^2 \geq \delta\|\mathbf{b}_i^*\|^2$ for all $1 \leq i < k \leq n$. (The case $i = k-1$ is just Lovász' condition.) Any δ-DeepLLL-reduced basis satisfies the below properties Yasuda and Yamaguchi (2019), Theorem 1:

- $\|\mathbf{b}_1\| \leq \alpha^{\frac{n-1}{2n}} \left(1 + \frac{\alpha}{4}\right)^{\frac{(n-1)(n-2)}{4n}} \text{vol}(L)^{\frac{1}{n}}$, where α is the same as in LLL.
- $\|\mathbf{b}_i\| \leq \sqrt{\alpha} \left(1 + \frac{\alpha}{4}\right)^{\frac{n-2}{2}} \lambda_i(L)$ for $1 \leq i \leq n$, and $\prod_{i=1}^{n} \|\mathbf{b}_i\| \leq \left(1 + \frac{\alpha}{4}\right)^{\frac{n(n-1)}{4}} \text{vol}(L)$.

These properties are strictly stronger than the case of LLL. The basic DeepLLL algorithm Schnorr and Euchner (1994) is given below (see also Bremner 2011, Fig. 5.1 or Cohen 2013, Algorithm 2.6.4).

Algorithm: The basic DeepLLL Schnorr and Euchner (1994)

Input: A basis $\mathbf{B} = (\mathbf{b}_1, \ldots, \mathbf{b}_n)$ of a lattice L, and a reduction parameter $\frac{1}{4} < \delta < 1$
Output: A δ-DeepLLL-reduced basis $\mathbf{B}$ of L
  1: Compute Gram–Schmidt information $\mu_{i,j}$ and $\|\mathbf{b}_i^*\|^2$ of the input basis $\mathbf{B}$
  2: $k \leftarrow 2$
  3: **while** $k \leq n$ **do**
  4:     Size-reduce $\mathbf{B}$ as in LLL
  5:     $C \leftarrow \|\mathbf{b}_k\|^2, i \leftarrow 1$
  6:     **while** $i < k$ **do**
  7:       **if** $C \geq \delta \|\mathbf{b}_i^*\|^2$ **then**
  8:         Compute $C \leftarrow C - \mu_{k,i}^2 \|\mathbf{b}_i^*\|^2$ and $i \leftarrow i + 1$ // $C = \|\pi_i(\mathbf{b}_k)\|^2$
  9:       **else**
 10:         $\mathbf{B} \leftarrow \sigma_{i,k}(\mathbf{B})$ // a deep insertion
 11:         Update the Gram–Schmidt information of $\mathbf{B}$, and $k \leftarrow \max(i, 2) - 1$
 12:       **end if**
 13:     **end while**
 14:     $k \leftarrow k + 1$
 15: **end while**

Compared with LLL, it is complicated to update the Gram–Schmidt information of $\mathbf{B}$ after every deep insertion. (See Yamaguchi and Yasuda 2017.) Every deep insertion does not always decrease the potential of an input basis, and thus the complexity of DeepLLL is no longer polynomial-time but potentially super-exponential in the lattice dimension. However, DeepLLL often finds much shorter lattice vectors than LLL in practice Gama and Nguyen (2008).

Block Korkine–Zolotarev (BKZ) Algorithm

Let us first introduce a strong notion of reduction: A basis $\mathbf{B} = (\mathbf{b}_1, \ldots, \mathbf{b}_n)$ of a lattice L is called *HKZ-reduced* if it is size-reduced and it satisfies $\|\mathbf{b}_i^*\| = \lambda_1(\pi_i(L))$ for all $1 \leq i \leq n$. For $1 \leq i \leq j \leq n$, denote by $\mathbf{B}_{[i,j]}$ the local projected block $(\pi_i(\mathbf{b}_i), \pi_i(\mathbf{b}_{i+1}), \ldots, \pi_i(\mathbf{b}_j))$, and by $L_{[i,j]}$ the lattice spanned by $\mathbf{B}_{[i,j]}$. The notion of BKZ-reduction is a local block version of HKZ-reduction Schnorr (1987), Schnorr (1992), Schnorr and Euchner (1994). For a blocksize $2 \leq \beta \leq n$, a basis $\mathbf{B} = (\mathbf{b}_1, \ldots, \mathbf{b}_n)$ of a lattice L is called *β-BKZ-reduced* if it is size-reduced and every local block $\mathbf{B}_{[j,j+\beta-1]}$ is HKZ-reduced for $1 \leq j \leq n - \beta + 1$. The second condition means $\|\mathbf{b}_j^*\| = \lambda_1(L_{[j,k]})$ for $1 \leq j \leq n - 1$ with $k = \min(j + \beta - 1, n)$. Every β-BKZ-reduced basis satisfies $\|\mathbf{b}_1\| \leq \gamma_\beta^{(n-1)/(\beta-1)} \lambda_1(L)$ Schnorr (1992). The

BKZ algorithm Schnorr and Euchner (1994) finds a β-BKZ-reduced basis, and it calls LLL to reduce every local block before finding the shortest vector over the block lattice. (As β increases, a shorter lattice vector can be found, but the running time is more costly.)

Algorithm: The basic BKZ Schnorr and Euchner (1994)

Input: A basis $\mathbf{B} = (\mathbf{b}_1, \ldots, \mathbf{b}_n)$ of a lattice L, a blocksize $2 \leq \beta \leq n$, and a reduction parameter $\frac{1}{4} < \delta < 1$ of LLL

Output: A β-DeepBKZ-reduced basis $\mathbf{B}$ of L

1: $\mathbf{B} \leftarrow \text{LLL}(\mathbf{B}, \delta)$ // Compute $\mu_{i,j}$ and $\|\mathbf{b}_j^*\|^2$ of the new basis $\mathbf{B}$ together

2: $z \leftarrow 0, j \leftarrow 0$

3: **while** $z < n - 1$ **do**

4: $j \leftarrow (j \bmod (n-1)) + 1, k \leftarrow \min(j + \beta - 1, n), h \leftarrow \min(k+1, n)$

5: Find $\mathbf{v} \in L$ such that $\|\pi_j(\mathbf{v})\| = \lambda_1(L_{[j,k]})$ by enumeration or sieve

6: **if** $\|\pi_j(\mathbf{v})\|^2 < \|\mathbf{b}_j^*\|^2$ **then**

7:   $z \leftarrow 0$ and call $\text{LLL}((\mathbf{b}_1, \ldots, \mathbf{b}_{j-1}, \mathbf{v}, \mathbf{b}_j, \ldots, \mathbf{b}_h), \delta)$ // Insert $\mathbf{v} \in L$ and remove the linear dependency to obtain a new basis

8: **else**

9:   $z \leftarrow z + 1$ and call $\text{LLL}((\mathbf{b}_1, \ldots, \mathbf{b}_h), \delta)$

10: **end if**

11: **end while**

It is customary to terminate the BKZ algorithm after a selected number of calls to an exact-SVP algorithm over block lattices. (See Hanrot et al. 2011 for analysis.) Efficient variants such as BKZ 2.0 Chen (2011) have been proposed, and some of them have been implemented in The FPLLL development team (2016). The *Hermite factor* is a good index to measure the practical output quality of a reduction algorithm. (See Gama and Nguyen 2008 for their experiments.) It is defined by $\gamma = \frac{\|\mathbf{v}\|}{\text{vol}(L)^{1/n}}$, where $\mathbf{v}$ is the shortest basis vector output by a reduction algorithm for a basis of a lattice L of dimension n. Under the Gaussian Heuristic and GSA, a limiting value of the root Hermite factor of BKZ with blocksize β is predicted in Chen (2013) as

$$\lim_{n \to \infty} \gamma^{\frac{1}{n}} = \left(v_\beta^{-\frac{1}{\beta}} \right)^{\frac{1}{\beta-1}} \sim \left(\frac{\beta}{2\pi e} (\pi \beta)^{\frac{1}{\beta}} \right)^{\frac{1}{2(\beta-1)}}.$$

There are experimental evidences to support this prediction for high blocksizes such as $\beta > 50$. (Note that the Gaussian Heuristic holds in practice for random lattices in high dimensions, but unfortunately it is violated in low dimensions.) In a simple form based on the Gaussian Heuristic, the GSA shape of a β-BKZ-reduced basis of volume 1 is predicted as $\|\mathbf{b}_i^*\| \approx \alpha_\beta^{\frac{n-1}{2} - i}$, where $\alpha_\beta = \left(\frac{\beta}{2\pi e} \right)^{1/\beta}$. This is reasonably accurate in practice for $\beta > 50$ and $\beta \ll n$. (See Chen 2013, 2011; Yu and Ducas 2017.) Other variants of BKZ have been proposed such as slide reduction Gama and Nguyen (2008), self-dual BKZ Micciancio and Walter (2016), and progressive-

BKZ Aono et al. (2016). As a mathematical improvement of BKZ, DeepBKZ was recently proposed in Yamaguchi and Yasuda (2017), in which DeepLLL is called a subroutine alternative to LLL. In particular, DeepBKZ finds a short lattice vector by smaller blocksizes than BKZ in practice. (Dual and self-dual variants of DeepBKZ were also proposed in Yasuda (2018), Yasuda et al. (2018).)

4 The SVP Challenge and Recent Strategies

To test algorithms solving SVP, sample lattice bases are presented in Darmstadt (2010) for dimensions from 40 up to 200. (The lattices are random in the sense of Goldstein and Mayer Goldstein and Mayer (2003).) For every lattice L, any non-zero lattice vector with (Euclidean) norm less than $1.05\mathrm{GH}(L)$ can be submitted to the hall of fame in the SVP challenge. To enter the hall of fame, the lattice vector is required to be shorter than a previous one in the same dimension (with possibly different seed). Note that not all lattice vectors in the hall of fame are necessarily the shortest. In this section, we introduce two recent strategies for solving the SVP challenge in high dimensions such as $n \geq 150$.

4.1 The Random Sampling Strategy

Early in 2017, a non-zero vector in a lattice L of dimension $n = 150$ with norm less than $1.05\mathrm{GH}(L)$ was first found by Teruya and Kashiwabara using many high-performance servers. (See Teruya et al. 2018 for their large-scale experiments.) Their strategy is based on the work of Fukase and Kashiwabara (2015), which is an extension of Schnorr's random sampling reduction (RSR) Schnorr (2003). Here we review random sampling (SA) and RSR. For a lattice L of dimension n, fix $1 \leq u < n$ to be a constant of search space bound. Given a basis $\mathbf{B} = (\mathbf{b}_1, \ldots, \mathbf{b}_n)$ of L, SA samples a vector $\mathbf{v} = \sum_{i=1}^{n} v_i \mathbf{b}_i^*$ in L satisfying $v_i \in (-1/2, 1/2]$ for $1 \leq i < n - u$, $v_i \in (-1, 1]$ for $n - u \leq i < n$ and $v_n = 1$. Let $S_{u,\mathbf{B}}$ denote the set of such lattice vectors. Since the number of candidates for v_i with $|v_i| \leq 1/2$ (resp. $|v_i| \leq 1$) is 1 (resp. 2), there are 2^u lattice vectors in $S_{u,\mathbf{B}}$. By calling SA up to 2^u times, RSR generates $\mathbf{v}$ satisfying $\|\mathbf{v}\|^2 < 0.99\|\mathbf{b}_1\|^2$ Schnorr (2003), Theorem 1. Two extensions are proposed in Fukase and Kashiwabara (2015) for solving the SVP challenge; the first one is to represent a lattice vector by a sequence of natural numbers via the Gram–Schmidt orthogonalization, and to sample lattice vectors on an appropriate distribution of the representation. The second one is to decrease the sum of the squared Gram–Schmidt lengths $\mathrm{SS}(\mathbf{B}) := \sum_{i=1}^{n} \|\mathbf{b}_i^*\|^2$ to make it easier to sample very short lattice vectors. The effectiveness of their extensions is guaranteed by their statistical analysis on lattices. Specifically, under the randomness assumption (RA),[2] they roughly estimate that the distribution of the squared length of a sampled vector $\|\mathbf{v}\|^2 = \sum_{i=1}^{n} v_i^2 \|\mathbf{b}_i^*\|^2$ follows the normal distribution $\mathcal{N}(\mu, \sigma^2)$ with

$$\mu = \frac{\sum_{i=1}^{n} \|\mathbf{b}_i^*\|^2}{12} \quad \text{and} \quad \sigma = \left(\frac{\sum_{i=1}^{n} \|\mathbf{b}_i^*\|^4}{180}\right)^{1/2}.$$

This implies that *shorter* lattice vectors are sampled as the squared-sum SS($\mathbf{B}$) becomes *smaller*. Then the basic strategy in Fukase and Kashiwabara (2015); Teruya et al. (2018) consists of the following two steps: (i) We reduce an input basis so that it decreases the sum of its squared Gram–Schmidt lengths as small as possible, by using LLL and insertion of sampled lattice vectors like BKZ. (See also Yasuda et al. 2017 for such procedure). (ii) With such reduced basis $\mathbf{B}$, we then find a short lattice vector by randomly sampling $\mathbf{v} = \sum_{i=1}^{n} v_i \mathbf{b}_i^*$.

As a sequential work, Aono and Nguyen (2017) introduced lattice enumeration with discrete pruning to generalize random sampling, and also provided a deep analysis of discrete pruning by using the volume of the intersection of a ball with a box. In particular, under RA, the expectation of the length of a short vector generated by lattice enumeration with discrete pruning from the so-called tag $\mathbf{t} = (t_1, \ldots, t_n) \in \mathbb{Z}^n$ is roughly given by $E(\mathbf{t}) = \sum_{i=1}^{n} \left(\frac{t_i^2}{4} + \frac{t_i}{4} + \frac{1}{12}\right) \|\mathbf{b}_i^*\|^2$, which is a generalization of the above mean μ. However, it is shown in Aono and Nguyen (2017) that the empirical correlation between $E(\mathbf{t})$ and the volume of ball-box intersection is negative. This is statistical evidence why decreasing SS($\mathbf{B}$) is important instead of increasing the volume of ball-box intersection. Furthermore, the calculation of the volume presented in Aono and Nguyen (2017) is much less efficient than the computation of SS($\mathbf{B}$). In 2018, Matsuda et al. (2018) investigated the strategy of Fukase and Kashiwabara (2015) by the Gram–Charlier approximation in order to precisely estimate the success probability of sampling short lattice vectors, and also discussed the effectiveness of decreasing SS($\mathbf{B}$) for sampling short lattice vectors.

4.2 The Sub-Sieving Strategy

Around the end of August 2018, many records for the SVP challenge in dimensions up to 155 had been found by the sub-sieving strategy of Ducas (2018). (See Albrecht et al. 2019 for their experiments report.) The basic idea is to reduce SVP in high dimensions to the *bounded distance decoding (BDD)* problem in low dimensions, a particular case of CVP, in which the target vector is known to be somewhat close to the lattice. It enforces us to find an enormous number of short vectors in projected

---

[2]RA states that the coefficients v_i of $\mathbf{v} = \sum_{i=1}^{n} v_i \mathbf{b}_i^*$ sampled by SA are uniformly distributed in $[-1/2, 1/2]$ for $1 \leq i < n - u$ and in $[-1, 1]$ for $n - u \leq i < n$. It does not hold strictly in practice.

lattices, and the sieve is useful to collect such vectors. In particular, the sieve is performed in projected lattices instead of the full lattice.

The specific strategy is as follows Ducas (2018), Section 3. Given a basis $\mathbf{B} = (\mathbf{b}_1, \ldots, \mathbf{b}_n)$ of a lattice L of high dimension n, we fix an integer d with $1 \leq d \leq n$, and perform the sieve in the projected lattice $\pi_d(L)$ to obtain a list of short lattice vectors

$$D := \left\{ \mathbf{v} \in \pi_d(L) \mid \mathbf{v} \neq \mathbf{0} \text{ and } \|\mathbf{v}\| \leq \sqrt{\frac{4}{3}} \mathrm{GH}\left(\pi_d(L) \right) \right\}.$$

We hope that the desired shortest non-zero vector $\mathbf{s}$ in the full lattice L projects to a vector in the above list D, that is, it satisfies $\pi_d(\mathbf{s}) \neq \mathbf{0}$ and $\|\pi_d(\mathbf{s})\| \leq \sqrt{\frac{4}{3}}\mathrm{GH}(\pi_d(L))$. (Note that $\pi_d(\mathbf{s}) = \mathbf{0}$ means that the vector $\mathbf{s}$ is in the sub-lattice $\mathcal{L}(\mathbf{b}_1, \ldots, \mathbf{b}_{d-1})$ of L. Here we do not care about the case.) Since $\|\pi_d(\mathbf{s})\| \leq \|\mathbf{s}\| \approx \mathrm{GH}(L)$, the condition

$$\mathrm{GH}(L) \leq \sqrt{\frac{4}{3}}\mathrm{GH}\left(\pi_d(L) \right) \tag{3}$$

is sufficient to satisfy our hope. This condition is not tight, since the projected vector $\pi_d(\mathbf{s})$ becomes shorter than the full vector $\mathbf{s}$ as the index d increases. By exhaustive search over the list D, we assume that the projected vector $\mathbf{s}_d := \pi_d(\mathbf{s}) \in D$ is known. We need to recover the full vector $\mathbf{s}$ from $\mathbf{s}_d$. Write $\mathbf{s} = \mathbf{Bx}$ for some $\mathbf{x} \in \mathbb{Z}^n$, and split $\mathbf{x}$ as $(\mathbf{x}_1 \mid \mathbf{x}_2)$ with $\mathbf{x}_1 \in \mathbb{Z}^{d-1}$ and $\mathbf{x}_2 \in \mathbb{Z}^{n-d+1}$. Then $\mathbf{s}_d = \pi_d(\mathbf{Bx}) = \mathbf{B}_d\mathbf{x}_2$ and hence $\mathbf{x}_2$ is known, where $\mathbf{B}_d = (\pi_d(\mathbf{b}_d), \ldots, \pi_d(\mathbf{b}_n))$. Now we need to recover $\mathbf{x}_1$ so that $\mathbf{s} = \mathbf{B}_1\mathbf{x}_1 + \mathbf{B}_2\mathbf{x}_2$ is small (or the shortest), where $\mathbf{B} = (\mathbf{B}_1 \mid \mathbf{B}_2)$. This is an easy BDD instance over the d-dimensional lattice spanned by $\mathbf{B}_1$ for the target vector $\mathbf{B}_2\mathbf{x}_2$. A sufficient condition to solve this problem using Babai's nearest plane algorithm Babai (1986) is that $|\langle \mathbf{b}_i^*, \mathbf{s}\rangle| \leq \frac{1}{2}\|\mathbf{b}_i^*\|^2$ for all $1 \leq i < d$. (See also Galbraith 2012, Chap. 18 for Babai's algorithms.) Since $|\langle \mathbf{b}_i^*, \mathbf{s}\rangle| \leq \|\mathbf{b}_i^*\|\|\mathbf{s}\|$, a further sufficient condition is that $\mathrm{GH}(L) \leq \frac{1}{2}\min_{i<d}\|\mathbf{b}_i^*\|$. This condition is far from tight, and it should not be a serious issue in practice. Indeed, even for a strongly reduced basis, the d first Gram–Schmidt lengths won't be much smaller than $\mathrm{GH}(L)$, say by more than a factor 2. (The BKZ-preprocessing with blocksize $\beta = \frac{n}{2}$ is assumed in Ducas (2018).) A concrete maximal value of d satisfying the condition (3) depends on the shape of a basis $\mathbf{B}$. It is estimated in Ducas (2018) that $d = \Theta(n/\log n)$ is suitable over a quasi-HKZ-reduced basis.

In 2019, Albrecht et al. (2019) proposed the General Sieve Kernel (G6K), an abstract stateful machine supporting a variety of advanced lattice reductions based on sieving algorithms. They have provided a highly optimized, multi-threaded, and tweakable implementation of G6K as an open-source C++ and Python library. A number of records in the hall of fame for the SVP challenge were found by the sub-sieving strategy on G6K. (In June 2019, the highest dimension to be solved in the SVP challenge is 157, using G6K.) Specifically, their experiments imply that in average $d = 11.46 + 0.0757n$ is a suitable free dimension of the sub-sieving strategy for the SVP challenge in high dimensions n. Furthermore, their solution for the SVP challenge in dimension 151 was found 400 times faster than the times reported for the SVP challenge in dimension 150, which was solved early in 2017 by the random sampling strategy.

Acknowledgements This work was supported by JST CREST Grant Number JPMJCR14D6, Japan. A part of this work was also supported by JSPS KAKENHI Grant Number JP16H02830.

References

M. Ajtai, Generating hard instances of lattice problems, in *Symposium on Theory of Computing (STOC 1996)* (ACM, 1996), pp. 99–108

M. Ajtai, R. Kumar, D. Sivakumar, A sieve algorithm for the shortest lattice vector problem, in *Symposium on Theory of Computing (STOC 2001)* (ACM, 2001), pp. 601–610

M. Albrecht, L. Ducas, G. Herold, E. Kirshanova, E.W. Postlethwaite, M. Stevens, The general sieve kernel and new records in lattice reduction. *Advances in Cryptology–EUROCRYPT 2019*, Lecture Notes in Computer Science, vol. 11477 (Springer, Berlin, 2019), pp. 717–746

M.R. Albrecht, B.R. Curtis, A. Deo, A. Davidson, R. Player, E.W. Postlethwaite, F. Virdia, T. Wunderer, Estimate all the LWE, NTRU schemes! *Security and Cryptography for Networks (SCN 2018)*, Lecture Notes in Computer Science, vol. 11035 (2018), pp. 351–367

Y. Aono, P.Q. Nguyen, Random sampling revisited: Lattice enumeration with discrete pruning. *Advances in Cryptology–EUROCRYPT 2017*, Lecture Notes in Computer Science, vol. 10211 (Springer, Berlin, 2017), pp. 65–102

Y. Aono, P.Q. Nguyen, T. Seito, J. Shikata, Lower bounds on lattice enumeration with extreme pruning. *Advances in Cryptology–CRYPTO 2018*, Lecture Notes in Computer Science, vol. 10992 (Springer, Berlin, 2018), pp. 608–637

Y. Aono, Y. Wang, T. Hayashi, T. Takagi, Improved progressive BKZ algorithms and their precise cost estimation by sharp simulator. *Advances in Cryptology–EUROCRYPT 2016*, Lecture Notes in Computer Science, vol. 9665 (Springer, Berlin, 2016), pp. 789–819. Progressive BKZ library is available from https://www2.nict.go.jp/security/pbkzcode/

L. Babai, On Lovász' lattice reduction and the nearest lattice point problem. Combinatorica **6**(1), 1–13 (1986)

S. Bai, T. Laarhoven, D. Stehlé, Tuple lattice sieving. LMS J. Comput. Math. **19**(A), 146–162 (2016)

M.R. Bremner, *Lattice Basis Reduction: An Introduction to the LLL Algorithm and Its Applications* (CRC Press, Bocca Raton, 2011)

M.S. Charikar, Similarity estimation techniques from rounding algorithms, in *Symposium on Theory of Computing (STOC 2002)* (ACM, 2002), pp. 380–388

Y. Chen, Réduction de réseau et sécurité concrete du chiffrement completement homomorphe. Ph.D. thesis, Paris 7 (2013)

Y. Chen, P.Q. Nguyen, BKZ 2.0: Better lattice security estimates. *Advances in Cryptology–ASIACRYPT 2011*, Lecture Notes in Computer Science, vol. 7073 (Springer, Berlin, 2011), pp. 1–20

H. Cohen, *A Course in Computational Algebraic Number Theory*, vol. 138, Graduate Texts in Mathematics (Springer Science & Business Media, Berlin, 2013)

T. Darmstadt, SVP challenge. (2010) https://www.latticechallenge.org/svp-challenge/

L. Ducas, Shortest vector from lattice sieving: a few dimensions for free. *Adavances in Cryptology–EUROCRYPT 2018*, Lecture Notes in Computer Science, , vol. 10820 (Springer, Berlin, 2018), pp. 125–145

U. Fincke, M. Pohst, Improved methods for calculating vectors of short length in a lattice, including a complexity analysis. Math. Comput. **44**(170), 463–471 (1985)

R. Fitzpatrick, C. Bischof, J. Buchmann, Ö. Dagdelen, F. Göpfert, A. Mariano, B.Y. Yang, Tuning Gauss Sieve for speed. *Progress in Cryptology–LATINCRYPT 2014*, Lecture Notes in Computer Science, vol. 8895 (Springer, 2014), pp. 288–305

M. Fukase, K. Kashiwabara, An accelerated algorithm for solving SVP based on statistical analysis. J. Inf. Process. (JIP) **23**(1), 67–80 (2015)

S.D. Galbraith, *Mathematics of Public Key Cryptography* (Cambridge University Press, Cambridge, 2012)

N. Gama, P.Q. Nguyen, Finding short lattice vectors within Mordell's inequality, in *Symposium on Theory of Computing (STOC 2008)* (ACM, 2008), pp. 207–216

N. Gama, P.Q. Nguyen, Predicting lattice reduction, *Advances in Cryptology–EUROCRYPT 2008*, Lecture Notes in Computer Science, vol. 4965 (Springer, Berlin, 2008), pp. 31–51

N. Gama, P.Q. Nguyen, O. Regev, Lattice enumeration using extreme pruning, *Advances in Cryptology–EUROCRYPT 2010*, Lecture Notes in Computer Science, vol. 6110 (Springer, Berlin, 2010), pp. 257–278

D. Goldstein, A. Mayer, *Forum Mathematicum*, vol. 15, On the equidistribution of Hecke points (De Gruyter, Berlin, 2003), pp. 165–190

G. Hanrot, X. Pujol, D. Stehlé, Analyzing blockwise lattice algorithms using dynamical systems, *Advances in Cryptology–CRYPTO 2011*, Lecture Notes in Computer Science, vol. 6841 (Springer, Berlin, 2011), pp. 447–464

C. Hermite, Extraits de lettres de M. Hermite à M. Jacobi sur différents objets de la théorie des nombres: Deuxième lettre. Journal für die Reine und Angewandte Mathematik (1850), pp. 279–315

G. Herold, E. Kirshanova, Improved algorithms for the approximate k-list problem in Euclidean norm, *Public Key Cryptography (PKC 2017)*, Lecture Notes in Computer Science, vol. 10174 (Springer, Berlin, 2017), pp. 16–40

R. Kannan, Improved algorithms for integer programming and related lattice problems, in *Symposium on Theory of Computing (STOC 1983)* (ACM, 1983), pp. 193–206

R. Kannan, Minkowski's convex body theorem and integer programming. Math. Oper. Res. **12**(3), 415–440 (1987)

A.K. Lenstra, H.W. Lenstra, L. Lovász, Factoring polynomials with rational coefficients. Mathematische Annalen **261**(4), 515–534 (1982)

J. Martinet, *Comprehensive Studies in Mathematics*, vol. 327, Perfect lattices in Euclidean spaces (Springer Science & Business Media, Berlin, 2013)

Y. Matsuda, T. Teruya, K. Kashiwabara, Estimation of the success probability of random sampling by the Gram-Charlier approximation. IACR ePrint 2018/815 (2018)

D. Micciancio, P. Voulgaris, Faster exponential time algorithms for the shortest vector problem, in *Symposium on Discrete Algorithms (SODA 2010)* (ACM-SIAM, 2010), pp. 1468–1480

D. Micciancio, M. Walter, Fast lattice point enumeration with minimal overhead, in *Symposium on Discrete algorithms (SODA 2014)* (ACM-SIAM, 2014), pp. 276–294

D. Micciancio, M. Walter, Practical, predictable lattice basis reduction, *Advances in Cryptology–EUROCRYPT 2016*, Lecture Notes in Computer Science, vol. 9665 (Springer, Berlin, 2016), pp. 820–849

P.Q. Nguyen, *The LLL Algorithm*, Hermite's constant and lattice algorithms (Springer, Berlin, 2009), pp. 19–69

P.Q. Nguyen, T. Vidick, Sieve algorithms for the shortest vector problem are practical. J. Math. Cryptol. **2**(2), 181–207 (2008)

M. Pohst, On the computation of lattice vectors of minimal length, successive minima and reduced bases with applications. ACM Sigsam Bull. **15**(1), 37–44 (1981)

M. Pohst, A modification of the LLL reduction algorithm. J. Symb. Comput. **4**(1), 123–127 (1987)

C.P. Schnorr, A hierarchy of polynomial time lattice basis reduction algorithms. Theor. Comput. Sci. **53**(2–3), 201–224 (1987)

C.P. Schnorr, Block Korkin-Zolotarev bases and successive minima. International Computer Science Institute (1992)

C.P. Schnorr, Lattice reduction by random sampling and birthday methods, *Symposium on Theoretical Aspects of Computer Science (STACS 2003)*, Lecture Notes in Computer Science, vol. 2607 (Springer, Berlin, 2003), pp. 145–156

C.P. Schnorr, M. Euchner, Lattice basis reduction: Improved practical algorithms and solving subset sum problems. Math. Program. **66**, 181–199 (1994)

Shor, P.W.: Algorithms for quantum computation: discrete logarithms and factoring, in *Symposium on Foundations of Computer Science (FOCS 1994)* (IEEE, 1994), pp. 124–134

C.C. Sims, *Computation with Finitely Presented Groups*, vol. 48 (Cambridge University Press, Cambridge, 1994)

T. Teruya, K. Kashiwabara, G. Hanaoka, Fast lattice basis reduction suitable for massive parallelization and its application to the shortest vector problem, *Public Key Cryptography (PKC 2018)*, Lecture Notes in Computer Science, vol. 10769 (Springer, Berlin, 018), pp. 437–460

The FPLLL development team: fplll, a lattice reduction library (2016), https://github.com/fplll/fplll

The National Institute of Standards and Technology (NIST): Post-quantum cryptography. (2016) https://csrc.nist.gov/projects/post-quantum-cryptography/post-quantum-cryptography-standardization

J. Yamaguchi, M. Yasuda, Explicit formula for Gram-Schmidt vectors in LLL with deep insertions and its applications, *Number-Theoretic Methods in Cryptology (NuTMiC 2017)*, Lecture Notes in Computer Science, vol. 10737 (Springer, Berlin, 2017), pp. 142–160

M. Yasuda, Self-dual DeepBKZ for finding short lattice vectors. J. Math. Cryptol. **14**(1), 84–94 (2020)

M. Yasuda, J. Yamaguchi, A new polynomial-time variant of LLL with deep insertions for decreasing the squared-sum of Gram-Schmidt lengths. Des. Codes Cryptogr. **87**, 2489–2505 (2019)

M. Yasuda, J. Yamaguchi, M. Ooka, S. Nakamura, Development of a dual version of DeepBKZ and its application to solving the LWE challenge, *Progress in Cryptology–AFRICACRYPT 2018*, vol. 10831, Lecture Notes in Computer Science (Springer, Berlin, 2018), pp. 162–182

M. Yasuda, K. Yokoyama, T. Shimoyama, J. Kogure, T. Koshiba, Analysis of decreasing squared-sum of Gram-Schmidt lengths for short lattice vectors. J. Math. Cryptol. **11**(1), 1–24 (2017)

Y. Yu, L. Ducas, Second order statistical behavior of LLL and BKZ, *Selected Areas in Cryptography (SAC 2017)*, Lecture Notes in Computer Science, vol. 10719 (Springer, Berlin, 2017), pp. 3–22

Recent Developments in Multivariate Public Key Cryptosystems

Yasufumi Hashimoto

Abstract The multivariate signature schemes UOV, Rainbow, and HFEv- have been considered to be secure and efficient enough under suitable parameter selections. In fact, several second round candidates of NIST's standardization project of Post-Quantum Cryptography are based on these schemes. On the other hand, there are few multivariate encryption schemes expected to be practical and despite that, various new schemes have been proposed recently. In the present paper, we summarize multivariate schemes UOV, Rainbow, and (variants of) HFE generating the second round candidates and study the practicalities of several multivariate encryption schemes proposed recently.

Keywords Multivariate public key cryptosystem (MPKC) · Post-quantum cryptography

1 Introduction

In 2016, NIST launched the standardization project of Post-Quantum Cryptography (NIST 2020). A lot of schemes were submitted to the first round of its project and 26 of them were chosen as the second round candidates in 2019 (NIST 2020). LUOV (Beullens et al. 2020), Rainbow (Ding et al. 2020) and GeMSS (Casanova et al. 2020) are multivariate signature schemes in the second round. These schemes are based on UOV (Kipnis et al. 1999; Patarin 1997), Rainbow (Ding et al. 2005), and HFEv- (Patarin et al. 2001), respectively, which were proposed before or around 2000 and have been still considered to be secure and efficient enough under suitable parameter

Y. Hashimoto (✉)
Department of Mathematical Sciences, University of the Ryukyus,
Nishihara-cho, Okinawa 903-0213, Japan
e-mail: hashimoto@math.u-ryukyu.ac.jp

© The Author(s) 2021
T. Takagi et al. (eds.), *International Symposium on Mathematics,*
Quantum Theory, and Cryptography, Mathematics for Industry 33,
https://doi.org/10.1007/978-981-15-5191-8_16

selections. On the other hand, there are few practical multivariate encryption schemes and despite that, various new schemes have been proposed in this decade.

The aim of this paper is to describe recent developments of multivariate public key cryptosystems, not yet presented in the previous paper (Hashimoto 2017). We first summarize in Sect. 2 the schemes UOV (Kipnis et al. 1999; Patarin 1997), Rainbow (Ding et al. 2005), and (variants of) HFE (Patarin 1996) with short surveys on the second round candidates LUOV (Beullens et al. 2020), Rainbow (Ding et al. 2020), and GeMSS (Casanova et al. 2020). Besides, we study in Sect. 3 the encryption schemes HFERP (Ikematsu et al. 2018), ZHFE (Porras et al. 2020), EFC (Szepieniec et al. 2016), and ABC (Tao et al. 2013) proposed recently, and show that the practicalities of these schemes are not much higher than the HFE variants for encryption, which are already known to be not too practical. Remark that MQDSS (Chen et al. 2016, 2020) is also a second round candidate and has been considered as a multivariate signature scheme since a set of randomly chosen multivariate quadratic forms is used in key generation, signature generation, and signature verification. However, it is based on Fiat–Shamir's transform of the 5-pass identification scheme (Sakumoto et al. 2011) and is far from other multivariate schemes. We then avoid to study MQDSS in this paper.

2 UOV, Rainbow, and Variants of HFE

In this section, we describe UOV (Kipnis et al. 1999; Patarin 1997), Rainbow (Ding et al. 2005), and variants of HFE (Patarin 1996) and give short surveys on the second round candidates LUOV (Beullens et al. 2020), Rainbow (Ding et al. 2020), and GeMSS (Casanova et al. 2020) of NIST's project (NIST 2020). We first propose the basic constructions of multivariate public key cryptosystems (MPKCS).

2.1 Basic Constructions of Multivariate Public Key Cryptosystems

Let $n, m \geq 1$ be integers, q a power of prime, and $\mathbf{F}_q$ a finite field of order q. Most MPKCs are described as follows.

Secret key. Two invertible affine maps $S : \mathbf{F}_q^n \to \mathbf{F}_q^n$, $T : \mathbf{F}_q^m \to \mathbf{F}_q^m$ and a quadratic map $G : \mathbf{F}_q^n \to \mathbf{F}_q^m$ to be inverted feasibly.

Public key. The quadratic map $F := T \circ G \circ S : \mathbf{F}_q^n \to \mathbf{F}_q^m$.

$$F : \mathbf{F}_q^n \xrightarrow{S} \mathbf{F}_q^n \xrightarrow{G} \mathbf{F}_q^m \xrightarrow{T} \mathbf{F}_q^m$$

Encryption scheme.

Encryption. For a plaintext $\mathbf{p} \in \mathbf{F}_q^n$, the ciphertext is $\mathbf{c} = F(\mathbf{p}) \in \mathbf{F}_q^m$.

Decryption. For a given ciphertext $\mathbf{c} \in \mathbf{F}_q^m$, compute $\mathbf{z} := T^{-1}(\mathbf{c})$ and find $\mathbf{y} \in \mathbf{F}_q^n$ with $G(\mathbf{y}) = \mathbf{z}$. Then the plaintext is $\mathbf{p} = S^{-1}(\mathbf{y})$.

Signature scheme.

Signature generation. For a message $\mathbf{m} \in \mathbf{F}_q^m$, compute $\mathbf{z} := T^{-1}(\mathbf{m})$ and find $\mathbf{y} \in \mathbf{F}_q^n$ with $G(\mathbf{y}) = \mathbf{z}$. Then the signature is $\mathbf{s} = S^{-1}(\mathbf{y})$.

Signature verification. The signature $\mathbf{s} \in \mathbf{F}_q^n$ is verified by $\mathbf{m} = F(\mathbf{s})$.

Efficiency. The encryption and signature verification are done by substituting $\mathbf{p}, \mathbf{s} \in \mathbf{F}_q^n$ into m quadratic forms of n variables. Their complexities are then $O(n^2 m)$ for most MPKCs under naive implementations. Furthermore, it is known (Hashimoto 2017) that the complexities of encrypting n plaintexts and of verifying n signatures simultaneously are $O(n^w m)$, where $2 \leq w < 3$ is a linear algebra constant. The complexities of decryption and signature generation depend mainly on how to invert G. We will discuss them in the individual schemes.

Security. There are two types of attacks on MPKCs. One is the *direct attack* to recover the plaintext $\mathbf{p}$ of a given ciphertext $\mathbf{c}$ directly by solving a system of m quadratic equations $F(\mathbf{x}) = (f_1(\mathbf{x}), \ldots, f_m(\mathbf{x})) = \mathbf{c}$ of n variables. The Gröbner basis attack is considered to be the most standard approach, and its complexity depends on the *degree d_{reg} of regularity* of the corresponding polynomial system $F(\mathbf{x}) - \mathbf{c}$. In general, d_{reg} is known to be smaller when the system is more over-defined ($m \gg n$) (Bardet et al. 2005). Furthermore, if q is small, the attacker will solve more efficiently by combining with the exhaustive search, which is called a *hybrid method* (Bettale et al. 2012). We also note that, if the system is massively under-defined ($n \gg m$), the attacker can find (at least) one of the solutions more effectively than the case of $n \sim m$ (Cheng et al. 2014; Kipnis et al. 1999; Miura et al. 2013; Tomae and Wolf 2012).

The other type is to recover partial information of the secret key (S, T) which is enough to invert F. In most known key recovery attacks on MPKCs, the attacker uses the property of the coefficient matrices of quadratic forms in G. Let $G_1, \ldots, G_m, F_1, \ldots, F_m$ be the coefficient matrices of $g_1(\mathbf{x}), \ldots, g_m(\mathbf{x}), f_1(\mathbf{x}), \ldots, f_m(\mathbf{x})$, respectively, i.e., $g_l(\mathbf{x}) = {}^t\mathbf{x}G_l\mathbf{x} + (\text{linear form})$ and $f_l(\mathbf{x}) = {}^t\mathbf{x}F_l\mathbf{x} + (\text{linear form})$ for $1 \leq l \leq m$. Since $F(\mathbf{x}) = T(G(S(\mathbf{x})))$, it holds

$$
\begin{pmatrix} F_1 \\ \vdots \\ F_m \end{pmatrix} = T \begin{pmatrix} {}^tSG_1S \\ \vdots \\ {}^tSG_mS \end{pmatrix}. \tag{1}
$$

This shows that, if $G_1, \ldots, G_m$ have special properties, partial information S, T will be recovered by the public information $F_1, \ldots, F_m$. How to recover and the complexity of the attack depend on $G_1, \ldots, G_m$, and then we discuss them in the individual schemes.

2.2   UOV

Let $o, v \geq 1$ be integers and put $n := o + v$, $m := o$. The quadratic map $G : \mathbf{F}_q^n \to \mathbf{F}_q^m$ is defined by

$$g_j(\mathbf{x}) = \sum_{1 \leq i \leq o} x_i \cdot (\text{linear form of } x_{o+1}, \ldots, x_n) \tag{2}$$
$$+ (\text{quadratic form of } x_{o+1}, \ldots, x_n),$$

for $1 \leq j \leq o$. UOV (Unbalanced Oil and Vinegar signature scheme, Patarin (1997), Kipnis et al. (1999) is constructed as follows.

Secret key. An invertible affine map $S : \mathbf{F}_q^n \to \mathbf{F}_q^n$ and the quadratic map $G : \mathbf{F}_q^n \to \mathbf{F}_q^m$ defined above.

Public key. The quadratic map $F := G \circ S : \mathbf{F}_q^n \to \mathbf{F}_q^m$.

Signature generation. For a message $\mathbf{m} = (m_1, \ldots, m_o) \in \mathbf{F}_q^m$, choose $u_1, \ldots, u_v \in \mathbf{F}_q$ randomly and find $y_1, \ldots, y_o \in \mathbf{F}_q$ such that

$$g_1(y_1, \ldots, y_o, u_1, \ldots, u_v) = m_1, \quad \ldots \quad , \quad g_o(y_1, \ldots, y_o, u_1, \ldots, u_v) = m_o. \tag{3}$$

The signature is $\mathbf{s} = S^{-1}(y_1, \ldots, y_o, u_1, \ldots, u_v)$.

Signature verification. The signature $\mathbf{s} \in \mathbf{F}_q^n$ is verified by $\mathbf{m} = F(\mathbf{s})$.

Complexity of signature generation. Since (3) is a system of o linear equations of o variables, we see that the complexity of signature generation of UOV is $O(n^3)$.

Security. The most important attack on UOV is Kipnis–Shamir's attack (Kipnis et al. 1999; Kipnis and Shamir 1998), which recovers an affine map S' such that $SS' = \begin{pmatrix} *_o & * \\ 0 & *_v \end{pmatrix}$ by using the fact that $G_1, \ldots, G_m$ are matrices having the forms of $\begin{pmatrix} 0_o & * \\ * & *_v \end{pmatrix}$. Its complexity is known to be $O(q^{\max(v-o,0)} \cdot n^4)$ (Kipnis et al. 1999), and then the parameter v must be sufficiently larger than o, namely n must be sufficiently larger than $2m$. This causes two inconveniences on UOV; one is that the sizes of keys are relatively large, and the other is that the approaches in Tomae and Wolf (2012), Cheng et al. (2014) weakens the security against the direct attacks a little. The later is easily covered by taking (n, m) a little larger. For the former, several approaches have been given until now. However, since some of key reduction approaches yield critical vulnerabilities (e.g., Hashimoto 2019; Peng and Tang 2018), the security of such UOVs must be studied quite carefully.

LUOV. LUOV (Beullens et al. 2020) is a signature scheme based on UOV and is a second round candidate of NIST's project. It is constructed over a finite field of even characteristic field and the components and coefficients in S, G, F are elements of $\mathbf{F}_2$. The size of keys is smaller and the security against the direct attack is not too less than the original UOV. Remark that the security against Kipnis–Shamir's attack is $O(2^{v-o} \cdot n^4)$ and a new attack on LUOV was quite recently proposed in Ding et al. (2013). Then the parameters o, v should be taken larger than the original version. See Beullens et al. (2020) for the latest version.

2.3 Rainbow

Rainbow (Ding et al. 2005) is a multi-layer version of UOV. We now describe the two-layer version. Let $o_1, o_2, v \geq 1$ be integers and put $n = o_1 + o_2 + v$, $m = o_1 + o_2$. Define the quadratic map $G : \mathbf{F}_q^n \to \mathbf{F}_q^m$ by

$$
\begin{aligned}
g_1(\mathbf{x}), \ldots, g_{o_1}(\mathbf{x}) &= \sum_{1 \leq i \leq o_1} x_i \cdot (\text{linear form of } x_{o_1+1}, \ldots, x_n) \\
&\quad + (\text{quadratic form of } x_{o_1+1}, \ldots, x_n), \\
g_{o_1+1}(\mathbf{x}), \ldots, g_m(\mathbf{x}) &= \sum_{o_1+1 \leq i \leq m} x_i \cdot (\text{linear form of } x_{m+1}, \ldots, x_n) \\
&\quad + (\text{quadratic form of } x_{m+1}, \ldots, x_n),
\end{aligned}
\tag{4}
$$

Rainbow is constructed as follows.

Secret key. Two invertible affine maps $S : \mathbf{F}_q^n \to \mathbf{F}_q^n$, $T : \mathbf{F}_q^m \to \mathbf{F}_q^m$ and the quadratic map $G : \mathbf{F}_q^n \to \mathbf{F}_q^m$ defined above.

Public key. The quadratic map $F := T \circ G \circ S : \mathbf{F}_q^n \to \mathbf{F}_q^m$.

Signature generation. For a message $\mathbf{m} \in \mathbf{F}_q^m$ to be signed, compute $\mathbf{z} = {}^t(z_1, \ldots, z_m) := T^{-1}(\mathbf{m})$ and choose $u_1, \ldots, u_v \in \mathbf{F}_q$ randomly. Find $y_{o_1+1}, \ldots, y_m \in \mathbf{F}_q$ such that

$$
g_{o_1+1}(y_1, \ldots, y_m, u_1, \ldots, u_v) = z_{o_1+1}, \quad \ldots, \quad g_m(y_1, \ldots, y_m, u_1, \ldots, u_v) = z_m.
\tag{5}
$$

After that, find $y_1, \ldots, y_{o_1} \in \mathbf{F}_q$ such that

$$
g_1(y_1, \ldots, y_m, u_1, \ldots, u_v) = z_1, \quad \ldots, \quad g_{o_1}(y_1, \ldots, y_m, u_1, \ldots, u_v) = z_{o_1}.
\tag{6}
$$

The signature is $\mathbf{s} = S^{-1}(y_1, \ldots, y_m, u_1, \ldots, u_v)$.

Signature verification. The signature $\mathbf{s} \in \mathbf{F}_q^n$ is verified by $\mathbf{m} = F(\mathbf{s})$.

Complexity of signature generation. Since (5) is a system of o_2 linear equations of o_2 variables and (6) is a system of o_1 linear equations of o_1 variables, we see that the complexity of signature generation is $O(n^3)$.

Security. Kipnis–Shamir's attack and rank attacks are major attacks on Rainbow. Since $G_1, \ldots, G_{o_1} = \begin{pmatrix} 0_{o_1} & * \\ * & *_{o_2+v} \end{pmatrix}$ and $G_{o_1+1}, \ldots, G_m = \begin{pmatrix} 0_{o_1} & 0 & 0 \\ 0 & 0_{o_2} & * \\ 0 & * & *_v \end{pmatrix}$, the complexity of Kipnis–Shamir's attack (Kipnis et al. 1999; Kipnis and Shamir 1998) on Rainbow is $O(q^{\max(o_2+v-o_1,0)} \cdot n^4)$. Furthermore, by checking the ranks of $G_1, \ldots, G_m$, we see that the complexities of min-rank attack and high-rank attack are $O(q^{o_2+v} \cdot n^4)$ and $O(q^{o_1} \cdot n^4)$, respectively (Yang and Chen 2005). Note that there have been several approaches to improve the efficiency of Rainbow. However, some of improvements are known to be insecure (e.g., Hashimoto 2019; Hashimoto et al. 2018; Peng and Tang 2018; Shim et al. 2017) and then the security of such efficient Rainbows must be studied carefully.

Rainbow on NIST's project. Rainbow (Ding et al. 2020) in the second round of NIST's project includes three versions; the standard Rainbow, the cyclic Rainbow, and the compressed Rainbow. The public keys and the numbers of arithmetics for signature verification for the later two Rainbows are smaller than the standard Rainbow. However, it is reported (Ding et al. 2020) that the verifications of the latter two versions are slower than the standard version. We consider that it is because the algorithms for verifications of the latter two versions are more complicated than the naive algorithm for the standard Rainbow. Better implementations are required for these arranged versions.

2.4 HFE

Let $n, m, d \geq 1$ be integers with $n = m, d < n$. Define $\mathscr{G} : \mathbf{F}_{q^n} \to \mathbf{F}_{q^n}$ by

$$\mathscr{G}(X) := \sum_{0 \leq i \leq j \leq d} \alpha_{ij} X^{q^i+q^j} + \sum_{0 \leq i \leq d} \beta_i X^{q^i} + \gamma,$$

where $\alpha_{ij}, \beta_i, \gamma \in \mathbf{F}_{q^n}$ and $G : \mathbf{F}_q^n \to \mathbf{F}_q^n$ by $G := \phi^{-1} \circ \mathscr{G} \circ \phi$ where $\phi : \mathbf{F}_q^n \to \mathbf{F}_{q^n}$ is an $\mathbf{F}_q$-isomorphism. HFE (Patarin 1996) is constructed as follows.

Secret key. Two invertible affine maps $S, T : \mathbf{F}_q^n \to \mathbf{F}_q^n$ and $\mathscr{G} : \mathbf{F}_{q^n} \to \mathbf{F}_{q^n}$ defined above.

Public key. The quadratic map $F := T \circ G \circ S = T \circ \phi^{-1} \circ \mathscr{G} \circ \phi \circ S : \mathbf{F}_q^n \to \mathbf{F}_q^n$.

Encryption. For a plaintext $\mathbf{p} \in \mathbf{F}_q^n$, the ciphertext is $\mathbf{c} := F(\mathbf{p}) \in \mathbf{F}_q^n$.

Decryption. For a given ciphertext $\mathbf{c}$, compute $\mathbf{z} := T^{-1}(\mathbf{c})$ and put $Z := \phi(\mathbf{z})$. Find $Y \in \mathbf{F}_{q^n}$ with $\mathscr{G}(Y) = Z$ and put $\mathbf{y} := \phi^{-1}(Y)$. The plaintext is $\mathbf{p} = S^{-1}(\mathbf{z})$.

Complexity of decryption. Since $\mathscr{G}(Y) = Z$ is a univariate polynomial equation of degree at most $2q^d$ over $\mathbf{F}_{q^n}$, the complexity of finding Y is

$$O((\deg \mathscr{G}(X))^3 + n(\deg \mathscr{G}(X))^2 \log q) = O(q^{3d} + nq^{2d} \log q)$$

by the Berlekamp algorithm (Berlekamp 1967, 1970). Then the parameter d should be $d = O(\log_q n)$.

Security. Let $\{\theta_1, \ldots, \theta_n\}$ be a basis of $\mathbf{F}_{q^n}$ over $\mathbf{F}_q$ and $\Theta := \left(\theta_j^{q^{i-1}}\right)_{1 \le i, j \le n}$. It is easy to see that $\Theta \mathbf{x} = {}^t(\phi(\mathbf{x}), \phi(\mathbf{x})^q, \ldots, \phi(\mathbf{x})^{q^{n-1}}) := {}^t(X, X^q, \ldots, X^{q^{n-1}})$. Since $F = (T \circ \phi^{-1}) \circ \mathscr{G} \circ (\phi \circ S)$, we have

$$\begin{pmatrix} F_1 \\ \vdots \\ F_n \end{pmatrix} = (T \cdot \Theta^{-1}) \begin{pmatrix} {}^t(\Theta S)\mathscr{G}^{(0)}(\Theta S) \\ \vdots \\ {}^t(\Theta S)\mathscr{G}^{(n-1)}(\Theta S) \end{pmatrix}, \qquad (7)$$

where $\bar{X} := \Theta \mathbf{x}$ and $\mathscr{G}^{(i)}$ is an $n \times n$ matrix over $\mathbf{F}_{q^n}$ such that $\mathscr{G}(X)^{q^i} = {}^t\bar{X}\mathscr{G}^{(i)}\bar{X} +$ (linear form of $\bar{X}$). This means that there exist $a_1, \ldots, a_n \in \mathbf{F}_{q^n}$ such that

$$a_1 F_1 + \cdots + a_n F_n = {}^t(\Theta S)\mathscr{G}^{(0)}(\Theta S) = {}^t(\Theta S) \begin{pmatrix} *_{d+1} \\ & 0_{n-d-1} \end{pmatrix} (\Theta S), \qquad (8)$$

and then $\mathrm{rank}(a_1 F_1 + \cdots + a_n F_n) \le d + 1$. The *min-rank attack* (Bettale et al. 2013; Kipnis and Shamir 1999) is an attack to recover such $(a_1, \ldots, a_n)$ and its complexity is estimated by $O\left(\binom{n+d+2}{d+2}^w\right) = O(n^{(d+2)w})$ under the assumption that a variant of Fröberg conjecture holds, where $2 \le w \le 3$ is a linear algebra constant. It is not difficult to check that the tuple $(a_1, \ldots, a_n)$ gives partial information of $T\Theta^{-1}$ and, once such a tuple is recovered, the attacker can recover partial information of ΘS, which is enough to decrypt arbitrary ciphertexts by elementary linear algebraic approaches. Since $d = O(\log_q n)$, the security of HFE is $n^{O(\log_q n)}$. Then the original HFE has been considered to be impractical. We also note that the security against Gröbner basis attack has been studied well (see e.g., Ding et al. 2011; Dubois and Gamma 2020; Faugère 2003; Granboulan et al. 2020; Huang et al. 2018). It is known that the rank condition (8) gives an upper bound of the degree d_{reg} of regularity of the polynomial system $F(\mathbf{x}) = \mathbf{c}$, in fact, $d_{\mathrm{reg}} \le \frac{1}{2}(q - 1)(d + 2)$ holds for HFE (Ding et al. 2011).

2.5 Variants of HFE

There have been various variants of HFE. In this subsection, we describe four major variants "plus (+)", "minus (−)", "vinegar (v)", and "projection (p)".

Plus (+). The "plus (+)" is a variant to add several polynomials on G. Let $r_+ \ge 1$ be an integer and $h_1(\mathbf{x}), \ldots, h_{r_+}(\mathbf{x})$ random quadratic forms of $\mathbf{x}$. For the map $G : \mathbf{F}_q^n \to \mathbf{F}_q^m$ of the original scheme, define $G_+ : \mathbf{F}_q^n \to \mathbf{F}_q^{m+r_+}$ by $G_+(\mathbf{x}) := {}^t(g_1(\mathbf{x}), \ldots, g_m(\mathbf{x}), h_1(\mathbf{x}), \ldots, h_{r_+}(\mathbf{x}))$. The public key $F_+ : \mathbf{F}_q^n \to \mathbf{F}_q^{m+r_+}$ of the plus is $F_+ := T_+ \circ G_+ \circ S$ where $T_+ : \mathbf{F}_q^{m+r_+} \to \mathbf{F}_q^{m+r_+}$ is an invertible affine map. It is mainly used for encryption when $m \geq n$. The decryption is as follows.

Decryption. For the ciphertext $\mathbf{c} \in \mathbf{F}_q^{m+r_+}$, compute $\mathbf{z} = (z_1, \ldots, z_{m+r_+}) := T_+^{-1}(\mathbf{c})$. Find $\mathbf{y} \in \mathbf{F}_q^n$ with $G(\mathbf{y}) = {}^t(z_1, \ldots, z_m)$ and verify whether ${}^t(h_1(\mathbf{y}), \ldots, h_{u_+}(\mathbf{y})) = {}^t(z_{m+1}, \ldots, z_{m+r_+})$. If it holds, the plaintext is $\mathbf{p} = S^{-1}(\mathbf{y})$. If not, try it again by another $\mathbf{y}$.

Complexity of decryption. If $m \geq n$, the number of $\mathbf{y}$ with $G(\mathbf{y}) = \mathbf{z}$ is (probably) small. Then the complexity of decryption of "plus" is not much larger than the original scheme.

Security. It is easy to see that an equation similar to (8) holds for the "plus" of HFE. Then the complexity of the min-rank attack on HFE+ is similar to the original HFE.

Minus (−). The "minus (−)" is to reduce several polynomials in F. Let $r_- \geq 1$ be an integer. For the public key $F : \mathbf{F}_q^n \to \mathbf{F}_q^m$ of the original scheme, the public key $F_- : \mathbf{F}_q^n \to \mathbf{F}_q^{m-r_-}$ of the minus is generated by $F_-(x) = {}^t(f_1(x), \ldots, f_{m-r_-}(x))$. It is mainly used for the signature scheme when $n \geq m$. The signature generation is as follows.

Signature generation. For a message $\mathbf{m} = {}^t(m_1, \ldots, m_{m-r_-}) \in \mathbf{F}_q^{m-r_-}$ to be signed, choose $u_1, \ldots, u_{r_-} \in \mathbf{F}_q$ randomly and let $\bar{\mathbf{m}} := {}^t(m_1, \ldots, m_{m-r_-}, u_1, \ldots, u_{r_-})$. Find $\mathbf{s} \in \mathbf{F}_q^n$ with $F(\mathbf{s}) = \bar{\mathbf{m}}$. If there exists such an $\mathbf{s}$, the signature is $\mathbf{s}$. If not, change $u_1, \ldots, u_{r_-}$ and repeat until such an $\mathbf{s}$ appears.

Complexities of signature generation. When $n \geq m$, the probability that $\mathbf{s}$ does not exist is considered to be not large. Then the complexity of the signature generation of the "minus" is not much larger than the original scheme.

Security. For the minus, it is easy to see that there exists an $(n - r_-) \times n$ matrix T_- such that

$$\begin{pmatrix} F_1 \\ \vdots \\ F_{n-r_-} \end{pmatrix} = (T_- \cdot \Theta^{-1}) \begin{pmatrix} {}^t(\Theta S)\mathscr{G}^{(0)}(\Theta S) \\ \vdots \\ {}^t(\Theta S)\mathscr{G}^{(n-1)}(\Theta S) \end{pmatrix}. \tag{9}$$

Then one can eliminate the contributions of $n - r_- - 1$ matrices in the right hand side by taking a linear combination of $F_1, \ldots, F_{n-r_-}$, namely there exist $a_1, \ldots, a_{n-r_-}$, $b_0, \ldots, b_{r_-} \in \mathbf{F}_{q^n}$ such that

$$a_1 F_1 + \cdots + a_{n-r_-} F_{n-r_-} = b_0 {}^t(\Theta S)\mathscr{G}^{(0)}(\Theta S) + \cdots + b_{r_-} {}^t(\Theta S)\mathscr{G}^{(r_-)}(\Theta S)$$

$$= {}^t(\Theta S) \begin{pmatrix} *_{d+r_-+1} & \\ & 0_{n-d-r_--1} \end{pmatrix} (\Theta S).$$

The min-rank attack is thus available on HFE- and its complexity can be estimated by $O(\binom{n+d+r_-+2}{d+r_-+2}^w) = O(n^{(d+r_-+2)w})$. This means that the "minus" enhances the security of HFE (see also Vates and Smith-Tone 2017).

Vinegar (v). The "vinegar (v)" is to add several variables on G. Let $r_v \geq 1$ be an integer. For the map $G : \mathbf{F}_q^n \to \mathbf{F}_q^m$ of the original scheme, define $G_v : \mathbf{F}_q^{n+r_v} \to \mathbf{F}_q^m$

such that $G_v(x_1, \ldots, x_n, u_1, \ldots, u_{r_v})$ is inverted similarly to $G(\mathbf{x})$ for any (or most) $u_1, \ldots, u_{r_v} \in \mathbf{F}_q$. For example, the map G_v of HFEv is given by $G_v := \phi_{-1} \circ \mathcal{G}_v \circ \phi_v$, where $\phi_v : \mathbf{F}_q^{n+r_v} \to \mathbf{F}_{q^n} \times \mathbf{F}_q^{r_v}$ is an $\mathbf{F}_q$-isomorphism and $\mathcal{G}_v : \mathbf{F}_{q^n} \times \mathbf{F}_q^{r_v} \to \mathbf{F}_{q^n}$ is the following polynomial map.

$$\mathcal{G}_v(X, x_{n+1}, \ldots, x_{n+r_v}) = \sum_{0 \le i,j \le d} \alpha_{ij} X^{q^i + q^j} + \sum_{0 \le i \le d} X^{q^i} \cdot (\text{linear form of } x_{n+1}, \ldots, x_{n+r_v})$$
$$+ (\text{quadratic form of } x_{n+1}, \ldots, x_{n+r_v}).$$

The public key $F_v : \mathbf{F}_q^{n+r_v} \to \mathbf{F}_q^n$ of the vinegar is $F_v := T \circ G_v \circ S_v$ where $S_v : \mathbf{F}_q^{n+r_v} \to \mathbf{F}_q^{n+r_v}$ is an invertible affine map. It is mainly used for signature when $n \ge m$. The signature generation is as follows.

Signature generation. For a message $\mathbf{m} \in \mathbf{F}_q^m$ to be signed, compute $\mathbf{z} := T^{-1}(\mathbf{m})$. Choose $u_1, \ldots, u_{r_v} \in \mathbf{F}_q$ randomly, and find $\mathbf{y} \in \mathbf{F}_q^n$ with $G_v(\mathbf{y}, u_1, \ldots, u_{r_v}) = \mathbf{z}$. If such an $\mathbf{y}$ does not exist, change $u_1, \ldots, u_{r_v}$ and try again. The signature is $\mathbf{s} = S_v^{-1}(\mathbf{y}, u_1, \ldots, u_{r_v})$.

Complexity of signature generation. Since $\mathbf{y}$ is found similarly to the original scheme, the complexity of finding $\mathbf{y}$ is almost the same as the original scheme. If $n \ge m$, the probability that $\mathbf{y}$ does not exist is considered to be not too large. Then the complexity of the "vinegar" is not too larger than the original scheme.

Security. For HFEv, we see that $\mathcal{G}_v(X, x_{n+1}, \ldots, x_{n+r_v}) = {}^t \bar{X}_v \left(\begin{array}{c|c} \begin{matrix} {}^{*d+1} & \\ & 0_{n-d-1} \end{matrix} & {}^* \\ \hline {}^* & {}_{*r_v} \end{array} \right) \bar{X}_v$

$+ (\text{linear form of } \bar{X}_v)$, where $\bar{X}_v = {}^t(X, \ldots, X^{q^{n-1}}, x_{n+1}, \ldots, x_{n+r_v})$. Then there exist $a_1, \ldots, a_n \in \mathbf{F}_{q^n}$ such that

$$a_1 F_1 + \cdots + a_n F_n = {}^t \left(\begin{pmatrix} \Theta & \\ & I_{r_v} \end{pmatrix} S_v \right) \left(\begin{array}{c|c} \begin{matrix} {}^{*d+1} & \\ & 0_{n-d-1} \end{matrix} & {}^* \\ \hline {}^* & {}_{*r_v} \end{array} \right) \left(\begin{pmatrix} \Theta & \\ & I_{r_v} \end{pmatrix} S_v \right).$$

Since the rank of the matrix in the right hand side above is at most $d + r_v + 1$, the security of HFEv against the min-rank attack is estimated by $O\left(\binom{n+d+r_v+2}{d+r_-+2}^w\right) = O(n^{(d+r_v+2)w})$.

Projection (p). The "projection" is to reduce several variables of the polynomials in F. Let $r_p \ge 1$ be an integer and $u_1, \ldots, u_{r_p} \in \mathbf{F}_q$. For the public key $F : \mathbf{F}_q^n \to \mathbf{F}_q^m$ of the original scheme, the public key $F_p : \mathbf{F}_q^{n-r_p} \to \mathbf{F}_q^m$ of the projection is generated by $F_p(x_1, \ldots, x_{n-r_p}) := F(x_1, \ldots, x_{n-r_p}, u_1, \ldots, u_{r_p})$. It is mainly used for encryption when $m \ge n$. The decryption is as follows.

Decryption. For the ciphertext $\mathbf{c} \in \mathbf{F}_q^m$, find $\mathbf{p} \in \mathbf{F}_q^n$ with $F(\mathbf{p}) = \mathbf{c}$ similarly to the original scheme. If $\mathbf{p} = (*, \ldots, *, u_1, \ldots, u_{r_p})$, the plaintext is $\tilde{\mathbf{p}} := (p_1, \ldots, p_{n-r_p}) \in \mathbf{F}_q^{n-r_p}$. If not, try it again by another $\mathbf{p}$.

Complexities of decryption. If $m \ge n$, the number of $\mathbf{p}$ with $F(\mathbf{p}) = \mathbf{c}$ is (probably) not too large. Then the complexity of decryption of the "projection" is not much larger than the original scheme.

Security. For the projection of HFE, we see that there exist $a_1, \ldots, a_n \in \mathbf{F}_{q^n}$ such that

$$a_1 F_1 + \cdots + a_n F_n = {}^t(\Theta \tilde{S}) \begin{pmatrix} *_{d+1} & \\ & 0_{n-d-1} \end{pmatrix} (\Theta \tilde{S}),$$

where $\tilde{S}$ is an $n \times (n - r_p)$ matrix with $S = (\tilde{S}, *)$. Then the min-rank attack is available and its complexity is almost the same as the original scheme.

The most successful variant of HFE is probably the signature scheme **HFEv-** (Patarin et al. 2001), a combination of "minus" and "vinegar" of HFE, since the security against the min-rank attack is enhanced drastically without slowing down the signature generation. In fact, **GeMSS** (Casanova et al. 2020) based on HFEv- was chosen as a second round candidate of NIST's project (NIST 2020). There are three kinds of GeMSS, called GeMSS, BlueGeMSS, and RedGeMSS, The major difference among these three GeMSSs is the degree of $\mathscr{G}_v$; the degrees are $513 (= 2^9 + 1)$, $129 (= 2^7 + 1), 17 (= 2^4 + 1)$, i.e., d's are 10, 8, 5, respectively. Of course, the signature generation of RedGeMSS is fastest and the BlueGeMSS is the next. Furthermore, the securities against the min-rank attack are enough if r_-, r_v are sufficiently large. On the other hand, as pointed out in Hashimoto (2018) for HMFEv (Petzoldt et al. 2017) (the vinegar of multi-HFE (Chen et al. 2020), the minus and the vinegar do not enhance the security against the high-rank attack. Though critical vulnerabilities of HFE variants against the high-rank attack have not been reported until now, we consider that an HFEv- with smaller d has a higher risk against the high-rank attack.

We recall that **Sflash** (Akkar et al. 2003) (a minus of Matsumoto–Imai's scheme (Matsumoto and Imai 1988) is a signature scheme selected by NESSIE (Preneel 2020) and broken by a differential attack (Fouque et al. 2005). Recently, its projections called **Pflash** (Cartor and Smith-Tone 2017; Smith-Tone et al. 2015) and **Eflash** (Cartor and Smith-Tone 2018) were proposed. Pflash is a signature scheme with $r_p < r_-$ and Eflash is an encryption scheme with $r_p > r_-$. The complexities of signature generation and decryption are about $q^{\min(r_p, r_-)}$ times of Matsumoto–Imai's scheme (Matsumoto and Imai 1988) and then we should take r_-, r_p by $\min(r_p, r_-) = O(\log_q n)$. It has been considered that the differential attack is not available on these schemes, and the security against the min-rank attack highly depends on r_-. The security of Eflash is thus $n^{O(\log_q n)}$. Similarly for the encryption scheme HFEp- with $r_p > r_-$, it is easy to see that the complexity of decryption is about q^{r_-} times of the original HFE and the complexity of the min-rank attack is roughly estimated by $O(n^{(3d+r_-+2)w})$. Since $3d + r_- = O(\log_q n)$, its security is also $n^{O(\log_q n)}$.

3   New Encryption Schemes

In this section, we study the encryption schemes HFERP (Ikematsu et al. 2018), ZHFE (Porras et al. 2020), EFC (Szepieniec et al. 2016), and ABC (Tao et al. 2013, 2015) proposed recently.

3.1   HFERP

HFERP (Ikematsu et al. 2018) is an encryption scheme constructed by a "plus" and "projection" of a combination of HFE and Rainbow. We first describe a one-layer version HFERP without "plus" and "projection".

Let $v, o, l, d_0 \geq 1$ be integers, $n := v + o$ and $m := v + o + l$. Define the map $\mathcal{G}_0 : \mathbf{F}_{q^v} \to \mathbf{F}_{q^v}$ by

$$\mathcal{G}_0(X) := \sum_{0 \leq i \leq j \leq d_0} \alpha_{ij} X^{q^i + q^j} + \sum_{0 \leq i \leq d_0} \beta_i X^{q^i} + \gamma,$$

where $\alpha_{ij}, \beta_i, \gamma \in \mathbf{F}_{q^v}$. The quadratic map $G : \mathbf{F}_q^n \to \mathbf{F}_q^m$ is given as follows.

$$^t(g_1(\mathbf{x}), \ldots, g_v(\mathbf{x})) = (\phi_0^{-1} \circ \mathcal{G}_0 \circ \phi_0)(\mathbf{x}_0),$$

$$g_{v+1}(\mathbf{x}), \ldots, g_m(\mathbf{x}) = \sum_{v+1 \leq i \leq n} x_i \cdot (\text{linear form of } \mathbf{x}_0) + (\text{quadratic form of } \mathbf{x}_0),$$

where $\phi_0 : \mathbf{F}_q^v \to \mathbf{F}_{q^v}$ is an $\mathbf{F}_q$-isomorphism and $\mathbf{x}_0 = {}^t(x_1, \ldots, x_v)$. HFERP (without "plus", "projection") is constructed as follows.

Secret key. Two invertible affine maps $S : \mathbf{F}_q^n \to \mathbf{F}_q^n$, $T : \mathbf{F}_q^m \to \mathbf{F}_q^m$ and the quadratic map $G : \mathbf{F}_q^n \to \mathbf{F}_q^m$.

Public key. The quadratic map $F := T \circ G \circ S : \mathbf{F}_q^n \to \mathbf{F}_q^m$.

Encryption. For a plaintext $\mathbf{p} \in \mathbf{F}_q^n$, the ciphertext is $\mathbf{c} = F(\mathbf{p}) \in \mathbf{F}_q^m$.

Decryption. For a given ciphertext $\mathbf{c}$, compute $\mathbf{z} = {}^t(z_1, \ldots, z_m) := T^{-1}(\mathbf{c})$. Let $Z_0 := \phi_0(z_1, \ldots, z_v) \in \mathbf{F}_{q^v}$ and find $Y_0 \in \mathbf{F}_{q^v}$ such that $\mathcal{G}_0(Y_0) = Z_0$. Put $(y_1, \ldots, y_v) := \phi_0^{-1}(Y_0) \in \mathbf{F}_q^v$ and find $y_{v+1}, \ldots, y_n \in \mathbf{F}_q$ with

$$g_{v+1}(y_1, \ldots, y_v, y_{v+1}, \ldots, y_n) = z_{v+1}, \quad \ldots, \quad g_m(y_1, \ldots, y_v, y_{v+1}, \ldots, y_n) = z_m. \tag{10}$$

The plaintext is $\mathbf{p} = S^{-1}(y_1, \ldots, y_n)$.

Complexity of decryption. Since the degree of $\mathcal{G}_0(X)$ is at most $2q^{d_0}$, the complexity of finding Y_0 is $O(q^{3d_0} + vq^{2d_0} \log q)$ by Berlekamp's algorithm. We see that (10) is a system of $o + l$ linear equations of o variables. We thus conclude that the total complexity of decryption is $O(q^{3d_0} + vq^{2d_0} \log q + n^3)$. The parameter d_0 should be taken by $d_0 = O(\log_q n)$.

Security. Let $\{\theta_1, \ldots, \theta_v\}$ be a basis of $\mathbf{F}_{q^v}$ over $\mathbf{F}_q$ and $\Theta_0 := \left(\theta_j^{q^{i-1}}\right)_{1 \le i, j \le v}$. By the definition of G, F, we see that

$$
\begin{pmatrix} F_1 \\ \vdots \\ F_m \end{pmatrix} = T \cdot \begin{pmatrix} \Theta_0^{-1} & \\ & I_{o+l} \end{pmatrix} \begin{pmatrix} {}^tS\left({}^t\Theta_0 \mathscr{G}_0^{(0)} \Theta_0 \quad 0_o \right)S \\ \vdots \\ {}^tS\left({}^t\Theta_0 \mathscr{G}_0^{(v-1)} \Theta_0 \quad 0_o \right)S \\ {}^tS\left(\begin{smallmatrix} *_v & * \\ * & 0_o \end{smallmatrix} \right)S \\ \vdots \\ {}^tS\left(\begin{smallmatrix} *_v & * \\ * & 0_o \end{smallmatrix} \right)S \end{pmatrix}
$$

and then there exist $a_1, \ldots, a_m \in \mathbf{F}_{q^v}$ such that

$$
a_1 F_1 + \cdots + a_m F_m = {}^tS\left({}^t\Theta_0 \mathscr{G}_0^{(0)} \Theta_0 \quad 0_o \right)S = {}^tS^t\begin{pmatrix} \Theta_0 \\ & I_o \end{pmatrix}\begin{pmatrix} *_{d_0+1} & \\ & 0_{n-d_0-1} \end{pmatrix}\begin{pmatrix} \Theta_0 \\ & I_o \end{pmatrix}S.
$$

The min-rank attack is thus available on HFERP and its complexity can be estimated by $O\left(\binom{m+d_0+2}{d_0+2}^w\right) = O(m^{(d_0+2)w})$ (Ikematsu et al. 2018). This situation is similar for its plus and projection. Since $d_0 = O(\log_q n)$, the security of HFERP is $n^{O(\log_q n)}$, which is almost the same as HFE. For the minus, we can easily check that the complexity of decryption is at most q^{r-} times of the original HFERP and the security against the min-rank attack is $O\left(\binom{m+d_0+2}{d_0+r_-+2}^w\right) = O(m^{(d_0+r_-+2)w})$. This means that the security of HFERP- is also $n^{O(\log_q n)}$.

3.2   ZHFE

ZHFE (Porras et al. 2020) is an encryption scheme constructed by two univariate polynomials over an extension field. In this subsection, we study the simplest version of ZHFE since the structure of the original version is not far from the simplest version.

Let $n, m, D \ge 1$ be integers with $m = 2n$ and define the quadratic forms $\mathscr{G}_1(X)$, $\mathscr{G}_2(X)$ of $\bar{X} = {}^t(X, X^q, \ldots, X^{q^{n-1}})$ such that the degree of $\Psi(X) := X^q \cdot \mathscr{G}_1(X) + X \cdot \mathscr{G}_2(X)$ is at most D. It is easy to see that the coefficient matrices $\mathscr{G}_1^{(0)}, \mathscr{G}_2^{(0)}$ of $\mathscr{G}_1(X), \mathscr{G}_2(X)$ as quadratic forms of $\bar{X}$ are

$$
\mathscr{G}_1^{(0)} = \left(\begin{array}{c|c} \begin{matrix} * & {\scriptstyle *\cdots *} \\ {\scriptstyle d+1} & {\scriptstyle 0\cdots 0} \\ & 0 \\ \hline * \ 0 & \\ \vdots \ 0 & 0 \\ * \ 0 & \end{matrix} & \begin{matrix} \\ \\ 0 \\ \hline \\ 0 \\ {\scriptstyle n-d-1} \end{matrix} \end{array}\right), \quad
\mathscr{G}_2^{(0)} = \left(\begin{array}{c|c} \begin{matrix} * & {\scriptstyle 0\cdots 0} \\ {\scriptstyle d+1} & {\scriptstyle * \ 0 \ *} \\ & 0 \\ \hline 0 \ * & \\ \vdots \ 0 & 0 \\ 0 \ * & \end{matrix} & \begin{matrix} \\ \\ 0 \\ \hline \\ 0 \\ {\scriptstyle n-d-1} \end{matrix} \end{array}\right), \tag{11}
$$

where $d := \lceil \log_q \frac{D-q}{2} \rceil$. Denote by $\phi_2 : \mathbf{F}_q^m \to \mathbf{F}_{q^n}^2$ an $\mathbf{F}_q$-isomorphism and $\mathscr{G}(X) := (\mathscr{G}_1(X), \mathscr{G}_2(X))$. ZHFE is constructed as follows.

Secret key. Two invertible affine maps $S : \mathbf{F}_q^n \to \mathbf{F}_q^n, T : \mathbf{F}_q^m \to \mathbf{F}_q^m$ and the quadratic map $G := \phi_2^{-1} \circ \mathscr{G} \circ \phi : \mathbf{F}_q^n \to \mathbf{F}_q^m$.

Public key. The quadratic map $F := T \circ G \circ S : \mathbf{F}_q^n \to \mathbf{F}_q^m$.

Encryption. For a plaintext $\mathbf{p} \in \mathbf{F}_q^n$, the ciphertext is $\mathbf{c} = F(\mathbf{p}) \in \mathbf{F}_q^m$.

Decryption. For a given ciphertext $\mathbf{c} \in \mathbf{F}_q^m$, compute $\mathbf{z} := T^{-1}(\mathbf{c})$. Let $(Z_1, Z_2) := \phi_2(z) \in \mathbf{F}_{q^n}^2$, and find $Y \in \mathbf{F}_{q^n}$ such that $\Psi(Y) - Y^q \cdot Z_1 - Y \cdot Z_2 = 0$. Verify whether $\mathscr{G}_1(Y) = Z_1, \mathscr{G}_2(Y) = Z_2$ hold and put $\mathbf{y} := \phi^{-1}(Y) \in \mathbf{F}_q^n$. The plaintext is $\mathbf{p} = S^{-1}(\mathbf{y})$.

Complexity of decryption. Since $\Psi(Y) - Y^q \cdot Z_1 - Y \cdot Z_2 = Y^q \cdot (\mathscr{G}_1(Y) - Z_1) + Y \cdot (\mathscr{G}_2(Y) - Z_2)$, at least one of Y satisfies $\mathscr{G}_1(Y) = Z_1, \mathscr{G}_2(Y) = Z_2$ if $\mathbf{z} \in G(\mathbf{F}_{q^n})$. The complexity of decryption is $O(D^3 + nD^2 \log q) = O(q^{3d} + nq^{2d} \log q)$ by Berlekamp's algorithm. The parameter d should be $d = O(\log_q n)$.

Security. Let $\{\theta_1, \ldots, \theta_n\}$ be a basis of $\mathbf{F}_{q^n}$ over $\mathbf{F}_q$ and $\Theta_2 := \left(\theta_j^{q^{i-1}} \cdot I_2\right)_{1 \le i, j \le n}$. We can easily check that

$$
\begin{pmatrix} F_1 \\ \vdots \\ F_m \end{pmatrix} = T\Theta_2^{-1} \begin{pmatrix} {}^t(\Theta S)\mathscr{G}_1^{(0)}(\Theta S) \\ {}^t(\Theta S)\mathscr{G}_2^{(0)}(\Theta S) \\ {}^t(\Theta S)\mathscr{G}_1^{(1)}(\Theta S) \\ \vdots \\ {}^t(\Theta S)\mathscr{G}_2^{(n-1)}(\Theta S) \end{pmatrix}
$$

and then there exist $a_1, \ldots, a_m \in \mathbf{F}_{q^n}$ such that

$$
a_1 F_1 + \cdots + a_m F_m = {}^t(\Theta S)\mathscr{G}_1^{(0)}(\Theta S).
$$

Since $\operatorname{rank}\mathscr{G}_1^{(0)} \le d + 2$ due to (11), the min-rank attack is available on ZHFE and its complexity can be estimated by $O(\binom{m+d+3}{d+3}^w) = O(m^{(d+3)w})$ (Cabarcas et al. 2017; Perlne and Smith-Tone 2016). Since $d = O(\log_q n)$, the security of ZHFE is also $n^{O(\log_q n)}$.

We note that the plus and projection do not enhance the security. For the minus, we see that there exist $a_1, \ldots, a_{m-r_-}, b_0, \ldots, b_{r_-} \in \mathbf{F}_{q^n}$ such that

$$a_1 F_1 + \cdots + a_{m-r_-} F_{m-r_-}$$

$$= b_0{}^t(\Theta S)\mathcal{G}_1^{(0)}(\Theta S) + b_1{}^t(\Theta S)\mathcal{G}_2^{(0)}(\Theta S) + \cdots + b_{r_-}{}^t(\Theta S)\mathcal{G}_{(r_- \bmod 2)+1}^{(\lfloor r_-/2 \rfloor)}(\Theta S)$$

$$= {}^t(\Theta S)\begin{pmatrix} *_{\lceil \frac{r_-}{2} \rceil + 1} & * & & * \\ * & *_{d-(r_- \bmod 2)} & 0 \\ * & 0 & 0 \end{pmatrix}(\Theta S).$$

Since the rank of the matrix above is $d + r_- + 2$, the complexity of the min-rank attack is $O(\binom{m+d+3}{d+r_-+3}^w) = O((2n)^{(d+r_-+3)w})$. However, the complexity of decryption is at most q^{r_-} times of the original ZHFE, and then the security of ZHFE- is also $n^{O(\log_q n)}$. Remark that (Perlne and Smith-Tone 2016) proposed a minus of ZHFE without slowing down the decryption by using a singular-type ZHFE. However, by studying the structure of such a ZHFE- carefully, we can easily check that such a minus does not enhance the security against the min-rank attack at all.

3.3   EFC

EFC (Szepieniec et al. 2016) is an encryption scheme constructed from the fact that an extension field can be expressed by a set of matrices.

Let $n, m \geq 1$ be integers with $m = 2n$, $h(t)$ an irreducible univariate polynomial over $\mathbf{F}_q$ and H an $n \times n$ matrix whose characteristic polynomial is $h(t)$. It is easy to see that $\mathcal{H} := \{a_0 I_n + a_1 H + \cdots + a_{n-1} H^{n-1} \mid a_0, \ldots, a_{n-1} \in \mathbf{F}_q\}$ is isomorphic to $\mathbf{F}_q[t]/\langle h(t) \rangle \simeq \mathbf{F}_{q^n}$. Choose $A_1, \ldots, A_m \in \mathcal{H}$ and define the map $G : \mathbf{F}_q^n \to \mathbf{F}_q^m$ by

$$\begin{aligned}{}^t(g_1(\mathbf{x}), g_3(\mathbf{x}), \ldots, g_{m-1}(\mathbf{x})) &= (x_1 A_1 + x_2 A_3 + \cdots + x_{m-1} A_n)\,\mathbf{x}, \\ {}^t(g_2(\mathbf{x}), g_4(\mathbf{x}), \ldots, g_m(\mathbf{x})) &= (x_1 A_2 + x_2 A_4 + \cdots + x_m A_n)\,\mathbf{x}.\end{aligned}$$

EFC (Szepieniec et al. 2016) is constructed as follows.

Secret key. Two invertible affine maps $S : \mathbf{F}_q^n \to \mathbf{F}_q^n$, $T : \mathbf{F}_q^m \to \mathbf{F}_q^m$ and the quadratic map $G : \mathbf{F}_q^n \to \mathbf{F}_q^m$ (i.e., the matrices $A_1, \ldots, A_m$) defined above.

Public key. The quadratic map $F := T \circ G \circ S : \mathbf{F}_q^n \to \mathbf{F}_q^m$.

Encryption. For a plaintext $\mathbf{p} \in \mathbf{F}_q^n$, the ciphertext is $\mathbf{c} = F(\mathbf{p}) \in \mathbf{F}_q^m$.

Decryption. For a given ciphertext $\mathbf{c}$, compute $\mathbf{z} = {}^t(z_1, \ldots, z_m) := T^{-1}(\mathbf{c})$. Solve a system of linear equations given by

$$\begin{aligned}(x_1 A_1 + x_2 A_3 &+ \cdots + x_n A_{m-1})\,{}^t(z_2, z_4, \ldots, z_m) \\ &= (x_1 A_2 + x_2 A_4 + \cdots + x_n A_m)\,{}^t(z_1, z_3, \ldots, z_{m-1}),\end{aligned} \tag{12}$$

and find a solution $\mathbf{y}$ of (12) satisfying $G(\mathbf{y}) = \mathbf{z}$. The plaintext is $\mathbf{p} = S^{-1}(\mathbf{y})$.

Complexity of decryption. Since $\mathcal{H}$ is commutative, it holds

$$(x_1 A_2 + x_2 A_4 + \cdots + x_n A_m)\,{}^t(g_1(\mathbf{x}), g_3(\mathbf{x}), \ldots, g_{m-1}(\mathbf{x}))$$
$$= (x_1 A_1 + x_2 A_3 + \cdots + x_n A_{m-1})\,{}^t(g_2(\mathbf{x}), g_4(\mathbf{x}), \ldots, g_m(\mathbf{x})).$$

Then at least one of solutions of (12) satisfies $G(\mathbf{y}) = \mathbf{z}$ if $\mathbf{z} \in G(\mathbf{F}_q^n)$. The equation (12) is written by $(z_1 B_1 + \cdots + z_m B_m)\mathbf{x} = 0$ with $n \times n$ matrices $B_1, \ldots, B_m$ are $n \times n$ derived from $A_1, \ldots, A_m$. The complexity of decryption is thus $O(n^3)$.

Note that, since the map G in EFC is over-defined, the complexity of the "plus" and the "projection" is almost the same as the original EFC and that of the "minus" is at most q^{r-} times of the original EFC.

Security. It is already known that the original EFC is insecure against the linearization attack (Szepieniec et al. 2016). We now study the security of EFC- against the min-rank attack. Let $\theta \in \mathbf{F}_{q^n}$ be a root of $h(t)$, choose a basis of $\mathbf{F}_{q^n}$ over $\mathbf{F}_q$ by $\{\theta_1, \ldots, \theta_n\} = \{1, \theta, \theta^2, \ldots, \theta^{n-1}\}$ and put $\Theta := \left(\theta_j^{q^{i-1}}\right)_{1 \le i, j \le n}$. Suppose that H is a companion matrix of $h(t)$. Since $A_1, \ldots, A_m \in \mathcal{H}$, there exist linear forms $L_1(\mathbf{x}), \ldots, L_m(\mathbf{x})$ of $\mathbf{x}$ over $\mathbf{F}_q$ such that

$$x_1 A_1 + x_2 A_3 + \cdots + x_n A_{m-1} = L_1(\mathbf{x})I_n + L_3(\mathbf{x})H + \cdots + L_{m-1}(\mathbf{x})H^{n-1},$$
$$x_1 A_2 + x_2 A_4 + \cdots + x_n A_m = L_2(\mathbf{x})I_n + L_4(\mathbf{x})H + \cdots + L_m(\mathbf{x})H^{n-1}.$$

Denote by

$$\mathscr{G}_1(X) := g_1(\mathbf{x})\theta_1 + g_3(\mathbf{x})\theta_2 + \cdots + g_{m-1}(\mathbf{x})\theta_n,$$
$$\mathscr{G}_2(X) := g_2(\mathbf{x})\theta_1 + g_4(\mathbf{x})\theta_2 + \cdots + g_m(\mathbf{x})\theta_n,$$
$$\mathscr{L}_1(X) := L_1(\mathbf{x})\theta_1 + L_3(\mathbf{x})\theta_2 + \cdots + L_{m-1}(\mathbf{x})\theta_n,$$
$$\mathscr{L}_2(X) := L_2(\mathbf{x})\theta_1 + L_4(\mathbf{x})\theta_2 + \cdots + L_m(\mathbf{x})\theta_n,$$

where $X := \phi(\mathbf{x}) = x_1\theta_1 + \cdots + x_n\theta_n$. It is easy to see that $\mathscr{G}_1(X), \mathscr{G}_2(X)$ are quadratic forms and $\mathscr{L}_1(X), \mathscr{L}_2(X)$ are linear forms of $\bar{X} = \Theta\mathbf{x} = {}^t(X, X^q, \ldots, X^{q^{n-1}})$. By the definition of G, we see that

$$\Theta^t(g_1(\mathbf{x}), g_3(\mathbf{x}), \ldots, g_{m-1}(\mathbf{x})) = \left(\sum_{1 \le i \le n} L_{2i-1}(\mathbf{x})(\Theta H \Theta^{-1})^{i-1}\right)(\Theta\mathbf{x}),$$

$$\Theta^t(g_2(\mathbf{x}), g_4(\mathbf{x}), \ldots, g_m(\mathbf{x})) = \left(\sum_{1 \le i \le n} L_{2i}(\mathbf{x})(\Theta H \Theta^{-1})^{i-1}\right)(\Theta\mathbf{x}).$$

$$(13)$$

Since $\Theta H \Theta^{-1} = \mathrm{diag}\left(\theta, \theta^q, \ldots, \theta^{q^{n-1}}\right)$ (e.g., Horn et al. 1985), we have $\mathscr{G}_1(X) = \mathscr{L}_1(X) \cdot X$, $\mathscr{G}_2(X) = \mathscr{L}_2(X) \cdot X$ due to (13). This means that the map G is written by $G = \phi_2^{-1} \circ \mathscr{G} \circ \phi$ where $\mathscr{G}(X) = (\mathscr{G}_1(X), \mathscr{G}_2(X)) = (\mathscr{L}_1(X) \cdot X, \mathscr{L}_2(X) \cdot X)$, and it holds

$$\begin{pmatrix} F_1 \\ \vdots \\ F_m \end{pmatrix} = T\Theta_2^{-1} \begin{pmatrix} {}^t(\Theta S)\mathscr{G}_1^{(0)}(\Theta S) \\ {}^t(\Theta S)\mathscr{G}_2^{(0)}(\Theta S) \\ {}^t(\Theta S)\mathscr{G}_1^{(1)}(\Theta S) \\ \vdots \\ {}^t(\Theta S)\mathscr{G}_2^{(n-1)}(\Theta S) \end{pmatrix}.$$

Then, for EFC-, there exist $a_1, \ldots, a_{m-r_-}, b_0, \ldots, b_{r_-} \in \mathbf{F}_{q^n}$ such that

$$a_1 F_1 + \cdots + a_{m-r_-} F_{m-r_-}$$
$$= b_0 {}^t(\Theta S)\mathscr{G}_1^{(0)}(\Theta S) + b_1 {}^t(\Theta S)\mathscr{G}_2^{(0)}(\Theta S) + \cdots + b_{r_-} {}^t(\Theta S)\mathscr{G}_{(r_- \bmod 2)+1}^{(\lfloor r_-/2 \rfloor)}(\Theta S)$$
$$= {}^t(\Theta S) \begin{pmatrix} *_{1+\lfloor \frac{r_-}{2} \rfloor} & * \\ * & 0 \end{pmatrix} (\Theta S).$$

Since the rank of the matrix above is at most $2\lfloor \frac{r_-}{2} \rfloor + 2$, the min-rank attack is available on EFC- and its complexity can be estimated by $O(\binom{2n-r_-+2\lfloor \frac{r_-}{2} \rfloor+3}{3+2\lfloor \frac{r_-}{2} \rfloor}^w) = O((2n)^{(r_-+3)w})$. Since $r_- = O(\log_q n)$, the security of EFC- is also $n^{O(\log_q n)}$. This situation is similar to the "plus" and "projection" of EFC-.

3.4 ABC

ABC (Tao et al. 2013, 2015) is an encryption scheme constructed by three polynomial matrices A, B, C. Let $r, n, m \geq 1$ be integers with $n = r^2, m = 2r^2$. For $\mathbf{x} = {}^t(x_1, \ldots, x_n)$, define the $r \times r$ matrices $A(\mathbf{x}), B(\mathbf{x}), C(\mathbf{x}), E_1(\mathbf{x}), E_2(\mathbf{x})$ by $A(\mathbf{x}) := \left(x_{j+r(i-1)}\right)_{1 \leq i,j \leq r}$, $B(\mathbf{x}) := \left(b_{ij}(\mathbf{x})\right)_{1 \leq i,j \leq r}$, $C(\mathbf{x}) := \left(c_{ij}(\mathbf{x})\right)_{1 \leq i,j \leq r}$, $E_1(\mathbf{x}) := A(\mathbf{x})B(\mathbf{x})$ and $E_2(\mathbf{x}) := A(\mathbf{x})C(\mathbf{x})$, where $b_{ij}(\mathbf{x}), c_{ij}(\mathbf{x})$ are linear forms of $\mathbf{x}$. The quadratic map $G : \mathbf{F}_q^n \to \mathbf{F}_q^m$ is generated by $E_1(\mathbf{x}) = \left(g_{j+r(i-1)}(\mathbf{x})\right)_{1 \leq i,j \leq r}$ and $E_2(\mathbf{x}) = \left(g_{n+j+r(i-1)}(\mathbf{x})\right)_{1 \leq i,j \leq r}$. The encryption scheme ABC (Tao et al. 2013) is constructed as follows.

Secret key. Two invertible affine maps $S : \mathbf{F}_q^n \to \mathbf{F}_q^n, T : \mathbf{F}_q^m \to \mathbf{F}_q^m$ and the quadratic map G defined above.

Public key. The quadratic map $F := T \circ G \circ S : \mathbf{F}_q^n \to \mathbf{F}_q^m$.

Encryption. For a plaintext $\mathbf{p} \in \mathbf{F}_q^n$, the ciphertext is $\mathbf{c} = F(\mathbf{p}) \in \mathbf{F}_q^m$.

Decryption. For a given ciphertext $\mathbf{c}$, compute $\mathbf{z} = {}^t(z_1, \ldots, z_m) := T^{-1}(\mathbf{c})$ and put $Z_1 := \left(z_{j+r(i-1)}\right)_{1 \leq i,j \leq r}$, $Z_2 := \left(z_{n+j+r(i-1)}\right)_{1 \leq i,j \leq r}$. Find $\mathbf{y} \in \mathbf{F}_q^n$ such that

$$B(\mathbf{y}) = C(\mathbf{y})Z_2^{-1}Z_1. \tag{14}$$

If Z_2 is not invertible, replace (14) into $B(\mathbf{y})Z_1^{-1}Z_2 = C(\mathbf{y})$. The plaintext is $\mathbf{p} = S^{-1}(\mathbf{y})$.

Complexity of decryption. The equation (14) yields a system of n linear equations of n variables. Then the complexity of decryption is $O(n^3)$. Remark that the decryption fails if $A(S(\mathbf{p}))$ is not invertible and its probability is about q^{-1}.

Security. It is easy to check that the coefficient matrix G_1 of the first polynomial $g_1(\mathbf{x})$ in $G(\mathbf{x})$ is $G_1 = \begin{pmatrix} *_r & * \\ * & 0_{n-r} \end{pmatrix}$. Then the min-rank attack is available and its complexity is $O(q^{2r} \cdot n^4)$ (Tao et al. 2013). Moody et. al. (Moody et al. 2014, 2017) proposed an asymptotically optimal attack with the complexity $O(q^{r+2} \cdot n^4)$ based on the structure of subspace differential invariants. Recently, Liu (Liu et al. 2018) proposed a key recovery attack by solving a system of linear equations derived from the construction of the polynomials, and extended its key recovery attack to the rectangular ABC (Tao et al. 2015) and Cubic ABC (Ding et al. 2014). They claimed that the complexities of these attacks are with the complexity $O(n^{2w})$, which is critical for the security of ABC schemes. On the other hand, one of the anonymous reviewers on the present paper claimed in his/her report that its attack seems doubtful. He/She may present his/her opinion somewhere in the near future.

Table 1 Signature schemes

	#{var.}	#{polyn.}	Sig. gen.	Security
UOV	$o + v$	o	n^3	$q^{v-o}n^4$ (KS)
Rainbow	$o_1 + o_2 + v$	$o_1 + o_2$	n^3	$q^{\min(o_2+v-o_1,o_1)}n^4$ (KS, HR)
HFEv-	$n + r_v$	$n - r_-$	q^{3d}	$n^{(d+r_-+r_v+2)w}$ (MR)

Table 2 Encryption schemes

	#{var.}	#{polyn.}	Decrypt.	Security
HFE var.	$n - r_p$	$n + r_+ - r_-$	q^{3d+r_-}	$n^{(d+r_-+2)w}$ (MR)
Eflash	$n - r_p$	$n - r_-$	$q^{r-}n^3$	$n^{(r_-+3)w}$ (MR)
HFERP var.	$n - r_p$	$n + l + r_+ - r_-$	$q^{3d_0+r_-}$	$(n + l)^{(d_0+r_-+2)w}$ (MR)
ZHFE var.	$n - r_p$	$2n + r_+ - r_-$	q^{3d+r_-}	$(2n)^{(d+r_-+3)w}$ (MR)
EFC var.	$n - r_p$	$2n + r_+ - r_-$	$q^{r-}n^3$	$(2n)^{(r_-+3)w}$ (MR)
ABC	r^2	$2r^2$	n^3	n^{2w} (Liu et. al.)

4 Conclusion

In Sect. 2, we describe the multivariate schemes UOV, Rainbow, HFE variants and the corresponding second round candidates of NIST's project. In Sect. 3, we discuss the practicalities of several new multivariate encryption schemes proposed recently. Tables 1 and 2 are rough sketches of the complexities of decryption/signature generation and the major attacks for the corresponding schemes. Remark that there are various other attacks concerned for implementations.

Table 1 shows that practical signature schemes can be implemented easily since signatures can be generated in polynomial time and the proposed attacks are in exponential time. On the other hand, Table 2 shows that the issues on the practicality of HFE variants have not been eliminated on the new encryption schemes. While selecting parameters for 80-, 100-, 120-bit securities on such encryption schemes might be possible, they will not be able to follow the future inflation of security levels. Further drastic approaches will be required to construct practical multivariate encryption schemes.

Acknowledgements The author would like to thank the anonymous reviewer(s) for reading the previous draft and giving helpful comments. He was supported by JST CREST no.JPMJCR14D6 and JSPS Grant-in-Aid for Scientific Research (C) no. 17K05181.

References

M.L. Akkar, N. Courtois, L. Goubin, R. Duteuil, A fast and secure implementation of Sflash, in *PKC'03*. LNCS, vol. 2567 (2003), pp. 267–278

M. Bardet, J.C. Faugère, B. Salvy, B.Y. Yang, Asymptotic expansion of the degree of regularity for semi-regular systems of equations, in *MEGA'05* (2005)

E.R. Berlekamp, Factoring polynomials over finite fields. Bell Syst. Tech. J. **46**, 1853–1859 (1967)

E.R. Berlekamp, Factoring polynomials over large finite fields. Math. Comput. **24**, 713–735 (1970)

L. Bettale, J.C. Faugère, L. Perret, Solving polynomial systems over finite fields: improved analysis of the hybrid approach. ISSAC **2012**, 67–74 (2012)

L. Bettale, J.C. Faugere, L. Perret, Cryptanalysis of HFE, multi-HFE and variants for odd and even characteristic. Designs, Codes Cryptogr. **69**, 1–52 (2013)

W. Beullens, B. Preneel, A. Szepieniec, F. Vercauteren, LUOV, an MQ signature scheme, https://www.esat.kuleuven.be/cosic/pqcrypto/luov/

D. Cabarcas, D. Smith-Tone, J.A. Verbel, Key recovery attack for ZHFE, in *PQCrypto'17*. LNCS, vol. 10346 (2017), pp. 289–308

R. Cartor, D. Smith-Tone, An updated security analysis of PFLASH, in *PQCrypto'17*. LNCS, vol. 10346 (2017), pp. 241–254

R. Cartor, D. Smith-Tone, EFLASH: a new multivariate encryption scheme, in *SAC'18*. LNCS, vol. 11349 (2018), pp. 281–299

A. Casanova, J.C. Faugère, G. Macario-Rat, J. Patarin, L. Perret, J. Ryckeghem, GeMSS: a great multivariate short signature, https://www-polsys.lip6.fr/Links/NIST/GeMSS.html

C.H.O. Chen, M.S. Chen, J. Ding, F. Werner, B.Y. Yang, Odd-char multivariate hidden field equations, http://eprint.iacr.org/2008/543

M.-S. Chen, A. Hülsing, J. Rijneveld, S. Samardjiska, P. Schwabe, From 5-pass MQ-based identification to MQ-based signatures, in *Asiacrypt'16*. LNCS, vol. 10032 (2016), pp. 135–165

M.-S. Chen, A. Hülsing, J. Rijneveld, S. Samardjiska, P. Schwabe, MQDSS, Post-quantum signature, http://mqdss.org/contact.html

C.M. Cheng, Y. Hashimoto, H. Miura, T. Takagi, A polynomial-time algorithm for solving a class of underdetermined multivariate quadratic equations over fields of odd characteristics, in *PQCrypto'14*. LNCS, vol. 8772 (2014), pp. 40–58

J. Ding, M.-S. Chen. A. Petzoldt, D. Schmidt, B.-Y. Yang, https://csrc.nist.gov/CSRC/media/Projects/Post-Quantum-Cryptography/documents/round-2/submissions/Rainbow-Round2.zip

J. Ding, T.J. Hodges, Inverting HFE systems is quasi-polynomial for all fields, in *Crypto'11*. LNCS, vol. 6841 (2011), pp. 724–742

J. Ding, A. Petzoldt, L.-C. Wang, The cubic simple matrix encryption scheme, in *PQCrypto'14*. LNCS, vol. 8772 (2014), pp. 76–87

J. Ding, D. Schmidt, Rainbow, a new multivariate polynomial signature scheme, in *ACNS'05*. LNCS, vol. 3531 (2005), pp. 164–175

J. Ding, Z. Zhang, J. Deaton, K. Schmidt, Vishakha, A new attack on the LUOV schemes, in *Second PQC Standardization Conference* (2019), https://csrc.nist.gov/events/2019/second-pqc-standardization-conference

V. Dubois, N. Gama, The degree of regularity of HFE systems, in *Asiacrypt'10*. LNCS, vol. 6477 (2010), pp. 557–576

J.C. Faugère, A new efficient algorithm for computing Grobner bases (F_4). J. Pure Appl. Algebra **139**, 61–88 (1999)

J.C. Faugère, A. Joux, Algebraic cryptanalysis of Hidden Field Equations (HFE) using Gröbner bases, in *Crypto'03*. LNCS, vol. 2729 (2003), pp. 44–60

P.A. Fouque, L. Granboulan, J. Stern, Differential cryptanalysis for multivariate schemes, in *Eurocrypt'05*. LNCS, vol. 3494 (2005), pp. 341–353

L. Granboulan, A. Joux, J. Stern, Inverting HFE is quasipolynomial, in *Crypto'06*. LNCS, vol. 4117 (2020), pp. 345–356

Y. Hashimoto, Multivariate public key cryptosystems, in *Mathematical Modelling for Next-Generation Cryptography* (Springer, berlin, 2017), pp. 17–42

Y. Hashimoto, High-rank attack on HMFEv. JSIAM Lett. **10**, 21–24 (2018)

Y. Hashimoto, Key recovery attack on Circulant UOV/Rainbow. JSIAM Lett. **11**, 45–48 (2019)

Y. Hashimoto, Y. Ikematsu, T. Takagi, Chosen message attack on multivariate signature ELSA at Asiacrypt, in *IWSEC'18*. LNCS, vol. 11049 (2018), pp. 3–18

R.A. Horn, Roger, C.R. Johnson, *Matrix Analysis* (Cambridge University Press, Cambridge, 1985)

M.-D.A. Huang, M. Kosters, Y. Yang, S.L. Yeo, On the last fall degree of zero-dimensional Weil descent systems. J. Symb. Comput. **87**, 207–226 (2018)

Y. Ikematsu, R.A. Perlner, D. Smith-Tone, T. Takagi, J. Vates, HFERP - a new multivariate encryption scheme, in *PQCrypto'18*. LNCS, vol. 10786 (2018), pp. 396–416

A. Kipnis, J. Patarin, L. Goubin, Unbalanced oil and vinegar signature schemes, in *Eurocrypt'99*. LNCS, vol. 1592 (1999), pp. 206–222, extended in http://www.goubin.fr/papers/OILLONG.PDF (2003)

A. Kipnis, A. Shamir, Cryptanalysis of the HFE public key cryptosystem by relinearization, in *Crypto'99*. LNCS, vol. 1666 (1999), pp. 19–30

A. Kipnis, A. Shamir, Cryptanalysis of the oil and vinegar signature scheme, in *Crypto'98*. LNCS, vol. 1462 (1998), pp. 257–267

J. Liu, Y. Yu, B. Yang, J. Jia, S. Wang, H. Wang, Structural key recovery of simple matrix encryption scheme family. Comput. J. **61**, 1880–1896 (2018)

T. Matsumoto, H. Imai, Public quadratic polynomial-tuples for efficient signature-verification and message-encryption, in *Eurocrypt'88*. LNCS, vol. 330 (1988), pp. 419–453

H. Miura, Y. Hashimoto, T. Takagi, Extended algorithm for solving underdefined multivariate quadratic equations, in *PQCrypto'13*. LNCS, vol. 7932 (2013), pp. 118–135

D. Moody, R. Perlner, D. Smith-Tone, An asymptotically optimal structural attack on the ABC multivariate encryption scheme, in *PQCrypto'14*. LNCS, vol. 8772 (2014), pp. 180–196

D. Moody, R. Perlner, D. Smith-Tone, Improved attacks for characteristic-2 parameters of the cubic ABC simple matrix encryption scheme, in *PQCrypto'17*. LNCS, vol. 10346 (2017), pp. 255–271

NIST, Post-quantum cryptography standardization, https://csrc.nist.gov/Projects/Post-Quantum-Cryptography/Post-Quantum-Cryptography-Standardization

NIST, Post-quantum cryptography, round 2 submissions, https://csrc.nist.gov/Projects/Post-Quantum-Cryptography/Round-2-Submissions

J. Patarin, Cryptanalysis of the Matsumoto and Imai public key scheme of Eurocrypt'88, in *Crypto'95*. LNCS, vol. 963 (1995), pp. 248–261

J. Patarin, Hidden fields equations (HFE) and isomorphisms of polynomials (IP): two new families of asymmetric algorithms, in *Eurocrypt'96*. LNCS, vol. 1070 (1996), pp. 33–48

J. Patarin, The oil and vinegar signature scheme, in *the Dagstuhl Workshop on Cryptography* (1997)

J. Patarin, N. Courtois, L. Goubin, Quartz, 128-bit long digital signatures, in *CT-RSA'01*. LNCS, vol. 2001 (2020), pp. 282–297

Z. Peng, S. Tang, Circulant UOV: a new UOV variant with shorter private key and faster signature generation. KSII Trans. Int. Inf. Syst. **12**, 1376–1395 (2018)

R. Perlner, D. Smith-Tone, Security analysis and key modification for ZHFE, in *PQCrypto'16*. LNCS, vol. 9606 (2016), pp. 197–212

A. Petzoldt, M.S. Chen, J. Ding, B.Y. Yang, HMFEv - An efficient multivariate signature scheme, in *PQCrypto 2017*. LNCS, vol. 10346 (2017), pp. 205–223

J. Porras, J. Baena, J. Ding, ZHFE, a new multivariate public key encryption scheme, in *PQCrypto'14*. LNCS, vol. 8772 (2014), pp. 229–245

B. Preneel, NESSIE project announces final selection of crypto algorithms, https://www.cosic.esat.kuleuven.be/nessie/deliverables/press_release_feb27.pdf

K. Sakumoto, T. Shirai, H. Hiwatari, Public-key identification schemes based on multivariate quadratic polynomials, in *Crypto'11*. LNCS, vol. 6841 (2011), pp. 706–723

K.-A. Shim, C.-M. Park, N. Koo, An existential unforgeable signature scheme based on multivariate quadratic equations, in *Asiacrypt'17*. LNCS, vol. 10624 (2017), pp. 37–64

D. Smith-Tone, M.-S. Chen, B.-Y. Yang, PFLASH - secure asymmetric signatures on smart cards, in *Lightweight Cryptography Workshop* (2015), http://csrc.nist.gov/groups/ST/lwc-workshop2015/papers/session3-smith-tone-paper.pdf

A. Szepieniec, J. Ding, B. Preneel, Extension field cancellation: a new central trapdoor for multivariate quadratic systems, in *PQC'16*. LNCS, vol. 9606 (2016), pp. 182–196

C. Tao, A. Diene, S. Tang, J. Ding, Simple matrix scheme for encryption, in *PQCrypto 2013*. LNCS, vol. 7932 (2013), pp. 231–242

C. Tao, H. Xiang, A. Petzoldt, J. Ding, Simple matrix - a multivariate public key cryptosystem (MPKC) for encryption. Finite Fields Their Appl. **35**, 352–368 (2015)

E. Tomae, C. Wolf, Solving underdetermined systems of multivariate quadratic equations revisited, in *PKC'12*. LNCS, vol. 7293 (2012), pp. 156–171

J. Vates, D. Smith-Tone, Key recovery attack for all parameters of HFE-, in *PQCrypto'17*. LNCS, vol. 10346 (2017), pp. 272–288

B.Y. Yang, J.M. Chen, Building secure tame-like multivariate public-key cryptosystems: the new TTS, in *ACISP'05*. LNCS, vol. 3574 (2005), pp. 518–531

Ramanujan Graphs for Post-Quantum Cryptography

Hyungrok Jo, Shingo Sugiyama, and Yoshinori Yamasaki

Abstract We introduce a cryptographic hash function based on expander graphs, suggested by Charles et al. '09, as one prominent candidate in post-quantum cryptography. We propose a generalized version of explicit constructions of Ramanujan graphs, which are seen as an optimal structure of expander graphs in a spectral sense, from the previous works of Lubotzky, Phillips, Sarnak '88 and Chiu '92. We also describe the relationship between the security of Cayley hash functions and word problems for group theory. We also give a brief comparison of LPS-type graphs and Pizer's graphs to draw attention to the underlying hard problems in cryptography.

Keywords Ramanujan graphs · Quaternion algebras · Cayley hash functions · Group word problem

1 Introduction

In the era of post-quantum cryptography, there exist four dominant research areas: Lattice-based, Code-based, Multivariate-based and Isogeny-based cryptography. Specifically, studies in the area of Isogeny-based cryptography have been numerous in the past decade, mainly due to the difficulty of finding a path in the Isogeny graph of supersingular elliptic curves.

H. Jo (✉)
Faculty of Engineering, Information and Systems, University of Tsukuba,
1-1-1 Tennodai, Tsukuba, Ibaraki 305-8573, Japan
e-mail: jo.hyungrok.gb@u.tsukuba.ac.jp

S. Sugiyama
Department of Mathematics, College of Science and Technology, Nihon University,
1-8-14 Suruga-Dai, Kanda, Chiyoda, Tokyo 101-8308, Japan
e-mail: s-sugiyama@math.cst.nihon-u.ac.jp

Y. Yamasaki
Graduate School of Science and Engineering, Ehime University,
2-5 Bunkyo-cho, Matsuyama, Ehime 790-8577, Japan
e-mail: yamasaki.yoshinori.mh@ehime-u.ac.jp

© The Author(s) 2021
T. Takagi et al. (eds.), *International Symposium on Mathematics,
Quantum Theory, and Cryptography*, Mathematics for Industry 33,
https://doi.org/10.1007/978-981-15-5191-8_17

231

In 2009, Charles et al. (2009a, 2009b) introduced cryptographic hash functions from expander graphs and explained the hardness of problems behind those schemes. They proposed two kinds of hash functions based on two families of Ramanujan graphs. One of their proposals is based on Ramanujan graphs by Lubotzky et al. (1988) (in short, LPS), which are Cayley graphs over the projective group with respect to well-chosen generating sets. The other is based on Ramanujan graphs by Pizer (1990), which are not (expected to be) Cayley graphs. So far, the variants of their proposal still survive against a quantum attack except the only known exponential complexity attack (Biasse et al. 2014).

In this article, we focus on not only the background of the families of LPS's graphs and their generalization (LPS-type Jo et al. 2020, 2018) with respect to the security of their Cayley-based hash functions, but also on the relationship between the families of LPS-type graphs and Pizer's graphs.

This article is organized as follows: In Sect. 2, we present some required preliminaries of expander graphs and Ramanujan graphs, and also of quaternion algebra theory. We summarize the security on Cayley hash functions and their cryptanalysis (variants of lifting attacks) related to solving word problems in group theory. In Sect. 3, we explain a way to generalize the explicit constructions of LPS and Chiu's Ramanujan graphs, and give a proof of the Ramanujan-ness of our graphs in the special case of "$P = 13$". In Sect. 4, we describe the relationship between the families of LPS-type graphs and Pizer's graphs. In Sect. 5, we summarize the arguments in this article and expound upon some unclarified problems and the relationships between explicit families of Ramanujan graphs.

2 Ramanujan Graphs and Their Cryptographic Applications

An expander graph is well known as a ubiquitous object in various research areas, especially in computer science for designing communication networks. It is said to be a sparse, but highly connected graph. The quality of the network on expander graphs is considered as the expanding ratio. Throughout this article, we assume that all graphs are finite, undirected, simple (i.e. no loops or multi-edges) and connected. Suppose that $X = (V, E)$ is a k-regular graph, composed of a vertex set $V = V(X)$ with n vertices and an edge set $E = E(X)$. For a subset T of V, the *boundary* ∂T of T is defined as

$$\partial T = \{(x, y) \in E | x \in T \text{ and } y \in V \setminus T\},$$

where $V \setminus T$ is the complement of T in V. The *expanding constant* $h(X)$ of X, which is defined as below, is a discrete analogue of the Cheeger constant in differential geometry (Lubotzky 1994):

$$h(X) = \min_{\substack{T \subset V \\ 0 < |T| \leq n/2}} \frac{|\partial T|}{|T|}.$$

We give the definition of an *expander graph*.

Definition 1 A family of k-regular graphs $(X_j)_{j \geq 1}$ such that $|V(X_j)| \to +\infty$ as $j \to +\infty$ is called an *expander family* if there is an $\epsilon > 0$ such that the expanding constant $h(X_j)$ satisfies $h(X_j) \geq \epsilon$ for all j.

For analysis of graphs, the *adjacency matrix* A of the graph X plays an important role; it is a square matrix indexed by pairs of vertices u, v whose (u, v)-entry $A_{u,v}$ is the number of edges between u and v. Since we assume that X has n vertices, A is an n-by-n, symmetric $(0, 1)$-matrix without diagonal entries (i.e. $A_{u,u} = 0$). For such a graph X, the adjacency matrix A of X has the spectrum $k = \lambda_0 > \lambda_1 \geq \cdots \geq \lambda_{n-1}$. It is known (Alon and Milman 1985; Dodziuk 1984) that

$$\frac{k - \lambda_1}{2} \leq h(X) \leq \sqrt{2k(k - \lambda_1)}.$$

If the spectral gap $k - \lambda_1$ is larger, the quality of the network of X is getting better as well. However, it is shown by Alon-Boppana as follows that it cannot be too large.

Theorem 1 *Let* $(X_j)_{j \geq 1}$ *be a family of* k-*regular graphs with* $|V(X_j)| \to +\infty$ *as* $j \to +\infty$. *Then*

$$\liminf_{j \to +\infty} \lambda_1(X_j) \geq 2\sqrt{k - 1}.$$

This fact motivates the definition of a *Ramanujan graph*.

Definition 2 A k-regular graph X is *Ramanujan* if, for every member λ of the spectrum of the adjacency matrix of X other than $\pm k$, one has $|\lambda| \leq 2\sqrt{k - 1}$. We call $2\sqrt{k - 1}$ the *Ramanujan bound* (RB).

For a more detailed exposition of the theory, see Davidoff et al. (2003), Lubotzky (1994), Terras (2010). In order to explain how to construct explicit Ramanujan graphs in the style of LPS, Chiu, LPS-type and Pizer, we recall basic facts and terminologies of quaternion algebras Vignéras (1980).

Let F be a field and $F^\times$ its unit group. Let $\mathcal{A} = \mathcal{A}_F$ be a *quaternion algebra* over F, i.e. a central simple algebra of dimension 4 over F. In this article, we always assume that F is not of characteristic 2. Then, there exist $a, b \in F^\times$ such that it can be written as $\mathcal{A} = \mathcal{A}_F(a, b) = \{\alpha = x + yi + zj + wk \mid x, y, z, w \in F\}$, where i, j, k satisfy $i^2 = a$, $j^2 = b$ and $ij = -ji = k$ (and hence $k^2 = -ab$). For $\alpha = x + yi + zj + wk \in \mathcal{A}$, its *conjugate*, the *reduced trace* and the *reduced norm* are defined by $\bar{\alpha} = x - yi - zj - wk$, $T(\alpha) = \alpha + \bar{\alpha} = 2x \in F$ and $N(\alpha) = \alpha\bar{\alpha} = \bar{\alpha}\alpha = x^2 - ay^2 - bz^2 + abw^2 \in F$, respectively.

Quaternion algebras over $\mathbb{F}_q$

Throughout this article, we denote by $\mathbb{P}$ the set of all prime numbers. For a prime $p \in \mathbb{P}$ and $d \in \mathbb{N}$, let $\mathbb{F}_{p^d}$ be the field of p^d elements. Let us fix $q \in \mathbb{P} \setminus \{2\}$. It is known that, for any $a, b \in \mathbb{F}_q^\times$, the quaternion algebra $\mathcal{A} = \mathcal{A}_{\mathbb{F}_q}(a, b)$ is isomorphic to the matrix algebra $M_2(\mathbb{F}_q)$ of the 2-by-2 matrices over $\mathbb{F}_q$. Let $\left(\frac{\cdot}{\cdot}\right)$ be the Kronecker symbol. When $\left(\frac{a}{q}\right) = \left(\frac{-b}{q}\right) = 1$, that is, $\sqrt{a}, \sqrt{-b} \in \mathbb{F}_q$, one has the following isomorphism.

Lemma 1 *Assume that* $\left(\frac{a}{q}\right) = \left(\frac{-b}{q}\right) = 1$. *Then, the map* $\psi_q : \mathcal{A} \to M_2(\mathbb{F}_q)$ *defined by*

$$\psi_q(x + yi + zj + wk) = \begin{bmatrix} x + y\sqrt{a} & \sqrt{-b}(z + w\sqrt{a}) \\ -\sqrt{-b}(z - w\sqrt{a}) & x - y\sqrt{a} \end{bmatrix}$$

is an isomorphism satisfying $\det(\psi_q(\alpha)) = N(\alpha)$ *and* $\psi_q(\overline{\alpha}) = \overline{\psi_q(\alpha)}$ *for* $\alpha \in \mathcal{A}$.

Here, $\overline{\begin{bmatrix} s & t \\ u & v \end{bmatrix}} = \begin{bmatrix} v & -t \\ -u & s \end{bmatrix}$ *for* $\begin{bmatrix} s & t \\ u & v \end{bmatrix} \in M_2(\mathbb{F}_q)$.

For a ring R, we denote by $R^\times$ the group of units of R. Let $GL_2(\mathbb{F}_q) = M_2(\mathbb{F}_q)^\times$ and $SL_2(\mathbb{F}_q) = \{A \in GL_2(\mathbb{F}_q) \mid \det A = 1\}$. Moreover, let $PGL_2(\mathbb{F}_q) = GL_2(\mathbb{F}_q)/Z(GL_2(\mathbb{F}_q))$ and $PSL_2(\mathbb{F}_q) = SL_2(\mathbb{F}_q)/Z(SL_2(\mathbb{F}_q))$. Here, for a group G, we denote by $Z(G)$ the *center* of G. We can naturally see that $PSL_2(\mathbb{F}_q)$ is a subgroup of $PGL_2(\mathbb{F}_q)$ of index 2 because now q is odd. Additionally, we remark that $|PGL_2(\mathbb{F}_q)| = q(q^2 - 1)$ and $|PSL_2(\mathbb{F}_q)| = \frac{q(q^2-1)}{2}$. Since $\mathcal{A} \simeq M_2(\mathbb{F}_q)$, we have $\mathcal{A}^\times \simeq GL_2(\mathbb{F}_q)$ via (the restriction of) ψ_q and hence obtain the isomorphism $\beta_q : \mathcal{A}^\times/Z(\mathcal{A}^\times) \to PGL_2(\mathbb{F}_q)$.

We need the following lemma later.

Lemma 2 (Davidoff et al. 2003, Chap. 3) *Assume that* $\left(\frac{a}{q}\right) = \left(\frac{-b}{q}\right) = 1$. *Let* $\alpha \in \mathcal{A}$ *with* $N(\alpha) = p \in \mathbb{P} \setminus \{q\}$, *which implies that* $\alpha \in \mathcal{A}^\times$. *Then,* $\beta_q(\alpha \mathbb{F}_q^\times) \in PSL_2(\mathbb{F}_q)$ *if and only if* $\left(\frac{p}{q}\right) = 1$.

Quaternion algebras over $\mathbb{Q}$

Let $a, b \in \mathbb{Z} \setminus \{0\}$ and $\mathcal{A} = \mathcal{A}_{\mathbb{Q}}(a, b)$ be a quaternion algebra over $\mathbb{Q}$. A place v of $\mathbb{Q}$ is said to be *split* in $\mathcal{A}$ if $\mathcal{A}_v := \mathcal{A} \otimes_{\mathbb{Q}} \mathbb{Q}_v \simeq M_2(\mathbb{Q}_v)$, where $\mathbb{Q}_v$ is the v-adic completion of $\mathbb{Q}$ and is said to be *ramified* if $\mathcal{A}_v$ is a division algebra. We denote by $\text{Ram}(\mathcal{A})$ the set of all places which are ramified in $\mathcal{A}$. Notice that $\text{Ram}(\mathcal{A})$ is a finite set, has an even cardinality, and determines an isomorphism class of quaternion algebras over $\mathbb{Q}$. The product of all primes (= finite places) in $\text{Ram}(\mathcal{A})$ is called the *discriminant* of $\mathcal{A}$ and is denoted by $\mathfrak{D}$. From now on, we assume that $\mathcal{A}$ is definite, that is, the infinite place ∞ is ramified in $\mathcal{A}$, whence there are an odd number of primes which are ramified in $\mathcal{A}$. Notice that $\mathcal{A} = \mathcal{A}_{\mathbb{Q}}(a, b)$ is definite if and only if $a < 0$ and $b < 0$.

A *lattice* $I \subset \mathcal{A}$ is a free $\mathbb{Z}$-submodule of $\mathcal{A}$ of rank 4. A lattice $O \subset \mathcal{A}$ is called an *order* if it is a ring with unity. In particular, it is called *maximal* if it is not properly contained in any other order. Notice that, if O is an order of $\mathcal{A}$, then $O \otimes_{\mathbb{Z}} \mathbb{Z}_p$ is an order of $\mathcal{A}_p$ for $p \in \mathbb{P}$. Here, $\mathbb{Z}_p$ is the ring of p-adic integers. Let O be an order of $\mathcal{A}$. We call a lattice I of $\mathcal{A}$ a *left* (resp. *right*) O-*ideal* if $O_L(I) = O$ (resp. $O_R(I) = O$), where $O_L(I) = \{\alpha \in \mathcal{A} \mid \alpha I \subset I\}$ (resp. $O_R(I) = \{\alpha \in \mathcal{A} \mid I\alpha \subset I\}$). We say that two left (resp. right) O-ideals I and J are equivalent, if there exists $\alpha \in \mathcal{A}^\times$ such that $I = J\alpha$ (resp. $I = \alpha J$). This is an equivalence relation. We denote by $H(O)$ the number of equivalence classes, which is shown to be finite, independent on left or right. We call $H(O)$ the *class number* of O.

We next give the definition of Eichler orders. To do that, we first recall the local situations. If $p \in \mathbb{P}$ is ramified in $\mathcal{A}$, then $\mathcal{A}_p$ is a division algebra which has a maximal order $O_p = \{\alpha \in \mathcal{A}_p \mid N(\alpha) \in \mathbb{Z}_p\}$. On the other hand if $p \in \mathbb{P}$ is split in $\mathcal{A}$, then $\mathcal{A}_p$ is isomorphic to $M_2(\mathbb{Q}_p)$ and a maximal order of $\mathcal{A}_p$ is isomorphic to a conjugate of the maximal order $M_2(\mathbb{Z}_p) = \begin{bmatrix} \mathbb{Z}_p & \mathbb{Z}_p \\ \mathbb{Z}_p & \mathbb{Z}_p \end{bmatrix}$ of $M_2(\mathbb{Q}_p)$ by an element of $\mathcal{A}_p^\times$.

Let $\mathfrak{D}$ be the discriminant of $\mathcal{A}$, and M be a positive square-free integer which is prime to $\mathfrak{D}$. An order O of $\mathcal{A}$ is called an *Eichler order* of level $(\mathfrak{D}, M)$ if the following local conditions are satisfied: For all $p \in \mathbb{P}$ being ramified in $\mathcal{A}$ (i.e., $p \mid \mathfrak{D}$), $O \otimes_{\mathbb{Z}} \mathbb{Z}_p = O_p$. On the other hand, for all $p \in \mathbb{P}$ being split in $\mathcal{A}$ (i.e. $p \nmid \mathfrak{D}$), $O \otimes_{\mathbb{Z}} \mathbb{Z}_p$ is isomorphic to a conjugate of the order $\begin{bmatrix} \mathbb{Z}_p & \mathbb{Z}_p \\ M\mathbb{Z}_p & \mathbb{Z}_p \end{bmatrix}$ of $M_2(\mathbb{Q}_p)$ by an element of $\mathcal{A}_p^\times$. Remark that an Eichler order is maximal when $M = 1$. If $p \mid M$, in this case we call p an *Eichler prime*. Notice that an Eichler order can be characterized as an order which is the intersection of two maximal orders. It is shown in Pizer (1976) that the class number of an Eichler order depends only on its level. Hence, we write $H(O)$ as $H(\mathfrak{D}, M)$ when O is of level $(\mathfrak{D}, M)$. Remark that $H(\mathfrak{D}, 1) = 1$ if and only if $\mathfrak{D} = 2, 3, 5, 7, 13$.

Let G be a group and S a generating set, which is symmetric (i.e. $S = S^{-1}$) and does not contain the identity of G. A *Cayley graph* over G with respect to S is a $|S|$-regular graph with a vertex set V and an edge set E, where $V = G$ and E consists of $(g_1, g_2) \in G \times G$ such that $g_1 = g_2 s$ for some $s \in S$.

The families of LPS's graphs Let p and q ($\gg 2\sqrt{p}$) be distinct primes congruent to 1 (mod 4). In Lubotzky et al. (1988), described how to construct a family of Ramanujan graphs of degree $p + 1$ having $O(q^3)$ vertices as $q \to +\infty$. These graphs are Cayley graphs over the groups $G = \mathrm{PGL}_2(\mathbb{F}_q)$ or $\mathrm{PSL}_2(\mathbb{F}_q)$ with respect to the generating set S_{LPS} defined as

$$S_{LPS} = \left\{ \begin{bmatrix} a_0 + ia_1 & a_2 + ia_3 \\ -a_2 + ia_3 & a_0 - ia_1 \end{bmatrix} \middle| a_0{}^2 + a_1{}^2 + a_2{}^2 + a_3{}^2 = p \right. \tag{1}$$

$$\left. \text{for odd } a_0 > 0 \text{ and even } a_1, a_2, a_3 \right\},$$

where $i \in \mathbb{Z}$ such that $i^2 \equiv -1$ (mod q). The diophantine Eq. (1) originally comes from the norm of their based-algebra $\mathcal{A}_{\mathbb{Q}}(-1, -1)$, where $i^2 = -1$, $j^2 = -1$ and $ij = -ji = k$, and is called the *Hamiltonian quaternion algebra*. By Jacobi's four-squares theorem Hirschhorn (1987), there are $8(p + 1)$ integer solutions $(a_0, a_1, a_2, a_3) \in \mathbb{Z}^4$ of (1). Since there are 8 units as $\pm 1, \pm i, \pm j, \pm k$, we see $|S_{LPS}| = p + 1$.

The families of Chiu's graphs In Margulis (1988), independently of LPS, alluded to the existence of essentially the same graphs as shown by LPS, but without an explicit description. In Chiu (1992), described how to construct a family of Ramanujan graphs, and explicitly covered the case of $p = 2$. Since the Hamiltonian quaternion algebra is not split at $p = 2$, Chiu chose a specific quaternion algebra $\mathcal{A}_{\mathbb{Q}}(-2, -13)$, which is split at 2 and has a maximal order of class number 1. Take a prime $q \in \mathbb{P} \setminus \{2, 13\}$ such that $\left(\frac{-2}{q}\right) = \left(\frac{13}{q}\right) = 1$. Chiu's cubic graphs are also Cayley graphs over the groups $G = \mathrm{PGL}_2(\mathbb{F}_q)$ or $\mathrm{PSL}_2(\mathbb{F}_q)$ with respect to the generating set S_C defined as

$$
S_C = \left\{ \begin{bmatrix} 1 & 0 \\ 0 & -1 \end{bmatrix}, \begin{bmatrix} 2+i' & i'j' \\ i'j' & 2-i' \end{bmatrix}, \begin{bmatrix} 2-i' & j'i' \\ j'i' & 2+i' \end{bmatrix} \right\},
$$

where $i', j' \in \mathbb{Z}$ such that $i'^2 \equiv -2$, $j'^2 \equiv 13 \pmod{q}$, respectively.

The families of Morgenstern's graphs   In Morgenstern (1994), described how to construct, for any prime power q, a family of Ramanujan graphs of degree $q + 1$. These graphs are given as Cayley graphs over the groups $G = \mathrm{PGL}_2(\mathbb{F}_{q^d})$ or $\mathrm{PSL}_2(\mathbb{F}_{q^d})$ for some $d \in \mathbb{N}$ with respect to the generating set $S_{M_{\mathrm{odd}}}$ when q is odd and $S_{M_{\mathrm{even}}}$ when q is even. For an odd prime power q, let ϵ be a non-square in $\mathbb{F}_q$. Let $g(x) \in \mathbb{F}_q[x]$ be irreducible of even degree d. We realize $\mathbb{F}_{q^d}$ as $\mathbb{F}_q[x]/g(x)\mathbb{F}_q[x]$. Let $i \in \mathbb{F}_{q^d}$ be such that $i^2 = \epsilon$. Then $S_{M_{\mathrm{odd}}}$ is defined as

$$
S_{M_{\mathrm{odd}}} = \left\{ \begin{bmatrix} 1 & a - ib \\ (a+ib)(x-1) & 1 \end{bmatrix} \ \middle| \ b^2\epsilon - a^2 = 1 \text{ for } a, b \in \mathbb{F}_q \right\}.
$$

For an even prime power q, let ϵ be a non-square in $\mathbb{F}_q$. Let $f(x) = x^2 + x + \epsilon$ be irreducible in $\mathbb{F}_q[x]$. Let $g(x) \in \mathbb{F}_q[x]$ be irreducible of even degree d. We also realize $\mathbb{F}_{q^d}$ as $\mathbb{F}_q[x]/g(x)\mathbb{F}_q[x]$. Let $i' \in \mathbb{F}_{q^d}$ be a root of $f(x)$. Then $S_{M_{\mathrm{even}}}$ is defined as

$$
S_{M_{\mathrm{even}}} = \left\{ \begin{bmatrix} 1 & a - i'b \\ (a+i'b+b)x & 1 \end{bmatrix} \ \middle| \ a^2 + ab + b^2\epsilon = 1 \text{ for } a, b \in \mathbb{F}_q \right\}.
$$

2.1 Security on Cayley Hashes and Word Problems

A *hash function* is a function that accepts a message as an arbitrarily long string of bits and outputs a hash value as a finite, fixed-length string of bits. An efficiency of the hashing process is a basic requirement in a practical point. Such a function should satisfy certain properties such as *collision resistant, second preimage resistant* and *preimage resistant*.

Let $n \in \mathbb{N}$ and let $\mathcal{H} : \{0, 1\}^* \to \{0, 1\}^n$; $m \mapsto h = \mathcal{H}(m)$, where $\{0, 1\}^*$ is the set of bit strings of arbitrary length and $\{0, 1\}^n$ is the set of bit strings of a fixed length n. The function $\mathcal{H}$ is said to be

- **Collision resistant** if it is *computationally infeasible* to find $m, m' \in \{0, 1\}^*$, $m \neq m'$, such that $\mathcal{H}(m) = \mathcal{H}(m')$,
- **Second preimage resistant** if $m \in \{0, 1\}^*$ is given, it is *computationally infeasible* to find $m' \in \{0, 1\}^*$, $m \neq m'$, such that $\mathcal{H}(m) = \mathcal{H}(m')$,

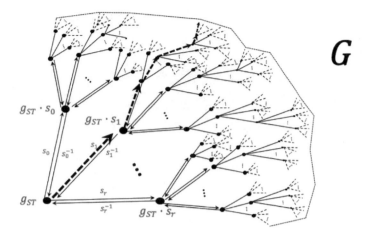

Fig. 1 Diffusion from the starting vertex g_{ST} along Cayley graphs over G with respect to $S = \{s_0, \ldots, s_r\}$

- **Preimage resistant** if $h \in \{0, 1\}^n$ is given, it is *computationally infeasible* to find $m \in \{0, 1\}^*$ such that $h = \mathcal{H}(m)$.

Let G be a non-commutative group and $S = \{s_0, \ldots, s_r\} \subset G$ be a generating set for the group G, symmetric and not having the identity. Charles et al. (2009a) and Petit et al. (2007), Petit and Quisquater (2010b) described a definition of Cayley hash functions, by which the input to hash is used as directions for walking around a graph, and the ending vertex is the output of the hash function as depicted in Fig. 1.

A message m is given as a string $m_1 \ldots m_\ell$, where $m_i \in \{0, \ldots, r\}$. Then the resulting hashing value h of m will be obtained as a group product

$$h := \mathcal{H}(m) = g_{ST} s_{m_1} s_{m_2} \cdots s_{m_\ell},$$

where g_{ST} is a fixed starting element in G. (We usually put g_{ST} as the identity in G.) To dispose a proper sequence of hashing bits inductively, we define a *choice function* π which assigns a next hashing bit with the bit of the message m and the previous hashing bit, while avoiding a back-tracking (i.e. ss^{-1} or $s^{-1}s$). We choose a function

$$\pi : \{0, \ldots, r\} \times S \to S \tag{2}$$

such that for any $s \in S$ the set $\pi(\{0, \ldots, r\} \times \{s\})$ is equal to $S \setminus \{s^{-1}\}$.

The security of Cayley hash functions lies on the hardness of solving *word problems* for group theory (Lubotzky 1994; Meier 2008; Petit and Quisquater 2010b), which are one of the most challenging open problems. There are three problems (*balance, representation and factorization problems*), which are related to the three properties of Cayley hash functions, respectively.

Let $L \in \mathbb{N}$ be small (approximately, $\log |G|$). We denote the product of group elements $s_{m_1}, s_{m_2}, \ldots, s_{m_\ell}$ by $\prod s_{m_i} = s_{m_1} s_{m_2} \cdots s_{m_\ell}$.

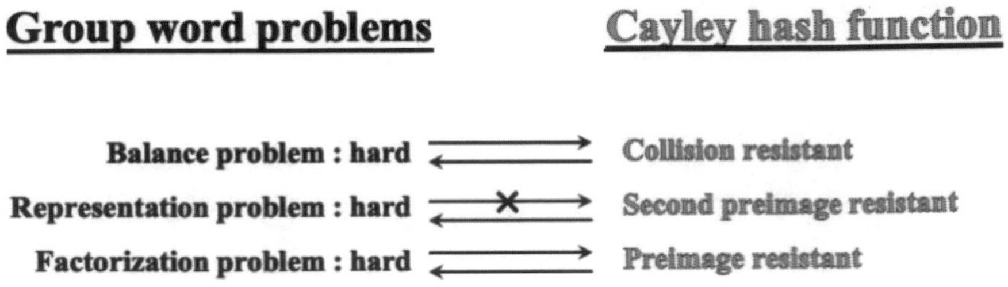

Group word problems   # Cayley hash function

Balance problem : hard $\longleftrightarrow$ Collision resistant

Representation problem : hard $\overset{\times}{\longleftrightarrow}$ Second preimage resistant

Factorization problem : hard $\longleftrightarrow$ Preimage resistant

Fig. 2 Relationship between the properties of Cayley hash functions and the hardness of Group word problems

- **Balance problem** : Find an "efficient" algorithm that returns two words $m_1 \ldots m_\ell$ and $m'_1 \ldots m'_{\ell'}$ with $\ell, \ell' < L$, $m_i, m'_i \in \{0, \ldots, r\}$ and $\prod s_{m_i} = \prod s_{m'_i}$.
- **Representation problem** : Find an "efficient" algorithm that returns a word $m_1 \ldots m_\ell$ with $\ell < L$, $m_i \in \{0, \ldots, r\}$ and $\prod s_{m_i} = 1$.
- **Factorization problem** : Find an "efficient" algorithm that given any element $g \in G$, returns a word $m_1 \ldots m_\ell$ with $\ell < L$, $m_i \in \{0, \ldots, r\}$ and $\prod s_{m_i} = g$.

A Cayley hash function is collision resistant if and only if the balance problem is hard; it is second preimage resistant only if the representation problem is hard; it is preimage resistant if and only if the corresponding factorization problem is hard (as described in Fig. 2).

The *diameter* of a Cayley graph over G with respect to S, which naturally came up from the problems above, is defined as the smallest ℓ such that every element of G can be expressed as a word of length at most ℓ in S. Babai and Seress (1992) conjectured that the diameter of any Cayley graph over any non-commutative simple group is polylogarithmic in the size of the group such as $\exp((|G| \log |G|)^{1/2}(1 + o(|G|)))$. Helfgott and Seress (2014) gave a quasipolynomial upper bound $\exp(\log \log |G|)^{O(1)}$, which is the best known upper bound for permutation groups.

Even after more than two decades of research in various areas (pure mathematics, computer sciences, cryptography, etc.), the hardness of the word problems is still difficult to break. For example, since suggested in Petit and Quisquater (2010b) as a challenge, it seems still open to solve the balance/representation/factorization problems for $G = \mathrm{SL}_2(\mathbb{F}_{2^n})$ with some specific generating set, which is tweaked from the generating set of Tillich and Zémor (1994). They also mentioned that it is an important challenge that we identify groups and their corresponding specific generating sets for the groups in which the balance, representation and factorization problems are difficult.

2.2 Lifting Attacks

In Zémor (1991), proposed the first scheme of hash functions from Cayley graphs upon SL_2 over a finite field having a large *girth*, which is the length of a shortest cycle. Right after the advent, Tillich and Zémor found a way to break Zémor's scheme by a *lifting attack* and suggested its improved version with SL_2 over a finite field of characteristic 2. Tillich–Zémor's scheme (Tillich and Zémor 1994) in resisted being cryptanalyzed for a decade and a half until Grassl et al. (2010) and Petit et al. (2009), Petit and Quisquater (2010a) found their collisions and even preimages in practical. A critical observation for both attacks is that the hardness of balance/representation/factorization problems does not change if we replace the generators for $SL_2(\mathbb{F}_{2^n})$ in order to use the Euclidean algorithm. Even Cayley hash functions based on LPS Ramanujan graphs proposed from Charles et al. (2009a) have been broken by Tillich and Zémor (2008) using a variant of a *lifting attack*.

In this subsection, we give a brief example of a lifting attack, which was used by Tillich and Zémor (2008). We have conditions on distinct prime numbers p and q that p and q satisfy $p \equiv q \equiv 1 \pmod 4$ and $\left(\frac{p}{q}\right) = 1$. First, the elements of $PSL_2(\mathbb{F}_q)$ are lifted to elements of $SL_2(\mathbb{Z}[i])$, where i is the imaginary unit. Even though the lifts of the generators do not generate the whole $SL_2(\mathbb{Z}[i])$ and only a subset Ω of $SL_2(\mathbb{Z}[i])$ with specific conditions shown in Tillich and Zèmor (2008), the lifting attack still works because Ω has a very simple nature as shown below.

$$\Omega = \left\{ \begin{bmatrix} x+iy & z+iw \\ -z+iw & x-iy \end{bmatrix} \middle| (x, y, z, w) \in E_\ell \text{ for some integer } \ell > 0 \right\},$$

where E_ℓ is the set of 4-tuples $(x, y, z, w) \in \mathbb{Z}^4$ such that

$$\begin{cases} x^2 + y^2 + z^2 + w^2 = p^\ell \\ x > 0, x \equiv 1 & \pmod 2 \\ y \equiv z \equiv w \equiv 0 & \pmod 2. \end{cases}$$

Tillich and Zémor solved the *representation problem* by lifting the identity to Ω, which amounts to solving the norm equation

$$(\lambda + xq)^2 + 4(yq)^2 + 4(zq)^2 + 4(wq)^2 = p^\ell \qquad (3)$$

with $\lambda, x, y, z, w \in \mathbb{Z}$ and $\ell \in \mathbb{N}$ (Once the identity is lifted, reduction by q and factoring become trivial). The equation is solved as follows: we arbitrarily fix $\ell = 2\ell'$ with $p^{\ell'} > mq^2$ and $\lambda + xq = p^{\ell'} - 2mq^2$ for some m. We substitute them for each variable in the norm Eq. (3). The norm equation can be deformed by $4q^2$, resulting in the equation of the form $y^2 + z^2 + w^2 = N := m(p^{\ell'} - mq^2)$.

The last equation is solved by generating random variables for w, checking the right parity to ensure that the resulting equation $y^2 + z^2 = N' := N - w^2$ has a solution, and we finally solve this equation with the continued fraction method (or with the advanced Euclidean algorithm, Cornacchia's algorithm, Pell's equation).

Subsequently, most of the existing Cayley hash functions based on explicit Ramanujan graphs Chiu (1988), Lubotzky (1994), Morgenstern (1992) have been broken by variants of a lifting attack Jo et al. (2008), Petit et al. (2008), Tillich and Zémor (2017) as lifting attacks are able to solve the factorization/representation problems for each case. As we can see in Table 1, when we attack Cayley hash func-

Table 1 Norm equations and N to Euclidean algorithm for Cryptanalysis on Cayley hashes

Ramanujan graphs	Norm equation and N for Euclidean algorithm
LPS's Ramanujan graph (Lubotzky 1988)	$x^2 + y^2 + z^2 + w^2 = p^\ell$ $N := p^\ell - z^2 - w^2$
Chiu's Ramanujan graph ($p = 2$) (Chiu 1992)	$x^2 + 2y^2 + 13z^2 + 26w^2 = 2^\ell$ $N := 2^\ell - 13z^2 - 26w^2$
LPS-type Ramanujan graph (Jo et al. 2020)	$x^2 + Py^2 + Qz^2 + PQw^2 = p^\ell$ $N := p^\ell - Qz^2 - PQw^2$

tions, we can apply a lifting attack, which corresponds to a norm equation of their base algebra with a Euclidean algorithm.

Thus, we want to make explicit Ramanujan graphs which have more various norm equations that use P and Q as coefficients ($P \in \{2, 3, 5, 7, 13\}$ and $Q \in \mathbb{P}$ satisfying $Q \equiv 3 \pmod 8$, $\left(\frac{-Q}{P}\right) = -1$ unless $P = 2$). At the very least, for applying variants of a lifting attack, we should set up an attack corresponding to each norm equation. It is also possible to put partial information (P, Q or both) unrevealed during the process of hashing as a private key. From this, we can build the digital signature schemes which mainly resist variants of a lifting attack. This motivates the study of a generalization of LPS's and Chiu's Ramanujan graphs.

3 The Families of LPS-Type Graphs

Now we recall Ibukiyama's construction (Ibukiyama 1982) of maximal orders of definite quaternion algebras over $\mathbb{Q}$ which is ramified at given primes.

Proposition 1 (Ibukiyama 1982) *Let r be an odd positive integer and $P_1, P_2, \ldots, P_r$ distinct prime numbers. Set $M = P_1 P_2 \cdots P_r$. Take a prime number Q such that $Q \equiv 3 \pmod 8$ and $\left(\frac{-Q}{P_i}\right) = -1$ for all i except for i with $P_i = 2$. Moreover, take an integer T such that $T^2 \equiv -M \pmod Q$. Then, $\mathcal{A}_\mathbb{Q}(-M, -Q)$ is a definite quaternion algebra which is ramified only at $\infty, P_1, P_2, \ldots, P_r$. Moreover, let*

$$\omega_1 = \frac{1 + j}{2}, \quad \omega_2 = \frac{i + k}{2} \quad and \quad \omega_3 = \frac{Tj + k}{Q}.$$

Then, $O_{-M, -Q} = \mathbb{Z} + \mathbb{Z}\omega_1 + \mathbb{Z}\omega_2 + \mathbb{Z}\omega_3$ is a maximal order of $\mathcal{A}_\mathbb{Q}(-M, -Q)$.

In Jo et al. (2020, 2018) a specific recipe for constructing LPS-type graphs is presented, and is shown below:

1. Fix a $p \in \mathbb{P}$.
2. Take $P \in \{2, 3, 5, 7, 13\}$ such that $P \neq p$.
3. We take a prime Q satisfying

$$Q \equiv 3 \pmod 8, \ \left(\frac{-Q}{P}\right) = -1 \text{ unless } P = 2$$

and an integer T satisfying $T^2 \equiv -P \pmod Q$. By Proposition 1, we have a definite quaternion algebra $\mathcal{A}_{\mathbb{Q}}(-P, -Q)$ (i.e., $i^2 = -P, j^2 = -Q, ij = -ji = k$) and its maximal order $O = O_{-P,-Q} = \mathbb{Z} + \mathbb{Z}\omega_1 + \mathbb{Z}\omega_2 + \mathbb{Z}\omega_3$ with class number 1, where

$$\omega_1 = \frac{1+j}{2}, \ \omega_2 = \frac{i+k}{2} \text{ and } \omega_3 = \frac{Tj+k}{Q}.$$

4. Find all elements in $O^\times = \{\alpha \in O \mid N(\alpha) = 1\}$.
5. Find all elements in $\{\alpha \in O \mid N(\alpha) = p\}$. Moreover, seek a suitable complete representative of $\{\alpha \in O \mid N(\alpha) = p\}/O^\times$. Define S by the suitable complete representative. Then $|S|$ is exactly equal to $p + 1$, which follows by the class number 1 condition Chiu (1992, Proposition 3.4).
6. Take a $q \in \mathbb{P} \setminus \{2\}$ satisfying $q \neq p$, $\left(\frac{-P}{q}\right) = \left(\frac{Q}{q}\right) = 1$ and $\left(\frac{p}{q}\right) = 1$.
7. Via the isomorphism ψ_q in Lemma 1 and using Lemma 2, we realize S as a subset of $\mathrm{PSL}_2(\mathbb{F}_q)$. Write S_{JSY} for the subset.
8. We have a Cayley graph $X_{P,Q}^{(p,q)} = \mathrm{Cay}(\mathrm{PSL}_2(\mathbb{F}_q), S_{JSY})$.

In Table 2, we present some numerical results by Magma and MATLAB which show the Ramanujan-ness of our constructions. Actually, we will show in the next subsection that our LPS-type graphs are Ramanujan when $P = 13$, which is the only choice of $P \in \{2, 3, 5, 7, 13\}$ such that $O^\times$ is equal to $\{\pm 1\}$. For the cases of $P \in \{2, 3, 5, 7\}$, at present, we have no ideas to prove or disprove the Ramanujan-ness of our graphs.

Table 2 Numerical results on the Ramanujan-ness of LPS-type graphs $X = X_{P,Q}^{(p,q)}$

p	Parameters (P, Q, q, T)	$\lambda_1(X)$	$2\sqrt{p}$ (RB)	$\lvert V(X) \rvert = q(q^2 - 1)/2$
2	$(13, 11, 7, 3)$	2.7253	2.8284	168
3	$(2, 3, 11, 1)$	3.3322	3.4641	660
5	$(2, 3, 11, 1)$	4.4718	4.4721	660
7	$(5, 67, 3, 14)$	3	5.2915	12
11	$(13, 11, 7, 3)$	6	6.6332	168

3.1 Proof of the Ramanujan-Ness of Graphs $X_{P,Q}^{(p,q)}$ when $P = 13$

We show that our graph $X_{P,Q}^{(p,q)}$ constructed as above is Ramanujan when $P = 13$. Let $O = \mathbb{Z} + \mathbb{Z}\omega_1 + \mathbb{Z}\omega_2 + \mathbb{Z}\omega_3$ be the maximal order we constructed as above for a fixed p, P, Q, T. Then, O has the class number 1.

Take a complete representative $S_{JSY} = \{\alpha_1, \ldots, \alpha_s\} \cup \{\bar{\alpha}_1, \ldots, \bar{\alpha}_s\} \cup \{\beta_1, \ldots, \beta_t\}$ of $\{\alpha \in O \mid N(\alpha) = p\}/O^\times$ so that $\bar{\beta}_j = \epsilon_j \beta_j$ for some $\epsilon_j \in O^\times$ for every j. In this case, $p + 1 = 2s + t$. In the same way as Coan and Perng (2012, Theorem 4.8) and Lubotzky (1988, Lemma 3.1), we have the following:

Lemma 3 *Any $\alpha \in O$ with $N(\alpha) = p^k$ for some $k \in \mathbb{N}$ is uniquely decomposed into the product*

$$\alpha = \epsilon p^r R(\alpha_1, \ldots, \alpha_s, \bar{\alpha}_1, \ldots, \bar{\alpha}_s, \beta_1, \ldots, \beta_t),$$

where $\epsilon \in O^\times$, $r \in \mathbb{N}$ and $R(\alpha_1, \ldots, \alpha_s, \bar{\alpha}_1, \ldots, \bar{\alpha}_s, \beta_1, \ldots, \beta_t)$ is a reduced word of $\alpha_1, \ldots, \alpha_s, \bar{\alpha}_1, \ldots, \bar{\alpha}_s, \beta_1, \ldots, \beta_t$ with length $m = k - 2r$.

The unit group $O^\times$ is $\{\pm 1\}$ only when $P = 13$. In such a case, we can prove the Ramanujan-ness of our graph $X_{P,Q}^{(p,q)}$ in the same way as Lubotzky (1988). For the variable $v = (x, y, z, w)$, we set

$$Q_q(v) = x^2 + qxy + q^2\left(\frac{1+Q}{4}\right)y^2 + q^2 Tyz$$

$$+ q^2 P\left(\frac{1+Q}{4}\right)z^2 + q^2 Pzw + q^2\left(\frac{P+T^2}{Q}\right)w^2.$$

It is a positive-definite quadratic form of order 4 corresponding to the reduced norm on O. Let A_q be the symmetric matrix such that $Q_q(v) = \frac{1}{2}{}^t v A_q v$, i.e.

$$A_q = \begin{bmatrix} 2 & q & 0 & 0 \\ q & \frac{q^2(1+Q)}{2} & 0 & q^2 T \\ 0 & 0 & \frac{q^2 P(1+Q)}{2} & q^2 P \\ 0 & q^2 T & q^2 P & 2q^2\frac{P+T^2}{Q} \end{bmatrix}.$$

Hence, A_q is an even matrix, i.e. $A_q \in M_4(\mathbb{Z})$ and every diagonal component is contained in $2\mathbb{Z}$. The *level* of Q_q is defined as the smallest positive integer N such that $N A_q^{-1}$ is an even matrix (cf. Schoeneberg 2012, Chap. IX). By $\det(A_q) = P^2 q^6$ and

$$A_q^{-1} = \frac{1}{P^2 q^6}\begin{bmatrix} q^6\frac{1+Q}{2}P(\frac{P+T^2}{Q}) & -q^5 P\left(\frac{P+T^2}{Q} + T^2\right) & -q^5 PT & q^5 PT\frac{1+Q}{2} \\ -q^5 P\left(\frac{P+T^2}{Q} + T^2\right) & 2q^4 P(\frac{P+T^2}{Q} + T^2) & 2q^4 PT & -q^4 PT(1+Q) \\ -q^5 PT & 2q^4 PT & 2q^4 P & -PQq^4 \\ q^5 PT\frac{1+Q}{2} & -q^4 PT(1+Q) & -PQq^4 & q^4 PQ\frac{(1+Q)}{2} \end{bmatrix},$$

the level of Q_q is equal to Pq^2.

Set $r_{Q_q}(n) := |\{\alpha \in O \mid N(\alpha) = n\}|$ for $n \in \mathbb{N}$. Then, the theta series $\Theta_{Q_q}(z) := \sum_{n=0}^{\infty} r_{Q_q}(n)e^{2\pi i n z} = \sum_{v \in \mathbb{Z}^4} e^{2\pi i Q_q(v)z}$ for $z \in \mathbb{C}$ with $\text{Im}(z) > 0$ is absolutely and locally uniformly convergent by Schoeneberg (2012, Chap. IX, Sect. 1.1). Referring to Schoeneberg (2012, Chap. IX, Theorem 4) and Schoeneberg (2012, Chap. IX, Theorem 5) for $\mathbf{h} = \mathbf{0}$, the theta series $\Theta_{Q_q}(z)$ is a holomorphic modular form of weight 2 and level $\Gamma_0(Pq^2)$ with trivial nebentypus. Here, $\Gamma_0(Pq^2)$ is the Hecke congruence subgroup of level Pq^2. We remark that Q_q, A_q, Θ_{Q_q}, are valid for a general $q \in \mathbb{N}$.

Assume $P = 13$. Let Λ' be the set of all $\alpha \in O$ such that $N(\alpha) = p^k$ for some $k \in \mathbb{N}$. We define an equivalence relation on Λ so that $\alpha \sim \beta$ means $\alpha = \epsilon p^n \beta$ for some $\epsilon \in O^\times$ and $n \in \mathbb{Z}$. Since $O^\times = \{\pm 1\}$ holds, the quotient set $\Lambda := \Lambda'/\sim = \{[\alpha] \mid \alpha \in \Lambda'\}$ has a natural group structure by $[\alpha][\beta] = [\alpha\beta]$. By Lemma 3, it is generated by S_{JSY}, a complete representative of $\{\alpha \in O \mid N(\alpha) = p\}/O^\times$, and $\text{Cay}(\Lambda, S_{JSY})$ is a $(p+1)$-regular tree. The homomorphism $\Lambda \to \text{PSL}_2(\mathbb{F}_q)$ as a restriction of ψ_q of Lemma 1 induces $\Lambda/\Lambda(q) \to \text{PSL}_2(\mathbb{F}_q)$ with $\Lambda(q) = \ker(\psi_q|_\Lambda)$. This homomorphism $\Lambda/\Lambda(q) \to \text{PSL}_2(\mathbb{F}_q)$ is surjective as in the theory of quadratic diophantine equations applied to the quadratic form Q_1 (cf. Lubotzky et al. 1988, p. 267; Malishev 1962). Then our graph $X_{13,Q}^{(p,q)} = \text{Cay}(\text{PSL}_2(\mathbb{F}_q), S_{JSY})$ is identified with $\Lambda/\Lambda(q)$ as a graph.

For proving Ramanujan-ness, let $\lambda_0 = p + 1 > \lambda_1 \geq \cdots \geq \lambda_{n-1}$ be the spectrum of the adjacency matrix of $X_{13,Q}^{(p,q)}$ (so we set $n = |X_{13,Q}^{(p,q)}| = |\text{PSL}_2(\mathbb{F}_q)|$). Then, we have only to show $\theta_j \in \mathbb{R}$ for all $j \in \{1, \ldots, n-1\}$, where $\theta_j \in \mathbb{C}$ is taken so that $\lambda_j = 2\sqrt{p}\cos\theta_j$ for each $j \in \{0, \ldots, n-1\}$. By the trace formula for a regular graph as in Lubotzky (1988, p. 270–272 and p. 274, Remark 2), we have the expression

$$r_{Q_q}(p^k) = \frac{2p^{k/2}}{n} \sum_{j=0}^{n-1} \frac{\sin(k+1)\theta_j}{\sin\theta_j}.$$

Recall that this is the p^k-th Fourier coefficient of the modular form Θ_{Q_q}. Since the theta series is a sum of a linear combination of cuspidal Hecke eigenforms and that of Eisenstein series of weight 2 and level $\Gamma_0(Pq^2)$, we may take a cusp form f_1 and a non-cusp form f_2 of weight 2 so that $\Theta_{Q_q} = f_1 + f_2$. Let $a(m)$ and $C(m)$ be the m-th Fourier coefficients of f_1 and f_2 at the cusp ∞ for $m \in \mathbb{N}$, respectively. Then, $r_{Q_q}(p^k)$ has the following expression:

$$C(p^k) + a(p^k) = r_{Q_q}(p^k) = \frac{2p^{k/2}}{n} \sum_{j=0}^{n-1} \frac{\sin(k+1)\theta_j}{\sin\theta_j}.$$

By Deligne's bound as a resolution of the Ramanujan–Petersson conjecture (Deligne 1969, 1974), we have $|a(p^k)| = O_\epsilon(p^{k(1/2+\epsilon)})$. Due to the explicit nature of Fourier coefficients of Eisenstein series, $C(m)$ can be described as $C(m) = \sum_{d \mid m} F(d)$ for a periodic function $F : \mathbb{N} \to \mathbb{C}$ (cf. Lubotzky 1988, p. 272). By $\left(\frac{p}{q}\right) = 1$ and $\theta_0 = i \log \sqrt{p}$, we have

$$C(p^k) = \frac{2}{n} \frac{p^{k+1} - 1}{p - 1} - a(p^k) + o(p^k) = \frac{2}{n} \frac{p^{k+1} - 1}{p - 1} + o(p^k).$$

By the Deligne bound of $a(p^k)$ and Lubotzky (1988, Lemma 4.4), we have $C(p^k) = \frac{2}{n} \frac{p^{k+1}-1}{p-1}$ because of $\left(\frac{p}{q}\right) = 1$. As a consequence, for any $\epsilon > 0$,

$$\frac{2}{n} \sum_{j=1}^{n-1} \frac{\sin(k+1)\theta_j}{\sin \theta_j} = \frac{1}{p^{k/2}} O_\epsilon(p^{k(1/2+\epsilon)}) = O_\epsilon(p^{k\epsilon}),$$

which leads us that every θ_j for $j \in \{1, \ldots, n-1\}$ is real. Therefore, we obtain $|\lambda_j| \le 2\sqrt{p}$ for all $j = 1, \ldots, n-1$, which implies that $X_{13,Q}^{(p,q)}$ is a Ramanujan graph.

We remark an adelic approach toward Ramanujan-ness. As we see Costache et al. (2018, Sect. 7.2) (see also Lubotzky 1994, Theorem 7.1.1), we can prove the Ramanujan-ness of $X_{P,Q}^{(p,q)}$ for $P = 13$ by using an adelic interpretation as well as by using the Jacquet–Langlands correspondence between automorphic representations of the adelic group $GL_2(\mathbb{A}_\mathbb{Q})$ and those of $\mathcal{A}^\times(\mathbb{A}_\mathbb{Q}) = (\mathcal{A} \otimes \mathbb{A}_\mathbb{Q})^\times$, which is the adelization of the anisotropic inner form $\mathcal{A}^\times$ of GL_2.

4 Relationship Between LPS-Type Graphs and Pizer's Graphs

While research in the field of Cayley-based cryptography has been declining, research in the field of Isogeny-based cryptography is quite robust, in part due to its key role in post-quantum cryptography.

However, it is also natural to investigate whether attacks on group word problems of Cayley hash functions based on LPS's graphs are related to the problem of finding a path in an isogeny graph of supersingular elliptic curves, which is explained in detail in Charles et al. (2009b).

Costache et al. (2018) described a wide range of usage of Ramanujan graphs in cryptography and also pointed out some different aspects of LPS's graphs and Pizer's graphs with specific features. They presented the construction of LPS's graphs as Cayley graphs, in terms of local double cosets. They used strong approximation (Costache et al. 2018, Sect. 7; Lubotzky 1994, Sect. 6.3) as a main tool to present the connection between local and adelic double cosets for LPS's and Pizer's graphs. They also compared the two types of graphs in an aspect of appearance by restricting the degree of the graphs (i.e. $p = 5$).

In this section, we give some comparisons between LPS-type graphs and Pizer's graphs as Costache et al. did. First, we describe Pizer's Ramanujan graphs referred to in Pizer (1990, 1998), Costache et al. (2018).

The families of Pizer's graphs Pizer (1990, 1998) showed how to construct the family of Ramanujan graphs as follows: Let $\mathcal{A}$ be the quaternion algebra over $\mathbb{Q}$ that is ramified exactly at odd $q \in \mathbb{P}$ and ∞. We shall consider special orders, which are generalizations of Eichler orders, of level $L = (q, M)$ and $L = (q^2, M)$. The vertex set of Pizer's graph $G(L, p)$ shall be in bijection with (a subset of) the isomorphism classes of left ideals of an order. Since the class number of the order depends only on its level, we may write $H(L)$ for it, which is equal to the size of such a graph. Notice that, by Pizer (1998, Proposition 4.4), we have

$$H(q, M) = \frac{q-1}{12} M \prod_{d|M}(1 + 1/d) + \begin{cases} \frac{1}{4}\left(1 - \left(\frac{-4}{q}\right)\right)\prod_{d|M}\left(1 + \left(\frac{-4}{d}\right)\right) & \text{if } 4 \nmid M \\ 0 & 4 \mid M \end{cases}$$
$$+ \begin{cases} \frac{1}{3}\left(1 - \left(\frac{-3}{q}\right)\right)\prod_{d|M}\left(1 + \left(\frac{-3}{d}\right)\right) & \text{if } 9 \nmid M \\ 0 & \text{if } 9 \mid M \end{cases}$$

and

$$H(q^2, M) = \frac{q^2-1}{12} M \prod_{d|M}(1 + 1/d) + \begin{cases} 0 & \text{if } q \geq 5 \\ \frac{4}{3}\prod_{d|M}\left(1 + \left(\frac{-3}{d}\right)\right) & \text{if } q = 3. \end{cases}$$

Here, the product is over all primes d dividing M.

We give a definition of a Brandt matrix. Let $\{I_1, I_2, \ldots, I_H\}$ with $H = H(L)$ be a complete representative of the left ideal classes of O. For each $i \in \{1, \ldots, H\}$, let O_i be the right order of the ideal I_i, and e_i be the number of $O_i^\times$. For $n \in \mathbb{N}$, the *Brandt matrix* $B(L; n) = \left[b_{i,j}^{(n)}\right]$ associated to an order of level L is a square matrix of size $H(L)$ having (i, j)-entry

$$b_{i,j}^{(n)} = e_j^{-1} \cdot |\{\alpha \in I_j^{-1}I_i \mid N(\alpha)N(I_j)/N(I_i) = n\}|,$$

where $N(I)$ is the norm of an ideal I defined as the greatest common divisor of the norms of its nonzero elements. Let p be a prime which is coprime to qM. If we restrict the parameters p and q, the edge set of $G(L, p)$ is given by a Brandt matrix $B(L; p)$, namely the adjacency matrix of $G(L, p)$ is given by $B(L; p)$. By Pizer (1998, Proposition 4.6), we see that $G(L, p)$ is undirected (i.e. $B(L; p)$ is symmetric) when $L = (q, M)$ with $q \equiv 1 \pmod{12}$ and $L = (q^2, M)$ with $q > 3$. Moreover, it has no loops if $\operatorname{tr} B(L; p) = 0$ and no multiple edges if $\operatorname{tr} B(L; p^2) = H(L)$ (Costache et al. 2018; Pizer 1998). The regularity $p + 1$

Table 3 The families of Pizer's graphs $G(L, p)$

Conditions \ Level	$L = (q, M)$	$L = (q^2, M)$	
Coprimality	$(p, qM) = 1$		
Bipartite-ness	non-bipartite	if $\left(\frac{p}{q}\right) = -1$ bipartite	if $\left(\frac{p}{q}\right) = 1$ non-bipartite
# of vertices	$H(L)$	$H(L)$	$H(L)/2$
Undirected-ness	$q \equiv 1 \pmod{12}$	$q > 3$	
No loops	$\operatorname{tr} B(L; p) = 0$		
No multiple edges	$\operatorname{tr} B(L; p^2) = H(L)$		
Regularity	$(p + 1)$-regular		

of $G(L, p)$ and its connectedness can be obtained from using $B(L; p)$ as the adjacency matrix, as shown in Pizer (1998, Proposition 5.1). We summarize the necessary properties of the families of Pizer's graphs $G(L, p)$ in Table 3.

4.1 Similarities and Differences

As Costache et al. (2018) argued, we explicate the similarities and differences among LPS, LPS-type and Pizer's graphs from a number-theoretic perspective. These families can be viewed as sets of local double cosets, i.e. as graphs of the form

$$\Gamma \backslash \mathrm{PGL}_2(\mathbb{Q}_p)/\mathrm{PGL}_2(\mathbb{Z}_p),$$

where Γ is a discrete cocompact subgroup.

Discrete local double cosets (LPS-type) Let p be a split prime in $\mathcal{A}$. For $N \in \mathbb{N}$, we set

$$\Gamma(N) := \ker(\mathcal{A}^\times(\mathbb{Z}[p^{-1}]) \to \mathbb{Z}[p^{-1}]^\times \backslash \mathcal{A}^\times(\mathbb{Z}[p^{-1}]/N\mathbb{Z})).$$

It is a discrete cocompact subgroup in $\mathcal{A}_p^\times$. We have

$$\mathrm{Cay}(\mathrm{PSL}_2(\mathbb{F}_q), S) \cong \Gamma(q) \backslash \mathrm{PGL}_2(\mathbb{Q}_p)/\mathrm{PGL}_2(\mathbb{Z}_p)$$

for some suitable S.

For LPS-type graphs, the local double cosets are also isomorphic to adelic double cosets, but in this case the corresponding set of adelic double cosets is smaller relative to the quaternion algebra and we do not have the same chain of isomorphisms as shown below. On the other hand, Pizer's graphs, via strong approximation (Costache et al. 2018; Lubotzky 1994), can be viewed as graphs on adelic double cosets which are in turn the set of classes of an order of $\mathcal{A}$ that is related to a discrete cocompact subgroup Γ. Moreover, the class set $\text{Cl}(O)$ of a maximal order O from Pizer's graph is in bijection with supersingular elliptic curves (Charles et al. 2009b, Sect. 5.3.1) and offers convincing evidence that an isomorphism is obtained with a supersingular isogeny graph (SSIG).

The chain of isomorphisms
(LPS)

$$\text{Cay}(\text{PSL}_2(\mathbb{F}_q), S_{LPS}) \cong \Gamma(2q)\backslash\text{PGL}_2(\mathbb{Q}_p)/\text{PGL}_2(\mathbb{Z}_p)$$

(LPS-type with $P = 13$)

$$\text{Cay}(\text{PSL}_2(\mathbb{F}_q), S_{JSY}) \cong \Gamma(q)\backslash\text{PGL}_2(\mathbb{Q}_p)/\text{PGL}_2(\mathbb{Z}_p)$$

(Pizer)

$$O[p^{-1}]^\times\backslash\text{GL}_2(\mathbb{Q}_p)/\text{GL}_2(\mathbb{Z}_p) \cong \text{Cl}(O) \cong \text{SSIG}$$

Each of the underlying quaternion algebras vary with their own choice of parameters. In the case of LPS's graphs, we use the Hamiltonian quaternion algebra, ramified at 2 and ∞ and split at p. In the case of LPS-type graphs, we use the definite quaternion algebra, ramified at 13 and ∞ and split at p. Varying the parameter q, we can have different Ramanujan graphs of LPS and LPS-type, depending on the congruence subgroup $\Gamma(2q)$ and $\Gamma(q)$, respectively, without changing each of their underlying quaternion algebras. On the other hand, in the case of Pizer's graphs, we use the definite quaternion algebra, ramified at q and ∞.

5 Open Problems

It is unknown whether the link exists between the hardness of the path-finding problem in Supersingular Isogeny (Pizer) graphs and the the hardness of group word problems in Cayley-type Ramanujan graphs. If it is possible to connect those two problems theoretically or schematically, there are some expected ways to analyze the hardness of the path-finding problem in Pizer's graphs by employing the approach previously used for Cayley graphs. As a part of these approaches, it is also important to investigate much more general versions of explicit constructions of Ramanujan graphs. It is in the process to construct the family of $(2p + 1)$-regular graphs, where p is an Eichler prime based on the quaternion algebra with an explicit construction of Eichler order having class number 1 in Jo et al. (2020). We now study the Ramanujan-ness of these graphs by similar arguments in LPS-type graphs.

Additionally, even though it is difficult to predict that Pizer's graph can be represented as a Cayley graph over a group with respect to a suitable generating set (actually, all graphs with a small number of vertices, suggested as examples in Pizer

1998 are not Cayley graphs), it is not clear whether a Pizer's graph with a sufficiently large number of vertices is a Cayley graph or not.

Acknowledgements This work was supported by JST CREST Grant Number JPMJCR14D6, Japan. The authors would like to thank Meghan Delaney for pointing out grammatical errors.

References

N. Alon, V. Milman, λ_1, isoperimetric inequalities for graphs, and superconcentrators. J. Comb. Theory. B. **38**(1), 73–88 (1985)

L. Babai, Á. Seress, On the diameter of permutation groups. European. J. Combin. **13**(4), 231–243 (1992)

J.F. Basilla, On the solution of $x^2 + dy^2 = m$. P. Jpn. Acad. A-Math **80**(5), 40–41 (2004)

J.F. Biasse, D. Jao, A. Sankar, A quantum algorithm for computing isogenies between supersingular elliptic curves. Indocrypt LNCS **8885**, 428–442 (2014)

D.X. Charles, E.Z. Goren, K.E. Lauter, Cryptographic hash functions from expander graphs. J. Cryptol. **22**(1), 93–113 (2009a)

D.X. Charles, E.Z. Goren, K.E. Lauter, Families of Ramanujan graphs and quaternion algebras. Groups and symmetries, in *CRM Proceedings and Lecture Notes*, vol. 47 (American Mathematical Society, Providence, RI, 2009b), 53–80

P. Chiu, Cubic Ramanujan graphs. Combinatorica **12**(3), 275–285 (1992)

B. Coan, C. Perng, Factorization of Hurwitz quaternions. Int. Math. Forum **7**(41–44), 2143–2156 (2012)

A. Costache, B. Feigon, K.E. Lauter, M. Massierer, A. Puskás, Ramanujan graphs in cryptography. arXiv:1806.05709 (2018)

G. Davidoff, P. Sarnak, A. Valette, *Elementary Number Theory, Group Theory and Ramanujan Graphs* (Cambridge University Press, Cambridge, 2003)

L. De Feo, D. Jao, J. Plût, Towards quantum-resistant cryptosystems from supersingular elliptic curve isogenies. J. Math. Cryptol. **8**(3), 209–247 (2014)

P. Deligne, Formes modulaires et représentations l-adiques, Séminaire N. Bourbaki, exp. n°, 139–172 (1968–1969)

P. Deligne, La conjecture de Weil. I, Inst. Hautes Études Sci. Publ. Math. **43**, 273–307 (1974)

J. Dodziuk, Difference equations, isoperimetric inequality and transience of certain random walks. T. Am. Math. Soc. **284**(2), 787–794 (1984)

M. Eichler, Zur Zahlentheorie der Quaternionen-Algebren. J. Reine Angew. Math. **195**(1955), 127–151 (1956)

M. Eichler, S. Sundaravaradan, Lectures on modular correspondences. Tata Institute of Fundamental Research (1956) Available via DIALOG. http://www.math.tifr.res.in/~publ/ln/tifr09.pdf

M. Eichler, The basis problem for modular forms and the traces of the Hecke operators, in *Modular Functions of One Variable*, vol. 320 ed. by W. Kuyk (Springer, Heidelberg, 1973), 75–152

M. Grassl, I. Ilić, S. Magliveras, R. Steinwandt, Cryptanalysis of the Tillich-Zémor Hash Function. J. Cryptol. **24**(1), 148–156 (2010)

O. Goldreich, *Foundations of Cryptography* (Cambridge University Press, Cambridge, 2004)

H.A. Helfgott, Á. Seress, On the diameter of permutation groups. Ann. Math. **179**, 611–658 (2014)

M. Hirschhorn, A simple proof of Jacobi's four-square theorem. P. Am. Math. Soc. **101**(3), 436–438 (1987)

H. Hoory, N. Linial, A. Wigderson, Expander graphs and their applications. B. Am. Math. Soc. **43**(4), 439–561 (2006)

T. Ibukiyama, A basis and maximal orders of quaternion algebras over the rational number (In Japanese). MSJ, Sugaku **24**(4), 316–318 (1972) https://core.ac.uk/download/pdf/38181256.pdf

T. Ibukiyama, On maximal orders of division quaternion algebras over the rational number field with certain optimal embeddings. Nagoya. Math. J. **88**, 181–195 (1982)

Y. Ihara, Discrete Subgroups of PL(2, $\mathfrak{k}_p$). Proc. Symp. Pure Math. **18**, 272–278 (1966)

H. Jo, C. Petit, T. Takagi, Full cryptanalysis of hash functions based on cubic ramanujan graphs. IEICE Trans. Fundam. Electron. Commun. Comput. Sci. **100**(9), 1891–1899 (2017)

H. Jo, S. Sugiyama, Y. Yamasaki, A general explicit construction of LPS-type Ramanujan graphs, in preparation

H. Jo, Y. Yamasaki, LPS-type Ramanujan graphs, in 2018 International Symposium on Information Theory and Its Applications, ISITA 2018, 399–403 (2018)

M. Kirschmer, J. Voight, Algorithmic enumeration of ideal classes for quaternion orders. SIAM J. Comput. **39**(5), 1714–1747 (2010)

A. Lubotzky, R. Phillips, P. Sarnak, Ramanujan graphs. Combinatorica **8**(3), 261–277 (1988)

A. Lubotzky, Discrete groups, expanding graphs and invariant measures (Springer Science Business Mediam, Berlin, 1994)

G. Margulis, Explicit group-theoretical constructions of combinatorial schemes and their application to the design of expanders and concentrators. Probl. Peredachi. Inf. **24**(1), 51–60 (1988)

A.I. Malishev, On the representation of integers by positive definite forms (in Russian). Trudy Mat. Inst. Steklov. **65**, 1–319 (1962)

J. Meier, *Groups, graphs and trees; an introduction to the geometry of infinite groups* (Cambridge University Press, Cambridge, 2008)

J.F. Mestre, La méthode des graphes. Exemples et applications, in *Proceedings of the International Conference on Class Numbers and Fundamental Units of Algebraic Number Fields (Katata)*, 217–242 (1986)

J.F. Mestre, T.A. Jorza, The Method of Graphs. Examples and Applications. Notes. (2011)

M. Morgenstern, Existence and explicit constructions of $q + 1$ regular Ramanujan graphs for every prime power q. J. Comb. Theory, Ser. B **62**(1), 44–62 (1994)

O. Parzanchevski, P. Sarnak, Super-golden-gates for PU(2). Adv. Math. **327**, 869–901 (2018)

C. Petit, K.E. Lauter, J.J. Quisquater, Cayley hashes: A class of efficient graph-based hash functions, preprint. (2007)

C. Petit, K.E. Lauter, J.J. Quisquater, Full cryptanalysis of LPS and Morgenstern hash functions. SCN LNCS **5229**, 263–277 (2008)

C. Petit, J.J. Quisquater, Preimages for the Tillich-Zémor hash function, in *International Workshop on Selected Areas in Cryptography*. (Springer, Berlin, Heidelberg, 2010), 282–301

C. Petit, J.J. Quisquater, Rubik's for cryptographers. IACR Cryptology ePrint Archive, vol. 638 (2010)

C. Petit, J.J. Quisquater, J.P. Tillich, G. Zémor, Hard and easy components of collision search in the Zémor-Tillich hash function: New attacks and reduced variants with equivalent security, in *Cryptographers' Track at the RSA Conference* (Springer, Berlin, Heidelberg, 2009), 182–194

A.K. Pizer, Type numbers of Eichler orders. J. Reine Angew. Math. **264**, 76–102 (1973)

A.K. Pizer, On the arithmetic of quaternion algebras. Acta Arith. **31**, 61–89 (1976)

A.K. Pizer, Ramanujan graphs and Hecke operators. B. Am. Math. Soc. **23**(1), 127–137 (1990)

A.K. Pizer, Ramanujan graphs. AMS/IP Stud. Adv. Math. **7**, 159–178 (1998)

H.J. Rosson, B.J. Ellison, J.B. Wilson, Trees, Hecke operators, and quadratic forms, preprint. https://www.math.colostate.edu/~jwilson/math/PrePrintTree.pdf

P. Sarnak, *Some Applications of Modular Forms* (Cambridge University Press, Cambridge, 1999)

B. Schoeneberg, Elliptic Modular Functions: An Introduction, vol. 203 (Springer, Berlin, 2012)

A. Terras, *Zeta functions of graphs; a stroll through the garden*, vol. 128 (Cambridge University Press, Cambridge, 2010)

J.P. Tillich, G. Zémor, Hashing with SL_2, in *Annual International Cryptology Conference* (Springer, Berlin, Heidelberg, 1994), 40–49

J.P. Tillich, G. Zèmor, Collisions for the LPS expander graph hash function. Eurocrypt LNCS **3027**, 254–269 (2008)

M.F. Vignéras, *Arithmétique des algèbres de quaternions*. Lecture Notes in Mathematical, vol. 800 (Springer, Berlin, 1980)

G. Zémor, Hash functions and graphs with large girths, in *Workshop on the Theory and Application of Cryptographic Techniques* (Springer, Berlin, Heidelberg, 1991), 508–511

Post-Quantum Constant-Round Group Key Exchange from Static Assumptions

Katsuyuki Takashima

Abstract We revisit a generic compiler from a two-party key exchange (KE) protocol to a group KE (GKE) one by Just and Vaudenay. We then give two families of GKE protocols from *static* assumptions, which are obtained from the general compiler. The first family of the GKE protocols is a constant-round GKE by using secure key derivation functions (KDFs). As special cases, we have such GKE from *static* Ring-LWE (R-LWE), where "static" means that the parameter size in the R-LWE does not depend on the number of group members, n, and also from the standard SI-DDH and CSI-DDH assumptions. The second family consists of two-round GKE protocols from isogenies, which are proven secure from *new* isogeny assumptions, the first (resp. second) of which is based on the SIDH (resp. CSIDH) two-party KE. The underlying new *static* assumptions are based on indistinguishability between *a product value of supersingular invariants* and a random value.

Keywords Post-quantum cryptography · Constant-round group key exchange · Static assumptions · Lattice-based cryptography · Isogeny-based cryptography

1 Introduction

1.1 Background

It is well known that widely deployed cryptographic schemes (e.g., RSA and ECC) can be broken by using a large-scale quantum computer (Shor 1997). Hence, we should develop new cryptosystems based on quantum-resistant mathematical problems (called post-quantum cryptography (PQC)).

Group key exchange (GKE) is an important cryptographic primitive, and has been studied for a long time (since the seminal two-party Diffie–Hellman key exchange). In GKE, the number of rounds is a crucial measure for evaluating the efficiency and to obtain a constant-round GKE protocol is considered as a minimum desirable require-

K. Takashima (✉)
Mitsubishi Electric, 5-1-1 Ofuna, Kamakura, Kanagawa 247-8501, Japan
e-mail: Takashima.Katsuyuki@aj.MitsubishiElectric.co.jp

© The Author(s) 2021
T. Takagi et al. (eds.), *International Symposium on Mathematics, Quantum Theory, and Cryptography*, Mathematics for Industry 33,
https://doi.org/10.1007/978-981-15-5191-8_18

ment. Traditionally, the Burmester and Desmedt (BD) KE protocol (Burmester and Desmedt 1994) has been widely known from its simplicity and small round complexity, just two rounds. Subsequently, Just and Vaudenay (JV) (1996) generalized the BD construction in which *any* two-party KE can be used for obtaining GKE. However, their description was sketchy and a *rigorous* security proof was not presented before (see Boyd and Mathuria 2003 also).

In the post-quantum setting, there exist two variants BD-type GKE protocols from lattices (Apon et al. 2019) and isogenies (Furukawa et al. 2018).[1] Apon et al. (2019) proposed a lattice-based BD-type GKE from the Ring-LWE (R-LWE) assumption (in the random oracle model), in which the authors elaborately adjusted the original security proof to their new post-quantum setting. However, since the underlying R-LWE assumption depends on the number of group members, n, the size of data also gets large depending on n. Furukawa et al. (2018) proposed an isogeny-based BD-type GKE protocol called SIBD. However, the security proof of SIBD (Theorem 4 in Furukawa et al. 2018) is imperfect, and several points remain unclear, for example, on how to simulate some public variables. Applying the JV-type compiler to a post-quantum two-party KE is also considered as a reasonable approach, however, we should give a rigorous treatment on its (post-quantum) security proof.

As a result, we lack a post-quantum constant-round GKE protocol with a rigorous and reasonable security proof. We next consider what are reasonable underlying assumptions. The size of a problem instance in the above R-LWE setting is linear in the number of group members, n. Traditionally, in pairing-based cryptography, such linear-sized assumptions are called "non-static", "dynamic", or "q-type", which are not desirable from efficiency and security viewpoints. And, in a line of researches, we succeeded to replace q-type ones to static ones (e.g., Kowalczyk and Wee 2019; Okamoto and Takashima 2010; Takashima 2014) in paring cryptography. Hence, we have the following problem as our target:

Can we obtain (provably secure) post-quantum constant-round group key exchange from static assumptions ?

Recent cryptography research also considers *tight* security reduction (from a static assumption). In fact, the original BD GKE is proven tightly secure from the standard DDH assumption (Theorem 6). For obtaining tight security proof, it is not enough to employ a general form of the JV-type transformation which includes a *general KDF* function to a cyclic group $\mathbb{G}$ (denoted $KDF_{\mathbb{G}}$). We need a construction without using (general) $KDF_{\mathbb{G}}$ functions *for tight security* since $KDF_{\mathbb{G}}$ breaks mathematical structures in the underlying two-party KE.

---

[1] Boneh et al. (2018) recently proposed a one-round GKE from isogenies. However, it has a crucial mathematical difficulty so that it cannot be realized yet.

1.2 Our Contributions

We revisit previous post-quantum BD-type GKE schemes (Apon et al. 2019; Furukawa et al. 2018 and the JV compiler for GKE Boyd and Mathuria 2003; Just and Vaudenay 1996, and reformulate them under a provably secure generic compiler. We have two families of GKE protocols from *static* assumptions.

The first family of GKE protocols obtained from the general compiler is a constant-round GKE (from a two-party KE protocol) by using a secure $KDF_{\mathbb{G}}$ (Theorem 3). As special cases, we have such GKE from *static* Ring-LWE (R-LWE), where "static" means that the parameter size in the R-LWE does not depend on the number of group members, n (Corollary 1) and the standard SI-DDH and CSI-DDH assumptions (Corollary 2). The first family has a limitation that they cannot have a tight security proof since a general $KDF_{\mathbb{G}}$ is used.

The second family consists of two-round GKE protocols, which are proven secure from *new* isogeny assumptions, the first (resp. second) of which is based on the SIDH (resp. CSIDH) KE (Theorem 4 (resp. Theorem 5)). They are called SI-PBD and CSI-PBD GKEs, respectively. The underlying new *static* assumptions are obtained from indistinguishability between a random *product value* of supersingular invariants and a random value (in some appropriate finite field), which seem to have independent interests. They are called DSJP (Decisional Supersingular j-invariants Product) and DSMP (Decisional Supersingular Montgomery coefficients Product) assumptions, respectively. As the second family needs no $KDF_{\mathbb{G}}$'s, it may have some merits for approaching to tightly secure GKE. (However, we do not yet succeed it.)

Note that we have the Katz–Yung (KY) generic compiler from KE to authenticated KE (AKE) (Katz and Yung 2007), in which a signature scheme is required. Very interestingly, the first *practical* isogeny-based signature scheme, CSI-FiSh, was recently proposed (Beullens et al. 2019). Therefore, we have a practical authenticated GKE (AGKE) by applying the KY compiler to our isogeny-based GKE and CSI-FiSh, both of which are post-quantum from isogenies. (Refer to Bernstein et al. 2019; Peikert 2019 for recent estimates on post-quantum security of CSIDH and CSI-FiSh.) Since we have several lattice-based signatures, e.g., Ducas et al. (2018), Fouque et al. (2017), Akleylek et al. (2017), we also have lattice-based AGKE from our lattice GKE.

1.3 Key Techniques

Hereafter, the user indices are taken in a cycle: for example, $h_{n+1} := h_1$ and $h_0 := h_n$. We first review the BD GKE protocol briefly. It is defined on a cyclic group $\mathbb{G}$ of a prime order q and a generator $g \in \mathbb{G}$ as follows:

Round-1.   Each user i generates $a_i \leftarrow_R \mathbb{Z}/q\mathbb{Z}$, $h_i := g^{a_i}$ and broadcasts h_i.

Round-2.   Each user i calculates $J_{i-1,i} := (h_{i-1})^{a_i}$, $J_{i,i+1} := (h_{i+1})^{a_i}$ and $u_i := J_{i,i+1} \cdot J_{i-1,i}^{-1}$. User i broadcasts u_i.

KeyComp.   User i calculates $K_i := J_{i-1,i}^n \cdot u_i^{n-1} \cdot u_{i+1}^{n-2} \cdots u_{i-2}$. Then, $K := K_i = J_{1,2} \cdot J_{2,3} \cdots J_{n,1}$ is the shared key among the n users.

In the (tight) security proof of the BD key exchange protocol from DDH on $\mathbb{G}$, we should simulate broadcast values $(h_i, u_i)_{i \in [n]}$ as well as embed the DDH challenge element into the challenge shared key K.

The SIBD protocol (Furukawa et al. 2018) is obtained from the above BD GKE by replacing (h_i, J_i) with invariants of supersingular elliptic curves. Since the invariants are given by elements in finite fields, we also have

$$u_i := J_{i,i+1} \cdot J_{i-1,i}^{-1}, \quad K := K_i := J_{i-1,i}^n \cdot u_i^{n-1} \cdot u_{i+1}^{n-2} \cdots u_{i-2}. \tag{1}$$

We revisit the JV construction (Just and Vaudenay 1996), whose original description was sketchy and the security proof was not given there. Hence, we first give a security proof for JV carefully. Based on the proof, we present our isogeny-based GKE from newly proposed assumptions. Then, as is shown in the proof of Theorem 3, if $J_{i-1,i}$'s are uniformly and independently distributed in $\mathbb{G}$, the n elements $K, u_1, \ldots, u_{i-1}, u_{i+1}, \ldots, u_n$ are also uniformly and independently distributed in $\mathbb{G}$ for $i \in [n]$ (and u_i is given as $u_i = (u_1 \cdots u_{i-1} \cdot u_{i+1} \cdots u_n)^{-1}$). It means that if $J_{i-1,i}$'s are distributed uniformly and independently, the target shared key K is changed to a random one *just by using an information-theoretic game transformation*. This is a key lemma on the BD-type encoding (Lemma 6).

However, for the SIBD protocol (Furukawa et al. 2018), since $J_{i-1,i}$ are given by supersingular j-invariants, we have an efficient algorithm for distinguishing between $J_{i-1,i}$ and a uniformly random element in the finite field (see Sutherland 2012). Hence, for fixing the situation, we introduce new decisional assumptions called d-DSJP and d-DSMP ones. For simplicity, here we just show the 2-DSJP assumption, in which a product of two j-invariants, $J_{i-1,i}^{(1)}$ and $J_{i-1,i}^{(2)}$, that is, $J_{i-1,i}^{(1)} \cdot J_{i-1,i}^{(2)}$, should be indistinguishable from a uniformly random variable. At present, we have *no* efficient algorithm for the problems, and considered them as plausible assumptions.

According to the above ideas, in Sect. 4.1, we give a JV-type generic transformation from KE to GKE based on the BD-type encoding of (u_i) and K from $(J_{i-1,i})$ given in Eq. (1). We then consider the following two approaches for obtaining uniformly random $J_{i-1,i}$'s:

1. Using a secure $\text{KDF}_\mathbb{G}$ function φ to obtain random $J_{i-1,i} := \varphi(\kappa_{i-1,i})$ where $\kappa_{i-1,i}$'s are shared keys by secure two-party KE: By this approach, we obtain a new GKE from the "static" R-LWE assumption (Sect. 4.2). We also obtain new GKE protocols from SI-DDH and CSI-DDH assumptions.
2. Using new assumptions on supersingular invariants: By using new DSJP and DSMP assumptions, the local outputs, $(J_{i-1,i})$ and $(M_{i-1,i})$, from two-party key exchange can be computationally changed to random ones, and we obtain new GKE from these post-quantum assumptions (Sects. 4.3 and 4.4) without $\text{KDF}_\mathbb{G}$.

1.4 Organization

In Sect. 2, we introduce several preliminary facts: definition of group key exchange, supersingular invariants and underlying assumptions for SIDH and CSIDH. In Sect. 3, our new assumptions on supersingular invariants are presented. In Sect. 4, we propose new PQ GKE, i.e., lattice-based and isogeny-based GKE from static assumptions.

Notations. When A is a set (resp. a random variable), $y \leftarrow_R A$ denotes that y is uniformly generated from A (resp. randomly generated from A according to its distribution). We denote the finite field of order q by $\mathbb{F}_q$. We denote the set $\{1, \ldots, n\}$ by $[n]$.

2 Preliminaries

2.1 Group Key Exchange

We give definitions of group key exchange, its correctness and security.

Definition 1 (*Group Key Exchange (GKE)*) An algorithm $\Pi := \Pi_{r,n}(\lambda)$ is called as a r-round n-party key exchange protocol if it is composed of probabilistic polynomial-time algorithms (Setup, (Round-r')$_{r'=1}^{r}$, KeyComp), where Setup takes a security parameter λ as input, and outputs public parameters params$_\Pi$, Round-r' for each user i takes previous all public variables and his/her own secrets and outputs (broadcasts) the r'th his/her public values, and KeyComp for each user i takes all public variables and his/her own secrets and outputs the shared secret value K_i.

We call Π is correct if all (shared) keys $K_1, \ldots, K_n$ are the same values, i.e., $K := K_1 = \cdots = K_n$. The key space (or key set) is denoted by $\mathbb{K} := \mathbb{K}(\lambda)$ whose cardinality #$\mathbb{K}$ is exponentially large in λ (or has enough entropy).

For a GKE protocol Π, we let $\mathsf{Exec}_\Pi(\lambda)$ denote an execution of the protocol, resulting in a transcript Ψ of all messages sent during the course of that execution, along with the shared key K computed by the parties. We let $\mathsf{Adv}_{\mathcal{A}}^\Pi(\lambda)$ denote the advantage of a polynomial-time quantum adversary $\mathcal{A}$ in distinguishing between the following two distribution ensembles:

$$\{ (\Psi, K) : (\Psi, K) \leftarrow_R \mathsf{Exec}_\Pi(\lambda) \}_{\lambda \in \mathbb{N}} \quad \text{and}$$
$$\{ (\Psi, K') : (\Psi, K) \leftarrow_R \mathsf{Exec}_\Pi(\lambda), \ K' \leftarrow_R \mathbb{K} \}_{\lambda \in \mathbb{N}}.$$

Protocol Π is post-quantumly secure if $\mathsf{Adv}_{\mathcal{A}}^\Pi(\lambda)$ is negligible in λ for any polynomial-time quantum $\mathcal{A}$.

2.2 SIDH and CSIDH Key Exchange

In this section, we introduce two efficient Diffie–Hellman-type key exchange protocols using isogenies of supersingular elliptic curves: SIDH (Feo et al. 2014) and CSIDH (Castryck et al. 2018).

2.2.1 Supersingular Isogenies and Invariants

We summarize facts about elliptic curves. For details, see Washington (2008), for example.

Let p be a prime greater than 3 and $\mathbb{F}_p$ be the finite field with p elements. Let $\overline{\mathbb{F}}_p$ be its algebraic closure. Here, an elliptic curve E over $\overline{\mathbb{F}}_p$ is given by the Montgomery normal form

$$E : \delta y^2 = x^3 + mx^2 + x \tag{2}$$

for m and $\delta \in \overline{\mathbb{F}}_p$, where the discriminant of the RHS of Eq. (2) and δ are nonzero. We denote the point at infinity on E by O_E. Elliptic curves are endowed with a unique algebraic group structure, with O_E as a neutral element. The j-invariant and Montgomery coefficient of E are given as $j(E) := \frac{256(m^2-3)^3}{m^2-4}$, $m(E) := m$. Two elliptic curves over $\overline{\mathbb{F}}_p$ are isomorphic if and only if they have the same j-invariant. For $j \in \overline{\mathbb{F}}_p$, $E(j)$ denotes an elliptic curve whose j-invariant is j. For $N \in \mathbb{Z}_{>0}$, the N-torsion points is $E[N] := \{P \in E(\overline{\mathbb{F}}_p) \mid NP = O_E\}$.

Given two elliptic curves E and E' over $\overline{\mathbb{F}}_p$, a homomorphism $\phi : E \to E'$ is a morphism of algebraic curves that sends O_E to $O_{E'}$. A nonzero homomorphism is called an isogeny, and a separable isogeny with the cardinality ℓ of the kernel is called ℓ-isogeny. We consider only separable isogenies in this paper. We compute the ℓ-isogeny by using Vélu's formulas (Vélu 1971) for a small prime $\ell = 2, 3, \ldots$. For explicit formulas, see Jao et al. (2017) for SIDH and see Castryck et al. (2018) for CSIDH.

An elliptic curve E over $\overline{\mathbb{F}}_p$ is called supersingular if there are no points of order p, i.e., $E[p] = \{O_E\}$. The j-invariants of supersingular elliptic curves lie in $\mathbb{F}_{p^2}$. We define two sets as below, for SI-DDH and CSI-DDH assumptions.

$$\mathbb{J}_{p^2} := \{j\text{-invariants of supersingular elliptic curves over } \mathbb{F}_{p^2}\}, \tag{3}$$

$$\mathbb{M}_p := \{\text{Montgomery coefficients of supersingular elliptic curves over } \mathbb{F}_p\}. \tag{4}$$

2.2.2 SIDH Key Exchange and SI-DDH Assumption (Feo et al. 2014)

The detailed description of SIDH key exchange, i.e., $\Pi := \mathsf{SIDH}$, is given in Appendix 3.1. Here, we summarize necessary facts on SIDH for later sections. Public parameters are given as $\mathsf{params}_{\mathsf{SIDH}} := (p, E; P_A, Q_A, P_B, Q_B)$. All the messages during an execution are also given as transcript $\Psi_{AB} := (\mathsf{params}_{\mathsf{SIDH}}, E_A, \phi_A(P_B), \phi_A(Q_B), E_B, \phi_B(P_A), \phi_B(Q_A))$. Alice's and Bob's shared keys, i.e., $K_A := j(E_{AB})$ and $K_B := j(E_{BA})$, are equal, and the value is denoted by K.

Definition 2 (*Supersingular Isogeny Decision Diffie–Hellman (SI-DDH) assumption* Feo et al. 2014; Fujioka et al. 2018) Let $(\Psi_{AB}, j(E_{AB})) \leftarrow_R \mathsf{Exec}_{\mathsf{SIDH}}(\lambda)$, where $\Psi_{AB} := (\text{params}_{\text{SIDH}}, \ E_A, \phi_A(P_B), \phi_A(Q_B), E_B, \phi_B(P_A), \phi_B(Q_A))$. An SI-DDH problem instance is given as (Ψ_{AB}, J_β), where

$$J_0 := j(E_{AB}), \qquad J_1 \leftarrow_R \mathbb{J}_{p^2}, \tag{5}$$

$\beta \leftarrow_R \{0, 1\}$, and $\mathbb{J}_{p^2}$ is defined in Eq. (3). If $| \Pr[\mathcal{A}(\Psi_{AB}, J_0) = 1] - \Pr[\mathcal{A}(\Psi_{AB}, J_1) = 1] | < \text{negl}(\lambda)$ holds for any polynomial-time quantum algorithm $\mathcal{A}$, we say that the SI-DDH assumption holds.

Theorem 1 (Feo et al. 2014) *The SIDH key exchange is post-quantumly secure under the SI-DDH assumption.*

2.2.3 CSIDH Key Exchange and CSI-DDH Assumption (Castryck et al. 2018)

The detailed description of CSIDH key exchange, i.e., $\Pi := \text{CSIDH}$, is given in Appendix 3.2. Here, we summarize necessary facts on CSIDH. Public parameters are given as $\text{params} := (p, E)$. All the messages during a execution are also given as transcript $\Psi_{AB} := (\text{params}_{\text{CSIDH}}, [\mathfrak{a}]E, [\mathfrak{b}]E)$. Alice's and Bob's shared keys, i.e., $K_A := m([\mathfrak{a}][\mathfrak{b}]E)$ and $K_B := m([\mathfrak{b}][\mathfrak{a}]E)$, are equal, and the value is denoted by K.

Definition 3 (*Commutative Supersingular Isogeny Decisional Diffie–Hellman (CSI-DDH) assumption*) Let $(\Psi_{AB}, m([\mathfrak{a}][\mathfrak{b}]E)) \leftarrow_R \text{Exec}_{\text{CSIDH}}(\lambda)$ where $\Psi_{AB} := (\text{params}_{\text{CSIDH}}, [\mathfrak{a}]E, [\mathfrak{b}]E)$. A CSI-DDH problem instance is given as (Ψ_{AB}, M_β), where

$$M_0 := m([\mathfrak{a}][\mathfrak{b}]E), \qquad M_1 \leftarrow_R \mathbb{M}_p,$$

$\beta \leftarrow_R \{0, 1\}$, and $\mathbb{M}_p$ is defined in Eq. (4). If $| \Pr[\mathcal{A}(\Psi_{AB}, M_0) = 1] - \Pr[\mathcal{A}(\Psi_{AB}, M_1) = 1] | < \text{negl}(\lambda)$ holds for any polynomial-time quantum algorithm $\mathcal{A}$, we say that the CSI-DDH assumption holds.

Theorem 2 (Castryck et al. 2018) *The CSIDH key exchange is post-quantumly secure under the CSI-DDH assumption.*

3 New Assumptions on Supersingular Invariants

3.1 New Assumptions on Supersingular j-Invariants

Definition 4 (*Decisional Supersingular j-Invariants Product (d-DSJP) Assumption*) Let $\left(\Psi_{AB}^{(\mu)}, j\left(E_{AB}^{(\mu)} \right) \right)_{\mu \in [d]}$ be transcripts of d-time executions of SIDH with the same $\text{params}_{\text{SIDH}}$, where $\Psi_{AB}^{(\mu)} := \left(\text{params}_{\text{SIDH}}, \left(E_A^{(\mu)}, \phi_A^{(\mu)}(P_B), \phi_A^{(\mu)}(Q_B), E_B^{(\mu)}, \right.\right.$

$\phi_B^{(\mu)}(P_A), \phi_B^{(\mu)}(Q_A)\big)\big)$ and $\Psi_{AB} := \left(\Psi_{AB}^{(\mu)}\right)_{\mu \in [d]}$. A d-DSJP problem instance is given as (Ψ_{AB}, J_β), where

$$J_0 := \prod_{\mu=1}^{d} j\left(E_{AB}^{(\mu)}\right), \qquad J_1 \leftarrow_R \mathbb{F}_{p^2} \tag{6}$$

and $\beta \leftarrow_R \{0, 1\}$. For any adversary $\mathcal{B}$, the advantage of $\mathcal{B}$ is defined as $\mathsf{Adv}_{\mathcal{B}}^{d\text{-DSJP}}(\lambda)$ $:= |\Pr[\mathcal{B}(\Psi_{AB}, J_0) = 1] - \Pr[\mathcal{B}(\Psi_{AB}, J_1) = 1]|$, and the d-DSJP assumption holds if $\mathsf{Adv}_{\mathcal{B}}^{d\text{-DSJP}}(\lambda)$ is negligible in λ for any polynomial-time quantum adversary $\mathcal{B}$.[2]

3.1.1 Progressive Weakness Among d-DSJP Assumptions

The next lemma shows that the $(d + 1)$-DSJP assumption is weaker than the d-DSJP one. In other words, a security proof from the $(d + 1)$-DSJP assumption is considered better than that from the d-DSJP one.

Lemma 1 *The d-DSJP assumption is reduced to the $(d + 1)$-DSJP assumption.*

For any adversary $\mathcal{A}$, there is a probabilistic machine $\mathcal{B}$, whose running time is essentially the same as that of $\mathcal{A}$, such that for any security parameter λ, $\mathsf{Adv}_{\mathcal{A}}^{(d+1)\text{-DSJP}}(\lambda) \leq \mathsf{Adv}_{\mathcal{B}}^{d\text{-DSJP}}(\lambda)$.

Proof $\mathcal{B}$ receives a d-DSJP tuple (Ψ_{AB}, J_β), where Ψ_{AB} is defined as in Definition 4. J_β is $\prod_{\mu=1}^{d} j\left(E_{AB}^{(\mu)}\right)$ when $\beta = 0$ or a random element in $\mathbb{F}_{p^2}$ when $\beta = 1$. $\mathcal{B}$ generates a new SIDH public key pair $\left(E_A^{(d+1)}, \phi_A^{(d+1)}(P_B), \phi_A^{(d+1)}(Q_B)\right), \left(E_B^{(d+1)}, \phi_B^{(d+1)}\right.$ $\left.(P_A), \phi_B^{(d+1)}(Q_A)\right)$ and SIDH shared key $j\left(E_{AB}^{(d+1)}\right)$, then constructs a new tuple $\Psi'_{AB} := \left(\mathsf{params}, \left(\left(E_A^{(\mu)}, \quad \phi_A^{(\mu)}(P_B), \phi_A^{(\mu)}(Q_B)\right), \left(E_B^{(\mu)}, \phi_B^{(\mu)}(P_A), \phi_B^{(\mu)}(Q_A)\right)\right)_{\mu \in [d+1]}\right)$, and $J'_\beta := J_\beta \cdot j\left(E_{AB}^{(d+1)}\right)$. $\mathcal{B}$ gives a $(d + 1)$-DSJP tuple (Ψ'_{AB}, J'_β) to $\mathcal{A}$, and outputs β' when $\mathcal{A}$ outputs β'. $\qquad \square$

In fact, we show the 1-DSJP problem is efficiently solved (Lemma 2 in Sect. 3.1.2) and the 2-DSJP problem has a specific approach for solving it via modular polynomials (Sect. 3.1.3).

3.1.2 Case $d = 1$: Relation Between SI-DDH and 1-DSJP Assumptions

While the value of J_0 for SI-DDH in Eq. (5) is the same as that of the 1-DSJP assumption in Eq. (6), the other J_1's in the two assumptions are distributed in different

---

[2]Its "sum" version (instead of "product"), Decisional Supersingular j-invariants Sum (d-DSJS) assumption, seems to be reasonable for $d \geq 2$, and can be used in security proofs for the "sum" version SI-SBD GKE scheme of SI-PBD GKE in Sect. 4.3. This footnote comment is also applied to the d-DSMP assumption and CSI-PBD GKE in Sect. 4.4 in a similar manner.

manners. Namely, the first (resp. the second) is the uniform distribution over $\mathbb{J}_{p^2}(\subsetneq \mathbb{F}_{p^2})$ (resp. $\mathbb{F}_{p^2}$). As is shown below, the difference is important.

Lemma 2 *The 1-DSJP problem can be solved in (deterministic) polynomial time except with a negligible error probability.*

Proof In the 1-DSJP problem, J_0 (resp. J_1) is uniformly distributed in $\mathbb{J}_{p^2}$ (resp. $\mathbb{F}_{p^2}$). Therefore, by applying supersingular identifying algorithm, e.g., Sutherland (2012), we can solve the problem. $\qquad\square$

From the above fact, the direct assumption, decisional $(1, 1)$-SI-PBD assumption in Definition 6 picks up the target key κ_1 ($\beta = 1$ instance) from a uniform distribution in $\mathbb{J}_{p^2}$ instead of $\mathbb{F}_{p^2}$.

3.1.3 Case $d = 2$: An Approach for 2-DSJP via Modular Polynomials

Lemma 1 shows the 2-DSJP assumption is the strongest among the d-DSJP assumptions for $d \geq 2$. In fact, we have some possible approaches for solving the problem as indicated below. But, the attack is not yet effective at present.

Here, we introduce modular polynomials $\Phi_N(X, Y) := \sum c_{ik} X^i Y^k$, which satisfy that $\Phi_N(j, j') = 0$ for two j-invariants j and j' such that there exists an N-isogeny between the associated elliptic curves $E(j)$ and $E(j')$. From the above defining property, it holds that $\Phi_N(X, Y)$ are symmetric polynomials w.r.t. X and Y. Hence, if we set $S := X + Y$ and $T := XY$, $\Phi_N(X, Y)$ are given as $\Phi_N(X, Y) = \Xi_N(S, T) := \sum \gamma_{ik} S^i T^k$ for a two-variable polynomial Ξ_N.

The output J_0 of the 2-DSJP problem is given by the product of two supersingular j-invariants, i.e., $\tau := j\left(E^{(1)}\right) j\left(E^{(2)}\right)$. We substitute $T := \tau$ into $\Xi_N(S, T)$, which we obtain a one-variable polynomial equation $\Xi_N(S, \tau) = 0$. If $E^{(1)}$ and $E^{(2)}$ are N-isogenous, then $\sigma := j\left(E^{(1)}\right) + j\left(E^{(2)}\right)$ satisfies the equation, i.e., $\Xi_N(\sigma, \tau) = 0$.

Based on this fact, we obtain a possible cryptanalysis for the 2-DSJP problem given as below. The input of the algorithm is a 2-DSJP instance (Ψ_{AB}, J_β).

1. Set a set of (small) integers $\mathbb{I} := \{N_1, \ldots, N_t\}$.
2. For each $N \in \mathbb{I}$, solve a one-variable polynomial equation $\xi_N(S) := \Xi_N(S, J_\beta) = 0$, and the set of zero points of ξ_N in $\mathbb{F}_{p^2}$ is denoted by $\mathcal{Z} \subset \mathbb{F}_{p^2}$. For each $z \in \mathcal{Z}$, solve the quadratic equation $W^2 - zW + J_\beta = 0$.

   a. If the roots $w_1 \notin \mathbb{F}_{p^2}$ or $w_2 \notin \mathbb{F}_{p^2}$, quit this loop.
   b. Check whether both of w_1 and w_2 are supersingular j-invariants or not. If yes, output $\beta' := 0$.

3. Output $\beta' := 1$.

The degree of isogenous curves $E^{(1)}$ and $E^{(2)}$ above is usually large, therefore, if the security parameter λ is set large, the attack is ineffective. But, the above scenario shows some possible approach to this problem using a specific property on modular polynomials when $d = 2$.

3.2 New Assumptions on Supersingular Montgomery Coefficients

Definition 5 (*Decisional Supersingular Montgomery Coefficients Product (d-DSMP) Assumption*) Let $\left(\Psi_{AB}^{(\mu)}, m\left(E_{AB}^{(\mu)}\right)\right)_{\mu \in [d]}$ be transcripts of d-time executions of CSIDH with the same $\mathsf{params}_{\mathsf{CSIDH}}$, where $\Psi_{AB}^{(\mu)} := (\mathsf{params}_{\mathsf{CSIDH}}, \left(E_A^{(\mu)}, E_B^{(\mu)}\right))$ and $\Psi_{AB} := \left(\Psi_{AB}^{(\mu)}\right)_{\mu \in [d]}$, where $E_A^{(\mu)} := [\mathfrak{a}^{(\mu)}] E$, $E_B^{(\mu)} := [\mathfrak{b}^{(\mu)}] E$ and $E_{AB}^{(\mu)} := [\mathfrak{a}^{(\mu)}][\mathfrak{b}^{(\mu)}] E$. A d-DSMP problem instance is given as (Ψ_{AB}, M_β), where

$$M_0 := \prod_{\mu=1}^d m\left(E_{AB}^{(\mu)}\right), \qquad M_1 \leftarrow_R \mathbb{F}_p,$$

and $\beta \leftarrow_R \{0, 1\}$. For any adversary $\mathcal{B}$, the advantage of $\mathcal{B}$ is defined as $\mathsf{Adv}_{\mathcal{B}}^{d\text{-DSMP}}(\lambda) := |\Pr[\mathcal{B}(\Psi_{AB}, M_0) = 1] - \Pr[\mathcal{B}(\Psi_{AB}, M_1) = 1]|$, and the d-DSMP assumption holds if $\mathsf{Adv}_{\mathcal{B}}^{d\text{-DSMP}}(\lambda)$ is negligible in λ for any polynomial-time quantum adversary $\mathcal{B}$.

For the DSMP assumptions, we have similar results for the DSJP. In particular, we have the following lemmas.

Lemma 3 *The d-DSMP assumption is reduced to the $(d + 1)$-DSMP assumption.*

Lemma 4 *The 1-DSMP problem can be solved in (deterministic) polynomial time except with a negligible error probability.*

4 Proposed Post-Quantum Group Key Exchange (GKE)

4.1 A Generic JV-Type Compiler for GKE from Two-Party KE (Just and Vaudenay 1996)

We describe a generic BD-type GKE compiler from a two-party KE protocol Π, and the obtained GKE protocol is denoted as Π^{BD}. Such a generic compiler was first proposed by Just and Vaudenay (1996), Boyd and Mathuria (2003), but, no formal proof was attached yet. By describing the security proof carefully, we also give a security proof for our proposal in Sects. 4.3 and 4.4, and we found a condition for the compiler to work correctly. The number of group members is assumed to be $n \geq 3$. Assume that we have two-party key exchange Π with shared keyspace $\mathbb{K}$. We need a map $\varphi : \mathbb{K} \to \mathbb{G}$ (called $\mathbb{G}$-embedding map), where $\mathbb{G}$ is a cyclic group of order q in the BD-type Encoding (**BDEnc**) as indicated below. We assume that $\gcd(n, q) = 1$ for the number of group members n and the cyclic group order q. (Note that we do not assume the intractability of discrete log in $\mathbb{G}$.)

Exec-Π. Each user i runs the protocol Π with users $i - 1$ and $i + 1$, respectively, and obtains keys $\kappa_{i-1,i}$ and $\kappa_{i,i+1}$.

BDEnc. User i sets $J_{i-1,i} := \varphi(\kappa_{i-1,i})$ and $J_{i,i+1} := \varphi(\kappa_{i,i+1})$, and broadcasts $u_i := J_{i,i+1} \cdot J_{i-1,i}^{-1} \in \mathbb{G}$.

KeyComp. User i calculates $K_i := J_{i-1,i}^n \cdot u_i^{n-1} \cdot u_{i+1}^{n-2} \cdots u_{i-2}$. Then, $K := K_i = J_{1,2} \cdot J_{2,3} \cdots J_{n,1}$ is the shared key among the n users.

The correctness is shown as the same as the original BD key exchange. The security depends on the map φ. Below, we show that it is proven secure assuming that φ is a secure KDF (see Appendix 2 for its definition) and the underlying protocol Π is secure.

Theorem 3 *The GKE protocol* Π^{BD} *is (post-quantumly) secure if* Π *is (post-quantumly) secure,* φ *is a (post-quantumly) secure KDF and* $\gcd(n, q) = 1$ *where* q *is the order of* $\mathbb{G}$.

For any (quantum) adversary $\mathcal{A}$, *there exist (quantum) machines* $\mathcal{B}_l$ *and* $\mathcal{C}_l$, *whose running times are essentially the same as that of* $\mathcal{A}$, *such that* $\mathsf{Adv}_{\mathcal{A}}^{\Pi^{\mathsf{BD}}}(\lambda) \leq \sum_{l \in [2n]} \left(\mathsf{Adv}_{\mathcal{B}_l}^{\Pi}(\lambda) + \mathsf{Adv}_{\mathcal{C}_l}^{\mathsf{KDF}}(\lambda) \right) + \varepsilon(\lambda)$, *where* $\varepsilon(\lambda)$ *is a negligible function in* λ.

Proof The view of $\mathcal{A}$ consists of $(u_1, \ldots, u_n, K)$. To prove Theorem 3, we consider the following $2n + 2$ games. An underlined part indicates a variable that is changed in a game from the previous one.

Game 0: Original game, which is the same as the first case in Definition 1. The values of $J_{i-1,i}, u_i, K$ are given as $J_{i-1,i} := \varphi(\kappa_{i-1,i})$,

$$u_i := J_{i,i+1} \cdot J_{i-1,i}^{-1} \text{ for } i \in [n], \quad K := J_{1,2} \cdot J_{2,3} \cdots J_{n-1,n} \cdot J_{n,1}, \tag{7}$$

where $\kappa_{i-1,i}$ is a shared key by running Π between users $i - 1$ and i.

Game l ($l \in [n]$): The lth output of φ is $\underline{J_{l-1,l} \leftarrow_R \mathbb{G}}$ (for both of users $l - 1$ and l), all the other $J_{i-1,i}$'s for $i \neq l$ are generated as in Game $l - 1$, and the view of $\mathcal{A}$, i.e., $(u_1, \ldots, u_n, K)$, are generated as in Eq. (7) from all the $J_{i-1,i}$'s for $i \in [n]$.

Game $n + 1$: Same as Game n except that the shared key is $\underline{K \leftarrow_R \mathbb{G}}$, and all the other variables are generated as in Game n. Note that K is independent of all the other variables.

Game $n + 1 + l$ ($l \in [n]$): The lth output of φ is $\underline{J_{l-1,l} := \varphi(\kappa_{l-1,l})}$ (for both of users $l - 1$ and l), all the other $J_{i-1,i}$'s for $i \neq l$ are generated as in Game $n + l$, and $(u_1, \ldots, u_n)$ are generated as in Eq. (7) from all the $J_{i-1,i}$'s for $i \in [n]$ and $K \leftarrow_R \mathbb{G}$. Here, note that Game $2n + 1$ is the same as the second case in Definition 1.

Let $\mathsf{Adv}_{\mathcal{A}}^{(l)}(\lambda)$ be the advantage of $\mathcal{A}$ in Game l, respectively.

We will show three lemmas (Lemmas 5–7) that evaluate the gaps between pairs of the advantages in Game 0, ..., Game $2n + 1$. From these lemmas, we obtain $\mathsf{Adv}_{\mathcal{A}}^{\Pi^{\mathsf{BD}}}(\lambda) \leq \sum_{l \in [2n+1]} \left| \mathsf{Adv}_{\mathcal{A}}^{(l-1)}(\lambda) - \mathsf{Adv}_{\mathcal{A}}^{(l)}(\lambda) \right| \leq \sum_{l \in [2n]} \left(\mathsf{Adv}_{\mathcal{B}_l}^{\Pi}(\lambda) + \mathsf{Adv}_{\mathcal{C}_l}^{\mathsf{KDF}}(\lambda) \right) + \varepsilon(\lambda)$ where $\varepsilon(\lambda) := \sum_{l \in [2n]} \varepsilon_l(\lambda)$ is a negligible function. This completes the proof of Theorem 3. $\qquad\square$

Lemma 5 *For any (quantum) adversary $\mathcal{A}$, there exist (quantum) machines $\mathcal{B}_l$ and $\mathcal{C}_l$, whose running times are essentially the same as that of $\mathcal{A}$, such that $|\mathsf{Adv}_{\mathcal{A}}^{(l-1)}(\lambda) - \mathsf{Adv}_{\mathcal{A}}^{(l)}(\lambda)| \leq \mathsf{Adv}_{\mathcal{B}_l}^{\Pi}(\lambda) + \mathsf{Adv}_{\mathcal{C}_l}^{\mathsf{KDF}}(\lambda) + \varepsilon_l(\lambda)$ for $l \in [n]$, where $\varepsilon_l(\lambda)$ are negligible functions.*

Proof For the proof, we define an intermediate game, i.e., Game $l - 1/2$, between Games $l - 1$ and l. In Game $l - 1/2$, $\kappa_{l-1,l} \leftarrow_R \mathbb{K}$ and $J_{l-1,l} := \varphi(\kappa_{l-1,l})$, and the rest of variables are all generated in the same manner as in Game $l - 1$.

By the definition of two-party KE, the difference of the advantages of Games $l - 1$ and $l - 1/2$ is bounded by the advantage against the KE protocol Π, i.e., $\mathsf{Adv}_{\mathcal{B}_l}^{\Pi}(\lambda)$ (except with negligible probability). Since the keyspace $\mathbb{K}$ has enough entropy, by the definition of KDF, the difference of the advantages of Games $l - 1/2$ and l is bounded by the advantage against KDF, i.e., $\mathsf{Adv}_{\mathcal{C}_l}^{\mathsf{KDF}}(\lambda)$ (except with negligible probability). This completes the proof of Lemma 5.   □

Lemma 6 (BDEnc Information-Theoretic Security) *For any (quantum) adversary $\mathcal{A}$, for any security parameter λ, $\mathsf{Adv}_{\mathcal{A}}^{(n+1)}(\lambda) = \mathsf{Adv}_{\mathcal{A}}^{(n)}(\lambda)$.*

Proof We can set $J_{i-1,i} := g^{\alpha_{i-1}}$ for $i \in [n]$, where $g \in \mathbb{G}$ is a generator and $\alpha_i \leftarrow_R \mathbb{Z}/q\mathbb{Z}$ (which are independent from each other). Then, $u_i := J_{i,i+1} \cdot J_{i-1,i}^{-1} = g^{\alpha_i - \alpha_{i-1}}$. First, we see that n elements (α_1, $\alpha_2 - \alpha_1$, $\alpha_3 - \alpha_2$, ..., $\alpha_n - \alpha_{n-1}$) are uniformly and independently distributed. Since $\alpha_1 + \cdots + \alpha_n = n\alpha_1 + (n-1)(\alpha_2 - \alpha_1) + (n-2)(\alpha_3 - \alpha_2) + \cdots + (\alpha_n - \alpha_{n-1})$ and $n \bmod q$ has an inverse element (from the assumption $\gcd(n, q) = 1$), n elements ($\alpha_1 + \cdots + \alpha_n$, $\alpha_2 - \alpha_1$, $\alpha_3 - \alpha_2$, ..., $\alpha_n - \alpha_{n-1}$) are also uniformly and independently distributed. Since $K = g^{\alpha_1 + \cdots + \alpha_n}$, K is independent of all the other variables, i.e., h_i, u_i. This completes the proof of Lemma 6.   □

Lemma 7 *For any (quantum) adversary $\mathcal{A}$, there exists (quantum) machines $\mathcal{B}_{n+l}$ and $\mathcal{C}_{n+l}$, whose running times are essentially the same as that of $\mathcal{A}$, such that for any security parameter λ, $|\mathsf{Adv}_{\mathcal{A}}^{(n+l)}(\lambda) - \mathsf{Adv}_{\mathcal{A}}^{(n+l+1)}(\lambda)| \leq \mathsf{Adv}_{\mathcal{B}_{n+l}}^{\Pi}(\lambda) + \mathsf{Adv}_{\mathcal{C}_{n+l}}^{\mathsf{KDF}}(\lambda) + \varepsilon_{n+l}(\lambda)$ for $l \in [n]$, where $\varepsilon_{n+l}(\lambda)$ are negligible functions.*

Lemma 7 is proven in a similar manner to Lemma 5.

4.2 Constant-Round GKE from Static Standard Assumptions

We instantiate the above generic GKE by Apon et al.'s ring LWE based GKE (Apon et al. 2019) by using a two-party KE Π and some SHA-2 (or SHA-3) based KDF φ, whose range is $\mathbb{G} := \mathbb{F}^*$ for some finite field $\mathbb{F}$. Therefore, we have the following corollary.

Corollary 1 *There exists a post-quantum constant-round GKE from two-party KE Π in Apon et al. (2019) and some standard KDF function φ under the static ring LWE assumption.*

Apon et al.'s original GKE is based on the "non-static" or "dynamic" R-LWE assumption. That is, the noise size depends on the number of group members n, then the scheme itself gets to large sizes.

Corollary 2 *There exists a post-quantum constant-round GKE from two-party KE SIDH (resp. CSIDH) and some standard KDF function φ under the SI-DDH (resp. CSI-DDH) assumption.*

4.3 Two-Round Product-BD (PBD) GKE from d-DSJP Assumption

We modify the SIBD Group Key Exchange proposed in Furukawa et al. (2018) to a provably secure one, called Supersingular Isogeny Product-BD $((n, d)$-SI-PBD) protocol for n-parties. In other words, our general (n, d)-SI-PBD protocol is obtained via our generic compiler (in Sect. 4.1) from two-party $(2, d)$-SI-PBD protocol, where a $\mathbb{G}$-embedding map φ is given by the identity map $\varphi := \mathrm{id}_{\mathbb{G}} : \mathbb{G} \to \mathbb{G}$.

4.3.1 Construction

We consider n-party key exchange. Each user is indexed by $1, 2, \ldots, n$, where n is supposed to be even for simplicity. Note that we can easily obtain the protocol for odd n. The user indices are taken in a cycle: so $R_{n+1} := R_1$ and $R_0 := R_n$. We introduce the map $\iota(i) := i \bmod 2$ and we will simply write ι instead of writing $\iota(i)$.

Setup. Takes a security parameter λ and the number of users n. The algorithm outputs $\mathsf{params}_{\mathsf{SIDH}} := (p(= f\ell_0^{e_0}\ell_1^{e_1} \pm 1), E, \{P_0, Q_0\}, \{P_1, Q_1\})$ for SIDH.

Round-1. Takes the user index i and params as input. User i randomly chooses $k_i^{(\mu)} \in \mathbb{Z}/\ell_\iota^{e_\iota}\mathbb{Z}$ and computes $R_i^{(\mu)} := P_\iota + k_i^{(\mu)}Q_\iota$. User i then computes the isogeny $\phi_i^{(\mu)}$ and elliptic curve $E_i^{(\mu)} := E/\langle R_i^{(\mu)}\rangle$ such that $\phi_i^{(\mu)} : E \to E_i^{(\mu)}$, where $\ker(\phi_i^{(\mu)})=\langle R_i^{(\mu)}\rangle$. The user i then sets $\mathsf{pk}_i^1 = \left(E_i^{(\mu)}, \phi_i^{(\mu)}(P_{1-\iota}),\right.$ $\left.\phi_i^{(\mu)}(Q_{1-\iota})\right)_{\mu\in[d]}$ and $\mathsf{sk}_i^1 := \left(k_i^{(\mu)}\right)_{\mu\in[d]}$. Finally, the user i broadcasts pk_i^1 to the other users.

Round-2. Takes the user index i, $\mathsf{params}_{\mathsf{SIDH}}$, $(\mathsf{pk}_{i-1}^1, \mathsf{pk}_{i+1}^1)$, and sk_i^1. User i executes SIDH key exchange with users $i - 1$ and $i + 1$ to obtain elliptic curves $E_{i-1,i}^{(\mu)}$ and $E_{i,i+1}^{(\mu)}$, respectively, and then computes

$$J_{i-1,i} := \prod_{\mu=1}^d j\left(E_{i-1,i}^{(\mu)}\right) \quad \text{and} \quad J_{i,i+1} := \prod_{\mu=1}^d j\left(E_{i,i+1}^{(\mu)}\right).$$

The user then computes $u_i := J_{i,i+1} \cdot J_{i-1,i}^{-1}$ and set $\mathsf{pk}_i^2 := u_i$. Finally, the user i broadcasts pk_i^2 to the other users.

KeyComp. User i collects $\left(\mathsf{pk}_{i'}^2\right)_{i'\in[n]}$ and sk_i^1 and computes $K_i := J_{i-1,i}^n \cdot u_i^{n-1} \cdot u_{i+1}^{n-2} \cdot \cdots \cdot u_{i-3}^2 \cdot u_{i-2}$.

We can easily verify that $K_i = J_{1,2} \cdot J_{2,3} \cdots J_{n-1,n} \cdot J_{n,1}$ holds for any i.

4.3.2 Warm-Up: Security from a Nonstatic Assumption

We rephrase security of the (n, d)-SI-PBD protocol based on Definition 1 as a form of the following assumption (see Lemma 8).

Definition 6 (*Decisional SI-PBD ((n,d)-SI-PBD) Assumption*) Let $(\Psi_{n,d}, K) \leftarrow_R$ $\mathsf{Exec}_{(n,d)\text{-SI-PBD}}(\lambda)$, where $J_{i-1,i} := \prod_{\mu=1}^{d} j\left(E_{i-1,i}^{(\mu)}\right)$, $J_{i,i+1} := \prod_{\mu=1}^{d} j\left(E_{i,i+1}^{(\mu)}\right)$,

$u_i := J_{i,i+1} \cdot J_{i-1,i}^{-1}$, $\Psi_{n,d} := \left(\mathsf{params}_{\mathsf{SIDH}}, \left(\left(E_i^{(\mu)}, \phi_i^{(\mu)}(P_{1-\iota}), \phi_i^{(\mu)}(Q_{1-\iota})\right), u_i\right)_{i \in [n], \mu \in [d]}\right)$,

and $K := \prod_{i=1}^{n} J_{i,i+1}$. An (n, d)-SI-PBD problem instance is given as $(\Psi_{n,d}, \kappa_\beta)$, where

$$\kappa_0 := K, \qquad \kappa_1 \leftarrow_R \mathbb{F}_{p^2},$$

and $\beta \leftarrow_R \{0, 1\}$. For any quantum algorithm $\mathcal{B}$, the advantage of $\mathcal{B}$ is defined as $\mathsf{Adv}_{\mathcal{B}}^{(n,d)\text{-SI-PBD}}(\lambda) := |\Pr[\mathcal{B}(\Psi_{n,d}, \kappa_0) = 1] - \Pr[\mathcal{B}(\Psi_{n,d}, \kappa_1) = 1]|$, and the (n, d)-SI-PBD assumption holds if $\mathsf{Adv}_{\mathcal{B}}^{(n,d)\text{-SI-PBD}}(\lambda)$ is negligible in λ for any polynomial-time quantum adversary $\mathcal{B}$.

Remark 1 We have better security proofs when $d \geq 2$ for the (n, d)-SI-PBD GKE (Theorem 4). However, the above gives only security proofs for the $d = 1$ case, which is based on nonstatic assumptions. Note that since $n \geq 3$ and the key K is a n-time product of j-invariants, then we have no efficient distinguishing algorithm between κ_0 and κ_1.

Lemma 8 *The (n, d)-SI-PBD key exchange among n-parties is post-quantumly secure under the (n, d)-SI-PBD assumption.*

Proof Lemma 8 is trivially obtained from Definitions 1 and 6.   □

If the (n, d)-SI-PBD problem is quantum resistantly hard, the SI-PBD key exchange among n-parties is also quantum resistant. Therefore, we should investigate the post-quantum security of the (n, d)-SI-PBD assumption in the next section.

Moreover, as is shown in Lemma 1 for the d-DSJP assumptions, the family of (n, d)-SI-PBD assumptions also has natural sequential reductions among them.

Lemma 9 *The (n, d)-SI-PBD assumption is reduced to the $(n, d + 1)$-SI-PBD assumption.*

For any adversary $\mathcal{A}$, there is a (quantum) machine $\mathcal{B}$, whose running time is essentially the same as that of $\mathcal{A}$, such that for any security parameter λ, $\mathsf{Adv}_{\mathcal{A}}^{(n,d+1)\text{-SI-PBD}}(\lambda) \leq \mathsf{Adv}_{\mathcal{B}}^{(n,d)\text{-SI-PBD}}(\lambda)$.

Proof The proof of Lemma 9 is similarly given to that of Lemma 1.   □

Lemma 9 shows that $(n, d + 1)$-SI-PBD group key exchange is more secure than (n, d)-SI-PBD one while the former is less efficient than the latter in terms of data sizes and execution times.

4.3.3 Security from d-DSJP Assumption for $d \geq 2$

Theorem 4 *The (n, d)-SI-PBD key exchange among n-parties is post-quantumly secure under the d-DSJP assumption when $d \geq 2$ and $\gcd(n, p^2 - 1) = 1$. (Note that $p^2 - 1$ is the order of cyclic group $\mathbb{G} := \mathbb{F}^*_{p^2}$.)*

For any quantum adversary $\mathcal{A}$, there exist quantum machines $\mathcal{B}_l$, whose running times are essentially the same as that of $\mathcal{A}$, such that $\mathsf{Adv}^{(n,d)\text{-SI-PBD}}_{\mathcal{A}}(\lambda) \leq \sum_{l \in [2n]} \mathsf{Adv}^{d\text{-DSJP}}_{\mathcal{B}_l}(\lambda)$ when $d \geq 2$.

Proof The view of $\mathcal{A}$ consists of $(u_1, \ldots, u_n, K)$. To prove Theorem 4, we consider the following $2n + 2$ games. An underlined part indicates a variable that is changed in a game from the previous one.

Game 0: Original game. That is, the values of $J_{i-1,i}$, u_i, K are given as $J_{i-1,i} := \prod_{\mu=1}^d j\left(E^{(\mu)}_{i-1,i}\right)$,

$$u_i := J_{i,i+1} \cdot J^{-1}_{i-1,i} \text{ for } i \in [n], \quad K := J_{1,2} \cdot J_{2,3} \cdots J_{n-1,n} \cdot J_{n,1}. \tag{8}$$

Game l ($l \in [n]$): The lth output of φ is: $\underline{J_{l-1,l} \leftarrow_R \mathbb{F}_{p^2}}$ (for both of users $l - 1$ and l), all the other $J_{i-1,i}$'s for $i \neq l$ are generated as in Game $l - 1$, and the view of $\mathcal{A}$, i.e., $(u_1, \ldots, u_n, K)$, are generated as in Eq. (8) from all the $J_{i-1,i}$'s for $i \in [n]$.

Game $n + 1$: Same as Game n except that the shared key is $\underline{K \leftarrow_R \mathbb{F}_{p^2}}$, and all the other variables are generated as in Game n. Note that K is independent of all the other variables.

Game $n + 1 + l$ ($l \in [n]$): The lth output of φ is: $\underline{J_{l-1,l} := \prod_{\mu=1}^d j\left(E^{(\mu)}_{l-1,l}\right)}$ (for both of users $l - 1$ and l), all the other $J_{i-1,i}$'s for $i \neq l$ are generated as in Game $n + l$, $(u_1, \ldots, u_n)$, are generated as in Eq. (8) from all the $J_{i-1,i}$'s for $i \in [n]$ and $K \leftarrow_R \mathbb{F}_{p^2}$. Here, note that Game $2n + 1$ is the same as the $\beta = 1$ case in Definition 6.

Let $\mathsf{Adv}^{(l)}_{\mathcal{A}}(\lambda)$ be the advantage of $\mathcal{A}$ in Game i, respectively.

We will show three lemmas (Lemmas 10–12) that evaluate the gaps between pairs of the advantages in Game 0, ..., Game $2n + 1$. From these lemmas, we obtain $\mathsf{Adv}^{(n,d)\text{-SI-PBD}}_{\mathcal{A}}(\lambda) \leq \sum_{l \in [2n+1]} \left| \mathsf{Adv}^{(l-1)}_{\mathcal{A}}(\lambda) - \mathsf{Adv}^{(l)}_{\mathcal{A}}(\lambda) \right| \leq \sum_{l \in [2n]} \mathsf{Adv}^{d\text{-DSJP}}_{\mathcal{B}_l}(\lambda)$. This completes the proof of Theorem 4. $\qquad\square$

Lemma 10 *For any quantum adversary $\mathcal{A}$, there exists a quantum machine $\mathcal{B}_l$, whose running time is essentially the same as that of $\mathcal{A}$, such that for any security parameter λ, $|\mathsf{Adv}^{(l-1)}_{\mathcal{A}}(\lambda) - \mathsf{Adv}^{(l)}_{\mathcal{A}}(\lambda)| \leq \mathsf{Adv}^{d\text{-DSJP}}_{\mathcal{B}_l}(\lambda)$ for $l \in [n]$.*

Proof $\mathcal{B}$ is given a d-DSJP instance (Ψ_{AB}, J_β), where

$$\Psi_{AB} := \left(\text{params}, \left(\left(E^{(\mu)}_A, \phi^{(\mu)}_A(P_B), \phi^{(\mu)}_A(Q_B) \right), \left(E^{(\mu)}_B, \phi^{(\mu)}_B(P_A), \phi^{(\mu)}_B(Q_A) \right) \right)_{\mu \in [d]} \right).$$

$\mathcal{B}$ (implicitly) sets user $l - 1$ A and user l B, and their public keys $\left(E_{l-1}^{(\mu)}, \right.$

$\left. \phi_{l-1}^{(\mu)}(P_l), \phi_{l-1}^{(\mu)}(Q_l) \right)_{\mu \in [d]} := \left(E_A^{(\mu)}, \phi_A^{(\mu)}(P_B), \phi_A^{(\mu)}(Q_B) \right)_{\mu \in [d]}$ and $\left(E_l^{(\mu)}, \phi_l^{(\mu)}(P_{l-1}), \right.$

$\left. \phi_l^{(\mu)}(Q_{l-1}) \right)_{\mu \in [d]} := \left(E_B^{(\mu)}, \phi_B^{(\mu)}(P_A), \phi_B^{(\mu)}(Q_A) \right)_{\mu \in [d]}$, respectively.

$\mathcal{B}$ generates randomly $J_{i-1,i} \leftarrow_R \mathbb{F}_{p^2}$ for $i < l$, and sets $(l - 1)$th j-invariants product as $J_{l-1,l} := J_\beta$. $\mathcal{B}$ generates secret keys $k_i^{(\mu)} \leftarrow_R \mathbb{Z}/\ell_\tau^{e_\tau}\mathbb{Z}$ for all $i \in [n] \setminus \{l - 1, l\}$ where $\tau := i \mod n$, and then his/her own public keys $\left(E_i^{(\mu)}, \phi_i^{(\mu)}(P_{\tau-1}), \right.$

$\left. \phi_i^{(\mu)}(Q_{\tau-1}) \right)_{\mu \in [d]}$. Since $\mathcal{B}$ has all secret keys except for users $l - 1, l$, he can compute all correct j-invariant products $J_{i-1,i}$ for $i > l$.

Using $J_{i-1,i}$ for $i \in [n]$ as defined above, $\mathcal{B}$ computes $u_i := J_{i,i+1} \cdot J_{i-1,i}^{-1}$ and $K := \prod_{i \in [n]} J_{i-1,i}$, and then sends $\mathcal{A}$ the public keys, $(u_i)_{i \in [n]}$, and the challenge value K.

If $\mathcal{A}$ outputs β', then $\mathcal{B}$ also outputs β'. We easily see that the distribution generated by $\mathcal{B}$ is that in Game $l - 1$ when $\beta = 0$ and that in Game i when $\beta = 1$.

This completes the proof of Lemma 10. $\square$

Lemma 11 *For any (quantum) adversary $\mathcal{A}$, for any security parameter λ,* $\mathsf{Adv}_{\mathcal{A}}^{(n+1)}(\lambda) = \mathsf{Adv}_{\mathcal{A}}^{(n)}(\lambda).$

Proof The proof of Lemma 11 is the same as that of Lemma 6 (**BDEnc** Information Theoretic Security Lemma). $\square$

Lemma 12 *For any quantum adversary $\mathcal{A}$, there exists a quantum machine $\mathcal{B} := \mathcal{B}_{n+l}$, whose running time is essentially the same as that of $\mathcal{A}$, such that for any security parameter λ, $|\mathsf{Adv}_{\mathcal{A}}^{(n+l)}(\lambda) - \mathsf{Adv}_{\mathcal{A}}^{(n+l+1)}(\lambda)| \leq \mathsf{Adv}_{\mathcal{B}_{n+l}}^{d\text{-DSJP}}(\lambda)$ for $l \in [n]$.*

Lemma 12 is proven in a similar manner to Lemma 10.

4.4 Two-Round PBD GKE from d-DSMP Assumption

Setup. Takes a security parameter λ and the number of users n. The algorithm outputs $\mathsf{params}_{\mathsf{CSIDH}} := (p(= 4 \cdot \ell_1 \cdots \ell_s - 1), E)$.

Round-1. Takes the user index i and $\mathsf{params}_{\mathsf{CSIDH}}$ as input. User i randomly chooses $\mathbf{e}_i^{(\mu)} := \left(e_{i,1}^{(\mu)}, \ldots, e_{i,s}^{(\mu)} \right)$ and defines $\left[\mathfrak{a}_i^{(\mu)} \right] := \left[\mathfrak{l}_1^{e_{i,1}^{(\mu)}} \cdots \mathfrak{l}_s^{e_{i,s}^{(\mu)}} \right]$. User i then computes elliptic curve $E_i^{(\mu)} := \left[\mathfrak{a}_i^{(\mu)} \right] E$ and sets $\mathsf{pk}_i^1 := \left(E_i^{(\mu)} \right)_{\mu \in [d]} := \left(\left[\mathfrak{a}^{(\mu)} \right] E \right)_{\mu \in [d]}$ and $\mathsf{sk}_i^1 := \left(\mathbf{e}^{(\mu)} \right)_{\mu \in [d]}$. Finally, the user i broadcast pk_i^1 to the other users.

Round-2. Takes the user index i, $\mathsf{params}_{\mathsf{CSIDH}}$, $(\mathsf{pk}_{i-1}^1, \mathsf{pk}_{i+1}^1)$, and sk_i^1. User i executes CSIDH key exchange with users $i - 1$ and $i + 1$ to obtain elliptic curves $E_{i-1,i}^{(\mu)}$ and $E_{i,i+1}^{(\mu)}$, respectively, and then computes

$$M_{i-1,i} := \prod_{\mu=1}^d m \left(E_{i-1,i}^{(\mu)} \right) \quad \text{and} \quad M_{i,i+1} := \prod_{\mu=1}^d m \left(E_{i,i+1}^{(\mu)} \right).$$

The user then computes $u_i := M_{i,i+1} \cdot M_{i-1,i}^{-1}$ and set $\mathsf{pk}_i^2 := u_i$. Finally, the user i broadcasts pk_i^2 to the other users.

KeyComp. User i collects $\left(\mathsf{pk}_{i'}^2\right)_{i' \in [n]}$ and sk_i^1 and computes $K_i := M_{i-1,i}^n \cdot u_i^{n-1} \cdot u_{i+1}^{n-2} \cdots \cdots u_{i-3}^2 \cdot u_{i-2}$.

We can easily verify that $K_i = M_{1,2} \cdot M_{2,3} \cdots M_{n-1,n} \cdot M_{n,1}$ holds for any i. We have the following lemma and theorem as in the case of the SI-PBD key exchange. The (n, d)-CSI-PBD assumption is defined in Definition 7 in Appendix 4.

Lemma 13 *The (n, d)-CSI-PBD key exchange among n-parties is secure under the (n, d)-CSI-PBD assumption.*

Theorem 5 *The (n, d)-CSI-PBD key exchange among n-parties is post-quantumly secure under the d-DSMP assumption when $d \geq 2$ and $\gcd(n, p - 1) = 1$. (Note that $p - 1$ is the order of cyclic group $\mathbb{G} := \mathbb{F}_p^*$.)*

For any quantum adversary $\mathcal{A}$, there exist quantum machines $\mathcal{B}_i$, whose running times are essentially the same as that of $\mathcal{A}$, such that for any security parameter λ,
$$\mathsf{Adv}_{\mathcal{A}}^{(n,d)\text{-CSI-PBD}}(\lambda) \leq \sum_{i \in [2n]} \mathsf{Adv}_{\mathcal{B}_i}^{d\text{-DSMP}}(\lambda).$$

Acknowledgements This research was partially supported by JST CREST Grant Number JPMJCR14D6, Japan. The author would like to thank Tatsuaki Okamoto for his valuable comments on the generic GKE construction given in Sect. 4.1.

Appendix 1: BD Group Key Exchange (Burmester and Desmedt 1994)

We describe the BD Key Exchange among n users on a cyclic group $\mathbb{G}$ of a prime order q and a generator g.

Round-1. Each user i generates $a_i \leftarrow_R \mathbb{Z}/q\mathbb{Z}$, $h_i := g^{a_i}$ and broadcasts h_i.

Round-2. Each user i calculates $J_{i-1,i} := (h_{i-1})^{a_i}$, $J_{i,i+1} := (h_{i+1})^{a_i}$ and $u_i := J_{i,i+1} \cdot J_{i-1,i}^{-1}$. User i broadcasts u_i.

KeyComp. User i calculates $K_i := J_{i-1,i}^n \cdot u_i^{n-1} \cdot u_{i+1}^{n-2} \cdots u_{i-2}$. Then, $K_i = J_{1,2} \cdot J_{2,3} \cdots J_{n,1}$ is the shared key among the n users.

Theorem 6 (Burmester and Desmedt 1994; Katz and Yung 2007) *The BD group key exchange is tightly secure under the DDH assumption. For any adversary $\mathcal{A}$, there is a probabilistic machine $\mathcal{B}$, whose running time is essentially the same as that of $\mathcal{A}$, such that for any security parameter λ, $\mathsf{Adv}_{\mathcal{A}}^{\mathrm{BD}}(\lambda) \leq \mathsf{Adv}_{\mathcal{B}}^{\mathrm{DDH}}(\lambda)$.*

Proof DDH solver $\mathcal{B}$ uses an attacker $\mathcal{A}$ against the BD protocol. Below, we prove the case n is even for simplicity. $\mathcal{B}$ receives a DDH tuple (g, g^a, g^b, T) where T is g^{ab} or g^c with random c, and should simulate public information $(h_i, u_i)_{i \in [n]}$ and the shared key K. $\mathcal{B}$ implicitly sets $a_1 := a$ and $a_2 := b$, and generates random $\tilde{a}_2, \tilde{a}_3, \ldots, \tilde{a}_{n-1} \leftarrow \mathbb{Z}/q\mathbb{Z}$. $\mathcal{B}$ also implicitly sets relations

$$\tilde{a}_2 = a_2 - a_n, \ \tilde{a}_3 = a_3 - a_1, \ldots, \tilde{a}_{n-2} = a_{n-2} - a_{n-4}, \ \tilde{a}_{n-1} = a_{n-1} - a_{n-3}, \quad (9)$$

which determines $a_3, \ldots, a_{n-1}$ as linear combinations of $a(= a_1), b(= a_2), \tilde{a}_3, \ldots,$ $\tilde{a}_{n-1},$ that is, $a_3 := a_1 + \tilde{a}_3, \ldots, a_{n-2} := a_{n-4} + \tilde{a}_{n-2} = b + \tilde{a}_4 + \cdots + \tilde{a}_{n-2},$ $a_{n-1} := a_{n-3} + \tilde{a}_{n-1} = a + \tilde{a}_3 + \cdots + \tilde{a}_{n-1}, \ a_n := a_2 - \tilde{a}_2.$

Therefore, $\mathcal{B}$ simulates h_i as follows: $h_1 := g^a, \ h_2 := g^b, \ h_3 := g^{a_1 + \tilde{a}_3} = g^a \cdot g^{\tilde{a}_3}, \ h_4 := g^{a_2 + \tilde{a}_4} = g^b \cdot g^{\tilde{a}_4}, \ldots, h_{n-2} := g^{b + \tilde{a}_4 + \cdots + \tilde{a}_{n-2}} = g^b \cdot g^{\tilde{a}_4 + \cdots + \tilde{a}_{n-2}}, \ h_{n-1} := g^{a + \tilde{a}_3 + \cdots + \tilde{a}_{n-1}} = g^a \cdot g^{\tilde{a}_3 + \cdots + \tilde{a}_{n-1}}, \ h_n := g^{a_2 - \tilde{a}_2} = g^b \cdot g^{-\tilde{a}_2},$ and $\mathcal{B}$ also simulates u_i as follows using relations (9), $u_i := h_i^{\tilde{a}_{i+1}}$ for $i = 1, \ldots, n-2, u_{n-1} := h_{n-1}^{-\sum_{i=1,3,\ldots,n-3} \tilde{a}_{i+1}},$ $u_n := h_n^{-\sum_{i=2,4,\ldots,n-2} \tilde{a}_{i+1}},$ where $a_n - a_{n-2} = (a_2 - \tilde{a}_2) - (a_2 + \tilde{a}_4 + \cdots + \tilde{a}_{n-2}) = -\sum_{i=1,3,\ldots,n-3} \tilde{a}_{i+1}$ and $a_1 - a_{n-1} = -\sum_{i=2,4,\ldots,n-2} \tilde{a}_{i+1}$ hold. Here, $\mathcal{B}$'s simulations of h_i and u_i are perfect.

Since the correct $K = K_2$ is $K_2 = J_{1,2}^n \cdot u_2^{n-1} \cdot u_3^{n-2} \cdots u_n$ with $J_{1,2} = g^{ab}$, $\mathcal{B}$ simulates shared key K as $K := T^n \cdot u_2^{n-1} \cdot u_3^{n-2} \cdots u_n$ where T is given in the DDH instance and u_i are calculated as above, and then $\mathcal{B}$ give it to $\mathcal{A}$. When $\mathcal{A}$ answers to the question whether K is correct or random, $\mathcal{B}$ answers to his problem as the same way as $\mathcal{A}$.

If $T = g^{ab}$, then the simulation is the same as the real game, and if $T = g^c$, then K is uniformly random and independently distributed from other variables. $\qquad \square$

Appendix 2: Key Derivation Function (KDF)

Let two-party key exchange denote Π with shared key space $\mathbb{K}$. A map $\varphi : \mathbb{K} \to \mathbb{G}$ is called key derivation function (with a range $\mathbb{G}$) if two distributions $\{ \varphi(\kappa) \mid \kappa \leftarrow_R \mathbb{K} \}$ and $\{ J \leftarrow_R \mathbb{G} \}$ are indistinguishable. Such a KDF function can be obtained from a standard hash function, e.g., SHA-2 or SHA-3. For the details, see Abe et al. (2005), for example.

Appendix 3: SIDH and CSIDH Key Exchange

Appendix 3.1: SIDH Key Exchange (Feo et al. 2014)

A supersingular elliptic curve E and generators of smooth order rank-2 torsion subgroups are taken as pubic parameters. Alice and Bob set random cyclic subgroups as secret keys, respectively, and calculate isogenies whose kernels are the secret keys by using Vélu's formulas. They publish their public keys, range curves of the isogenies, and images of the generators, respectively. Finally, they calculate isogenies from public keys. The range curves of the isogenies are isomorphic; therefore their j-invariants become the same. The detailed protocol is given as follows.

Setup. Let $e_A, e_B \in \mathbb{Z}$, and ℓ_A, ℓ_B be small primes (e.g., 2, 3), where $\ell_A^{e_A}$ and $\ell_B^{e_B}$ are close. Let p be a prime which satisfies that $p = \ell_A^{e_A} \ell_B^{e_B} f \pm 1$ where f is a small positive integer. Let $E : \delta y^2 = x^3 + \alpha x^2 + x$ be a supersingular elliptic curve defined over $\mathbb{F}_{p^2}$, where the cardinality of $E(\mathbb{F}_{p^2})$ is $(\ell_A^{e_A} \ell_B^{e_B} f)^2$. Let P_A, Q_A be generators of $E[\ell_A^{e_A}]$, and P_B, Q_B are generators of $E[\ell_B^{e_B}]$. Let public parameters be $\mathsf{params}_{\mathsf{SIDH}} := (p, E, P_A, Q_A, P_B, Q_B)$.

Round-1. Alice chooses random numbers $k_A \in (\mathbb{Z}/\ell_A^{e_A}\mathbb{Z})^\times$, and calculates $R_A = P_A + k_A Q_A$. Here, an order of R_A is $\ell_A^{e_A}$. Alice calculates an $\ell_A^{e_A}$-isogeny $\phi_A : E \to E_A : = E/\langle R_A \rangle$ and $\phi_A(P_B), \phi_A(Q_B)$ by using Vélu formulas.

Similarly, Bob chooses random numbers $k_B \in (\mathbb{Z}/\ell_B^{e_B}\mathbb{Z})^\times$, and calculates $R_B = P_B + k_B Q_B$. Here, an order of R_B is $\ell_B^{e_B}$. Bob calculates an $\ell_B^{e_B}$-isogeny $\phi_B : E \to E_B : = E/\langle R_B \rangle$ and $\phi_B(P_A), \phi_B(Q_A)$ by using Vélu formulas.

Alice sends $E_A, \phi_A(P_B), \phi_A(Q_B)$ to Bob, and Bob sends $E_B, \phi_B(P_A), \phi_B(Q_A)$ to Alice.

KeyComp. Alice calculates $R_A' = \phi_B(P_A) + k_A \phi_B(Q_A)$. Here, an order of R_A' is $\ell_A^{e_A}$. Alice calculates an $\ell_A^{e_A}$-isogeny $\phi_A' : E_B \to E_{AB} : = E_B/\langle R_A' \rangle$ and $K_A = j(E_{AB})$ by using Vélu formulas.

Bob calculates $R_B' = \phi_A(P_B) + k_B \phi_A(Q_B)$. Here, an order of R_B' is $\ell_B^{e_B}$. Bob calculates an $\ell_B^{e_B}$-isogeny $\phi_B' : E_A \to E_{BA} : = E_A/\langle R_B' \rangle$ and $K_B = j(E_{BA})$ by using Vélu formulas.

It holds that $\ker (\phi_A' \circ \phi_B) = \phi_B^{-1}(\langle R_A' \rangle) = \langle R_A \rangle \oplus \langle R_B \rangle$ and $\ker (\phi_B' \circ \phi_A) = \phi_A^{-1}(\langle R_B' \rangle) = \langle R_B \rangle \oplus \langle R_A \rangle$. Hence, $K_A = K_B$ holds; therefore, SIDH is correct.

The SI-DDH assumption is defined in Definition 2.

Theorem 1 (Feo et al. 2014) *The SIDH key exchange is post-quantumly secure under the SI-DDH assumption.*

Appendix 3.2: CSIDH Key Exchange (Castryck et al. 2018)

CSIDH (Commutative Supersingular Isogeny Diffie–Hellman) was proposed by Castryck et al. in 2018 (Castryck et al. 2018).

Let a prime $p := 4 \cdot \ell_1 \cdots \ell_s - 1$, where $\ell_1, \ldots, \ell_s$ are small distinct odd primes. Let $\mathcal{O}$ be an order in an imaginary quadratic field, $\pi \in \mathcal{O}$, π_p the pth power Frobenius endomorphism and $\mathcal{E}\ell\ell_p(\mathcal{O}, \pi)$ the set of $\mathbb{F}_p$-isomorphism classes of $\mathbb{F}_p$-rational supersingular elliptic curves whose $\mathbb{F}_p$-endomorphism ring is equal to $\mathcal{O}$ and the Frobenius π_p is given by $\pi \in \mathcal{O}$. For CSIDH, we only consider the case that $\mathcal{O} \cong \mathbb{Z}[\pi_p]$. CSIDH is based on the action of the ideal class group $\mathrm{cl}(\mathcal{O})$ on $\mathcal{E}\ell\ell_p(\mathcal{O}, \pi)$. Alice and Bob generate random elements in $\mathrm{cl}(\mathcal{O})$ for their secret keys, and calculate the actions on $E/\mathbb{F}_p : y^2 = x^3 + x$. They publish the obtained elliptic curves as public keys. Finally, they calculate their secret key actions on the public keys, respectively. The obtained elliptic curves are isomorphic over $\mathbb{F}_p$, and the Montgomery coefficients are the same. The detailed protocol is given as follows.

Setup.   Let p be a prime as $p = 4 \cdot \ell_1 \cdots \ell_s - 1$, where the $\ell_1, \ldots, \ell_s$ are small distinct odd primes. Let E be the supersingular elliptic curve $y^2 = x^3 + x$ and public parameters $\mathsf{params_{CSIDH}} := (p, E)$.

Round-1.   One randomly chooses an integer vector $(e_1, \ldots, e_s)$ from $\{-\eta, \ldots, \eta\}^s$. Define $[\mathfrak{a}] = \left[\mathfrak{l}_1^{e_1} \cdots \mathfrak{l}_s^{e_s} \right] \in cl(\mathcal{O})$, where $\mathfrak{l}_i = (\ell_i, \pi_p - 1)$, $\mathfrak{l}_i^{-1} = (\ell_i, \pi_p + 1)$, and η is the smallest integer which satisfies that $2\eta + 1 \geq \sqrt[s]{\#cl(\mathcal{O})}$. One calculates the action of $[\mathfrak{a}]$ on E and the Montgomery coefficient $m \in \mathbb{F}_p$ of $[\mathfrak{a}]E : y^2 = x^3 + mx^2 + x$. Let the integer vector $(e_1, \ldots, e_s)$ (or $[\mathfrak{a}]$) be the secret key, and $m \in \mathbb{F}_p$ be the public key.

KeyComp.   Alice (resp. Bob) has her (resp. his) secret key, $[\mathfrak{a}]$ (resp. $[\mathfrak{b}]$). Alice calculates the action $[\mathfrak{a}]E_B = [\mathfrak{a}][\mathfrak{b}]E$, where $E_B : y^2 = x^3 + m_B x^2 + x$. Bob calculates the action $[\mathfrak{b}]E_A = [\mathfrak{b}][\mathfrak{a}]E$, where $E_A : y^2 = x^3 + m_A x^2 + x$. Define shared keys $K_A := m([\mathfrak{a}][\mathfrak{b}]E)$, and $K_B := m([\mathfrak{b}][\mathfrak{a}]E)$.

By commutativity of $cl(\mathcal{O})$ and the uniqueness of the Montgomery coefficient, it holds that $K_A = K_B$; therefore, CSIDH is correct.

The CSI-DDH assumption is defined in Definition 3.

Theorem 2 (Castryck et al. 2018) *The CSIDH key exchange is post-quantumly secure under the CSI-DDH assumption.*

Appendix 4: Decisional CSI-PBD ((n, d)-CSI-PBD) Assumption

Definition 7 (*Decisional CSI-PBD ((n, d)-CSI-PBD) Assumption*)

Let $(\Psi_{n,d}, K) \leftarrow_R \mathsf{Exec}_{(n,d)\text{-CSI-PBD}}(\lambda)$, where $M_{i-1,i} := \prod_{\mu=1}^{d} m\left(E_{i-1,i}^{(\mu)} \right)$, $M_{i,i+1} := \prod_{\mu=1}^{d} m\left(E_{i,i+1}^{(\mu)} \right)$, $u_i := M_{i,i+1} \cdot M_{i-1,i}^{-1}$, $\Psi_{n,d} := (\mathsf{params_{CSIDH}}, \left(E_i^{(\mu)}, u_i \right)_{i \in [n], \mu \in [d]})$, and $K := \prod_{i=1}^{n} M_{i,i+1}$. An (n, d)-CSI-PBD problem instance is given as $(\Psi_{n,d}, \kappa_\beta)$ where $\kappa_0 := K$, $\kappa_1 \leftarrow_R \mathbb{F}_p$, and $\beta \leftarrow_R \{0, 1\}$. For any quantum algorithm $\mathcal{B}$, the advantage of $\mathcal{B}$ is defined as $\mathsf{Adv}_{\mathcal{B}}^{(n,d)\text{-CSI-PBD}}(\lambda) := |\Pr[\mathcal{B}(\Psi_{n,d}, \kappa_0) = 1] - \Pr[\mathcal{B}(\Psi_{n,d}, \kappa_1) = 1]|$, and the (n, d)-CSI-PBD assumption holds if $\mathsf{Adv}_{\mathcal{B}}^{(n,d)\text{-CSI-PBD}}(\lambda)$ is negligible in λ for any polynomial-time quantum adversary $\mathcal{B}$.

References

M. Abe, R. Gennaro, K. Kurosawa, V. Shoup, Tag-KEM/DEM: a new framework for hybrid encryption and a new analysis of Kurosawa-Desmedt KEM, in *EUROCRYPT 2005* (2005), pp. 128–146

S. Akleylek, E. Alkim, P.S. Barreto, J. Buchmann, E. Eaton, G. Gutoski, J. Krämer, P. Longa, H. Polat, J.E. Ricardini, G. Zanon, qTESLA. Submission to NIST PQC Standardization (2017)

D. Apon, D. Dachman-Soled, H. Gong, J. Katz, Constant-round group key exchange from the ring-LWE assumption, in *PQCrypto 2019* (2019), pp. 189–205

D.J. Bernstein, T. Lange, C. Martindale, L. Panny, Quantum circuits for the CSIDH: optimizing quantum evaluation of isogenies, in *EUROCRYPT 2019*, Part II (2019), pp. 409–441

W. Beullens, T. Kleinjung, F. Vercauteren, CSI-FiSh: efficient isogeny based signatures through class group computations. IACR Cryptol. ePrint Arch. **2019**, 498 (2019)

D. Boneh, D. Glass, D. Krashen, K. Lauter, S. Sharif, A. Silverberg, M. Tibouchi, M. Zhandry, Multiparty non-interactive key exchange and more from isogenies on elliptic curves, in *MATHCRYPT 2018* (2018), https://eprint.iacr.org/2018/665

C. Boyd, A. Mathuria, *Protocols for Authentication and Key Establishment*, Information Security and Cryptography (Springer, Berlin, 2003)

M. Burmester, Y. Desmedt, A secure and efficient conference key distribution system (extended abstract), in *EUROCRYPT'94* (1994), pp. 275–286

W. Castryck, T. Lange, C. Martindale, L. Panny, J. Renes, CSIDH: an efficient post-quantum commutative group action, in *ASIACRYPT 2018*, Part III (2018), pp. 395–427

L. Ducas, E. Kiltz, T. Lepoint, V. Lyubashevsky, P. Schwabe, G. Seiler, D. Stehlé, Crystals-dilithium: a lattice-based digital signature scheme. IACR Trans. Cryptogr. Hardw. Embed. Syst. **2018**(1), 238–268 (2018)

L.D. Feo, D. Jao, J. Plût, Towards quantum-resistant cryptosystems from supersingular elliptic curve isogenies. J. Math. Cryptol. **8**(3), 209–247 (2014)

P.A. Fouque, J. Hoffstein, P. Kirchner, V. Lyubashevsky, T. Pornin, T. Prest, T. Ricosset, G. Seiler, W. Whyte, Z. Zhang, Falcon: fast-Fourier lattice-based compact signatures over ntru. Submission to NIST PQC Standardization (2017)

A. Fujioka, K. Takashima, S. Terada, K. Yoneyama, Supersingular isogeny Diffie-Hellman authenticated key exchange, in *ICISC 2018* (2018), pp. 177–195

S. Furukawa, N. Kunihiro, K. Takashima, Multi-party key exchange protocols from supersingular isogenies, in *ISITA 2018* (IEEE Xplore, 2018), pp. 208–212

D. Jao, R. Azarderakhsh, M. Campagna, C. Costello, L.D. Feo, B. Hess, A. Jalali, B. Koziel, B. LaMacchia, P. Longa, M.N.J. Renes, V. Soukharev, D. Urbanik, Sike: supersingular isogeny key encapsulation. Submission to NIST PQC Standardization (2017)

M. Just, S. Vaudenay, Authenticated multi-party key agreement, in *ASIACRYPT'96* (1996), pp. 36–49

J. Katz, M. Yung, Scalable protocols for authenticated group key exchange. J. Cryptol. **20**(1), 85–113 (2007)

L. Kowalczyk, H. Wee, Compact adaptively secure ABE for NC^1 from k-lin, in *EUROCRYPT 2019*, Part I (2019), pp. 3–33

T. Okamoto, K. Takashima, Fully secure functional encryption with general relations from the decisional linear assumption, in *CRYPTO 2010* (2010), pp. 191–208. Full version is available as an online first article in J. Cryptol

C. Peikert, He gives c-sieves on the CSIDH. IACR Cryptol. ePrint Arch. **2019**, 725 (2019)

P.W. Shor, Polynomial-time algorithms for prime factorization and discrete logarithms on a quantum computer. SIAM J. Comput. **26**(5), 1484–1509 (1997)

A. Sutherland, Identifying supersingular elliptic curves. LMS J. Comput. Math. **15**, 317–325 (2012)

K. Takashima, Expressive attribute-based encryption with constant-size ciphertexts from the decisional linear assumption, in *SCN 2014* (2014), pp. 298–317

J. Vélu, Isogénies entre courbes elliptiques. Comptes Rendus Acad. Sci. Paris, Sér. A. **273**, 238–241 (1971)

L. Washington, *Elliptic Curves: Number Theory and Cryptography*, 2nd edn. (CRC Press, Boca Raton, 2008)

Correction to: International Symposium on Mathematics, Quantum Theory, and Cryptography

Tsuyoshi Takagi, Masato Wakayama, Keisuke Tanaka,
Noboru Kunihiro, Kazufumi Kimoto, and Yasuhiko Ikematsu

Correction to:
T. Takagi et al. (eds.), *International Symposium
on Mathematics, Quantum Theory, and Cryptography*,
Mathematics for Industry 33,
https://doi.org/10.1007/978-981-15-5191-8

The original version of the book was published with incorrect part names in the website for second part FM now updated with correct title. This has been corrected in the updated version.

The updated version of the book can be found at
https://doi.org/10.1007/978-981-15-5191-8

Index

© The Author(s) 2021
T. Takagi et al. (eds.), *International Symposium on Mathematics,
Quantum Theory, and Cryptography*, Mathematics for Industry 33,
https://doi.org/10.1007/978-981-15-5191-8

Printed in the United States
by Baker & Taylor Publisher Services